PROBABILITY AND CAUSALITY

SYNTHESE LIBRARY

STUDIES IN EPISTEMOLOGY,

LOGIC, METHODOLOGY, AND PHILOSOPHY OF SCIENCE

VOLUME 192

WESLEY C. SALMON

(*Photograph by Barbara Boylan*)

PROBABILITY AND CAUSALITY

Essays in Honor of Wesley C. Salmon

With an Annotated Bibliography by Wesley C. Salmon

Edited by

JAMES H. FETZER

University of Minnesota, Duluth

D. REIDEL PUBLISHING COMPANY

A MEMBER OF THE KLUWER ACADEMIC PUBLISHERS GROUP

DORDRECHT / BOSTON / LANCASTER / TOKYO

Library of Congress Cataloging-in-Publication Data CIP

Probability and causality.

(Synthese library ; v. 192)
Bibliography: p.
Includes indexes.
1. Science — Philosphy. 2. Probabilities. 3. Causation. I. Salmon, Wesley C. II. Fetzer, James H., 1940– . III. Title. IV. Series.
Q175.3.P758 1987 501 87–28689
ISBN 90–277–2607–8
1–55608–052–2 (U.S. : pbk.)

Published by D. Reidel Publishing Company,
P.O. Box 17, 3300 AA Dordrecht, Holland.

Sold and Distributed in the U.S.A. and Canada
by Kluwer Academic Publishers,
101 Philip Drive, Norwell, MA 02061, U.S.A.

In all other countries, sold and distributed
by Kluwer Academic Publishers Group,
P.O. Box 322, 3300 AH Dordrecht, Holland.

Printed in the Netherlands

To
Wesley C. Salmon

TABLE OF CONTENTS

EPILOGUE

FOREWORD

The contributions to this special collection concern issues and problems discussed in or related to the work of Wesley C. Salmon. Salmon has long been noted for his important work in the philosophy of science, which has included research on the interpretation of probability, the nature of explanation, the character of reasoning, the justification of induction, the structure of space/time and the paradoxes of Zeno, to mention only some of the most prominent. During a time of increasing preoccupation with historical and sociological approaches to understanding science (which characterize scientific developments as though they could be adequately analysed from the perspective of political movements, even mistaking the phenomena of conversion for the rational appraisal of scientific theories), Salmon has remained steadfastly devoted to isolating and justifying those normative standards distinguishing science from non-science — especially through the vindication of general principles of scientific procedure and the validation of specific examples of scientific theories — without which science itself cannot be (even remotely) adequately understood.

In this respect, Salmon exemplifies and strengthens a splendid tradition whose most remarkable representatives include Hans Reichenbach, Rudolf Carnap and Carl G. Hempel, all of whom exerted a profound influence upon his own development. It has appeared to be quite fitting, therefore, to have a volume devoted to some (though by no means all) of the most central and significant problems and issues with which he has recently been concerned, namely, those involving connections between probability and causality. For the ability to grasp the fundamental difference between probabilistic and non-probabilistic scientific laws, between causal and non-causal scientific explanations, and between deterministic and non-deterministic scientific theories presupposes the possession of adequately defined and appropriately justified conceptions of probability and causality that are capable of fulfilling their roles within these contexts — problems whose importance is equalled by their difficulty.

Accordingly, the contributors to this volume were invited to focus

James H. Fetzer (ed.) Probability and Causality. xi—xvii.

upon this problem complex of probability and causality to which Salmon himself has made stimulating contributions. The result was quite gratifying, issuing in a sequence of twelve studies falling into the broad categories of probability, causality and modality, on the one hand, and probability, causality and decision, on the other. In addition, Salmon was invited to contribute a study of his own (relating his current thinking on these problems and taking the opportunity, should he be so inclined, of commenting on the other contributions to this volume) as well as an annotated bibliography of his publications to date that would afford an occasion for him to discuss his own research of the past. Indeed, this bibliography, which provides the insights associated with an intellectual autobiography, is among the most fascinating features of this work.

Our collection commences with a study in which Salmon explores the relations between objective and subjective probabilities, contending that efforts that take subjective probabilities ("degrees of belief" or "strengths of conviction") as fundamental and treat objective probabilities as derivative have their relationship backward. In an inquiry of broad scope, Salmon contrasts his views with those of Ramsey, Mellor and Lewis, among others, substantially clarifying his conception of the respective roles of frequencies, propensities and probabilities, maintaining (a) that there are two kinds of probabilities, personal probabilities and relative frequencies; (b) that propensities cannot technically qualify as interpretations of the calculus of probability; but (c) that observed frequencies function as evidence for estimates of propensities, which, in turn, generate expectations that serve as our "very guide of life".

"Probability, Causality, and Modality" thus begins with a thoughtful reconstruction of Hume's critique of induction, in which William ("Ted") Morris persuasively argues that understanding Hume's position is not quite so easily done as some assume. A systematic analysis of earlier efforts by J. L. Mackie and by D. C. Stove discloses substantial evidence that Hume's position is more complicated and less conclusive than other scholars have had reason to suppose. By a demonstration that the principles of the probability calculus and of estimates of relative frequencies can both be derived by assuming the definition of "probabilities" as relative frequencies, Abner Shimony substantially strengthens the conception of (what might be described as) "Humean rationality". Results such as these are especially significant, since Hume would tend to view universal laws as constant conjunctions and statistical laws as relative frequencies.

Ilkka Niiniluoto advances a searching exploration of frequency and propensity conceptions of probability against the background of the Principle of Plenitude, which asserts that no genuine possibility can remain forever unrealized. He offers reasons for believing that frequency interpretations satisfy this requirement, while propensity interpretations do not; nevertheless, the arguments he elaborates support propensity conceptions — even at the expense of plenitude! In his examination of attempts to analyze causality in terms of probability, on the one hand, and probability in terms of causality, on the other, James H. Fetzer pursues these issues further. He contends that limiting frequency accounts (both actual and hypothetical) are incapable of fulfilling the desiderata appropriate to "indeterministic causation" and that long-run propensity concepts are hopelessly deficient, defending the virtues of single-case propensities instead.

Were those the only alternatives, therefore, arguments such as these might carry the day. The studies by Wayne Davis and by Brian Skyrms, however, suggest that there are enormously varied views on these subjects. Davis, in particular, examines the "probabilistic theories of causation" advanced by Suppes, by Giere and by Cartwright, assessing their respective strengths and weaknesses in great detail, arriving at the conclusion that their major virtue appears to be the absence of viable competitors. Skyrms similarly explores the extent to which work by Suppes and by Good clarifies the nature of "probabilistic causation" as well as the degree to which work by Adams and by Stalnaker illuminates the "pragmatics of conditionals". While conceding that difficult obstacles confront these conceptions, he argues that a Bayesian theory of conditional chance provides an intriguing bridge between alternative — and sometimes conflicting — points of view.

"Probability, Causality, and Decision" begins with an exploration by Nancy Cartwright of the principle of the common cause, which was introduced by Reichenbach and which Salmon has endorsed. She contends that this principle can be formulated in a fashion that copes with counterexamples van Fraassen has advanced, which supports the conclusion that different statistical criteria are necessary for common causes that happen to operate in distinctively different ways. While Cartwright considers conjunctive forks, Ellery Eells focuses upon a pair of problems that threaten to undermine current understanding of "probabilistic causality", namely: the problem of interacting causal factors and the problem of disjunctive causal factors. He suggests that these problems can be overcome within the context of a revised theory of probabilistic causality that reflects a broader variety of ways in which probabilistic causal interactions can occur.

The adequacy of these positions can be tested against the views advocated in Elliott Sober's study of the common cause principle, where its tenability is measured against troublesome cases drawn, not from quantum mechanics, but from evolutionary biology, which have to do with inferences related to common ancestry. These considerations lead him to recommend that the principle needs to be elaborated conditionally by specifying when common cause explanations are to be preferred to separate cause explanations. D. H. Mellor lends weight to probabilistic theories of causality by endorsing the principle that "causes" must at least raise the probability of their effects; but he supports his position with a novel argument viewing causes as "means" toward their effects as "ends", within a broadly decision-theoretical framework. This study, moreover, concerns the analysis of indeterminstic causation and bears comparison with others in Part I.

Many readers will be intrigued to discover that Richard Jeffrey reverts to the defense of the evidentiary version of the first edition of *Logic of Decision* rather than the ratificationist revisions of the second edition in his new analysis of the Prisoner's Dilemma (this time for clones). Not the least significant feature of this presentation, moreover, is his employment of a 'two-level model" in which judgmental probabilities serve as estimates of objective probabilities. A fascinating critique of the evidentiary version of Bayesian decision theory is advanced by Paul Humphreys, finally, who maintains that, for evidential versions to preserve their integrity against Newcomb-style counterexamples, they must resort to the self-defeating strategy of declaring certain sorts of situations as altogether beyond the bounds of rational decision making, where these maneuvers are not justifiable as features of rationality or as idealizations of practice.

The issues at stake in these studies, it should be observed, are not merely those whose specific features are delineated above. On the contrary, a variety of significant problems lie just beneath the surface and only arise when certain views are juxtaposed with certain others, including, for example, this sampler:

First and foremost, the reader should beware that, even though (or perhaps because) these are technical philosophical problems, different authors may use the same phrases and expressions with very different meanings. Those who talk about "probabilistic theories of causation" may or may not mean the same thing by that phrase as do those who discuss "theories of probabilistic causality", and there are those who would insist upon important differences as matters of principle. The expression, "statistical causality", again, may or may not be used as a synonym for "indeterministic causation", which, in turn, may or may not be synonymous with "probabilistic causation". And there are authors for whom "indeterministic causation" is an incoherent concept, because causation (for them) has to be deterministic. It is therefore important here — more so than with most philosophical writing — that meanings not be taken for granted.

Second, while specific interpretations of probability (such as limiting and hypothetical frequency constructions) receive consideration from various authors, whether or not they have the same understanding of these interpretations should be viewed as subject to debate. Thus, while endorsing the long-run propensity conception in application to traditional games of chance (tosses of dice, draws of cards and the like), Niiniluoto reports that he no longer thinks that these long-run dispositions are of universal strength. Fetzer, by comparison, not only presumes that long-run propensities must be of universal strength but also rejects them, where the divergence between their positions rests upon complex issues concerning whether sequences of "chance events" of the traditional kind properly qualify as deterministic or as indeterministic, which ultimately entails distinguishing between ontic and epistemic concepts.

Third, the work of some figures, such as that of Patrick Suppes, receives consideration in several of the studies mentioned above; yet the authors ascribe to him very different positions. Davis, for example, views his work as in the relative frequency tradition and Fetzer treats it as in the limiting frequency tradition, but Skyrms claims that, "Suppes intends his account to be neutral among interpretations of probability". Such a position, alas, is very difficult to reconcile with the discovery that, while frequencies are standard probabilities (which satisfy ordinary axioms for conditional probabilities, including Bayes' theorem), propensities are not. A construction of probabilistic causation that is supposed to be neutral among interpretations of probability, therefore, does not even satisfy syntactical conditions for definitional adequacy, much less any other relevant theoretical desiderata. Given this result, frequency constructions of his views may be appropriate.

Fourth, the positions of other authors may entail subtle (even buried) semantical problems. Skyrms, for example, discusses the principle of simplification of disjunctive antecedents for subjunctive conditionals, where this principle maintains that the truth of 'If *p* or *q* were the case, then *r* would be the case' entails the truth of 'If *p* were the case, then *r* would be the case; and if *q* were the case, then *r* would be the case'. While such a principle initially may seem appealing, it appears to be satisfiable, in principle, only when the *p*-phenomenon and the *q*-phenomenon are sufficient conditions (to bring about) the *r*-phenomenon, which is a condition that cannot be fulfilled when causation is indeterministic. Indeed, its acceptability — to which Eells' study of disjunctive causal factors has relevance — may ultimately depend upon the respective merits of maximal-change formal semantics as opposed to minimal-change formal semantics within these contexts.

Fifth, during the course of his commentary on his book, *Scientific Explanation and the Causal Structure of the World*, Salmon reiterates the idea that he is not concerned with "the logic of scientific explanation", especially since he does not believe that any such logic exists: "What constitutes a satisfactory pattern for scientific explanation depends crucially upon the kind of world in which in fact we live, and perhaps upon the domain of science with which we are concerned". But this view seems to suggest that aspects of logic itself are world-specific and perhaps even domain-specific, contrary to the world and content independence that principles of logic are ordinarily supposed to possess. Salmon's position thus raises some exceptionally interesting questions, not only about the structure of models of explanation, but about principles of reasoning as well.

Sixth and finally, during the course of his commentary on his paper, "Propensities: A Discussion Review", Salmon contends that, although his point of departure is a book by Mellor, he provides "an extended critique of the so-called propensity interpretation of probability". This claim, however, is subject to dispute, since he does not consider the full range of accounts advanced by propensity theoreticians such as Peirce, Popper, Hacking, Giere and Fetzer, for example, whose views depart considerably from those of Mellor (whom he does discuss at length). Given his recent endorsement of propensities as "just [the sort of] dispositions that seem to me to lie at the foundation of probabilistic causality" (*Scientific Explanation and the Causal Structure of the World*, p. 204), therefore, many readers will appreciate his further reflections in "Dynamic Rationality".

What this means, of course, is that this volume should prove to be a bonanza for graduate students in search of a thesis as well as for philosophers who want to sharpen their wits. And this is very much as it should be in a collection of essays dedicated to this man. For as Humphreys has remarked in the prefatory note to his contribution, whether or not Salmon would agree with the questions that have been raised is almost beside the point: "I do not know if he would agree with what follows but even if he does not, it is irrelevant, because Wes is one of those rare people whose interest lies in getting things right rather than in insisting that *he* has got it right. Would that we were all as careful and honest." And in this spirit, I take the greatest pleasure in dedicating this volume to someone who has made and who continues to make an important difference to us all.

30 June 1987 J.H.F.

PROLOGUE

WESLEY C. SALMON

DYNAMIC RATIONALITY: PROPENSITY, PROBABILITY, AND CREDENCE

Since the time of Bishop Butler we have known that *probability is the very guide of life* — at least it should be if we are to behave rationally. The trouble is that we have had an extraordinarily difficult time trying to figure out what kind of thing — or things — probability is. I shall argue that causality is a key piece in the puzzle, and consequently, an indispensable aspect of rationality.

1. THE PROBLEM

EXAMPLE 1. In the autumn of 1944 I went as a student to the University of Chicago. The dormitory in which I lived was not far from Stagg Field, and I often walked by the entrance to the West Stands where a sign was posted saying "Metallurgical Laboratories." Nothing ever seemed to be going on there. Later, when I got a job as a technician at the Met Labs, I learned that this was the place where, about two years previously, the first humanly-created self-sustaining nuclear chain reaction had taken place. An atomic pile had been assembled and the decision was taken to make a trial run on 2 December 1942. The control rods were gradually withdrawn and the chain reaction occurred as predicted. It was an undertaking fraught with high risk; as one commentator has put it, "They were facing either the successful birth of the atomic age — or a catastrophic nuclear disaster within a packed metropolis."[1] I was glad I had not been there at the time. No one could be absolutely certain beforehand that the chain reaction would not spread, involving other substances in the vicinity and engulfing Chicago — or, perhaps, the entire earth — in a nuclear holocaust. On second thought, it did not much matter where one was in case of such an eventuality.

I once read that before this pile was put into operation, the probability of the spreading of the chain reaction had been calculated. The result was that this probability was no more than three in a million. I do not recall the source of this bit of information, but I do remember wondering how the probability had been calculated and how it had

James H. Fetzer (ed.) Probability and Causality. 3—40.

been decided that the risk was acceptable. I was even more curious about the meaning of the probability concept in this context.[2]

EXAMPLE 2. The foregoing perplexity was brought vivid back to mind by an article in *Science* in June of 1986 about the Rogers Commission Report on the Challenger space shuttle disaster.[3] In particular, the article reports on a press conference held by Richard Feynman, the famous and colorful physicist who was a member of that commission. Although Feynman did not refuse to sign the commission's report, he did make a supplemental statement to the press in which he sharply criticized NASA's methods of assessing risks.

Feynman objects most strongly to NASA's way of calculating risks. Data collected since the early days of the program . . . show that about one in every 25 solid rocket boosters has failed. About 2900 have been launched with 121 losses. Feynman says it is reasonable to adjust the anticipated crash rate a bit lower (to 1 in 50) to take account of today's better technology. He would even permit a little more tinkering with the numbers (to 1 in 100) to take credit for exceptionally high standards of part selection and inspection. In this way, the Challenger accident, the first solid rocket failure in 25 shuttle launches (with two boosters each), fits perfectly into Feynman's adjusted rate of one crash per 50 to 100 rocket firings.[4]

But Feynman was stunned to learn that NASA rejects the historical data and claims the actual risk of crash is only 1 in 100 000. This is the official figure as published in "Space Shuttle Data for Planetary Mission RTG Safety Analysis" on 15 February 1985.[5] . . . . Feynman searched for the origin of this optimism and found that it was "engineering judgment," pure and simple. Feynman concluded that NASA "for whatever purpose . . . exaggerates the reliability of its product to the point of fantasy."

The article goes on to report a conversation with NASA's chief engineer, Milton Silveira, who stated that NASA does not "use that number as a management tool."

The 1 in 100 000 figure was hatched for the Department of Energy (DOE), he says, for use in a risk analysis DOE puts together on radioactive hazards on some devices carried aboard the shuttle. . . .

DOE and General Electric, supplier of the power units, write up a detailed risk analysis before launch. They are accustomed to expressing risk in statistical terms. NASA is not, but it must help them prepare the analysis. To speak in DOE's language, NASA translates its "engineering judgment" into numbers. How does it do this? One NASA official said, "They get all the top engineers together down at Marshall Space Flight Center and ask them to give their best judgment of the reliability of all the components involved." The engineers' adjectival descriptions are then converted to numbers. For example . . . "frequent" equals 1 in 100; "reasonably probable" equals 1 in 1000; "occasional" equals 1 in 10 000; and "remote" equals 1 in 100 000.

> When all the judgments were summed up and averaged, the risk of a shuttle booster explosion was found to be 1 in 100 000. That number was then handed over to DOE for further processing. . . .

Among the things DOE did with the number was to conclude that "the overall risk of a plutonium disaster was found to be terribly, almost inexpressibly low. This is, 1 in 10 000 000, give or take a syllable."

> "The process," says one consultant who clashed with NASA, "is positively medieval." He thinks Feynman hit the nail exactly on the head. . . . Unless the risk estimates are based on some actual performance data, he says, "it's all tomfoolery."

Feynman is quoted as saying, "When playing Russian roulette, the fact that the first shot got off safely is little comfort for the next."

EXAMPLE 3. Then there is the old joke about precipitation probabilities. We have seen them in weather forecasts in newspapers and on TV; sometimes they serve as some sort of guide of life. They are always given in multiples of 10 — e.g., a 30 percent chance of showers. How are they ascertained and what do they mean? Well, there are 10 meteorologists, and they take a vote. If 3 of the 10 vote for showers, that makes a 30 percent chance.

2. GRADES OF RATIONALITY

When one speaks of rationality, it may refer to rational action or to rational belief. Rational action must somehow take account of probabilities of various outcomes. I am inclined to think that the policy of maximizing expected utilities — though perhaps not precisely correct — is satisfactory in most practical situations.[6] Expected utilities are defined in terms of probabilities. I want to discuss the nature of these probabilities. There is a long tradition that identifies them with degrees of belief. I am not going to deny that answer, but I want to look at it with some care. First, however, a matter of terminology.

In a recent article, Patrick Maher has argued — quite cogently, in my opinion — that belief has nothing to do with rational *behavior*.[7] He takes belief to be an all or nothing affair; one either believes a given proposition or does not believe it. Beliefs have the purely cognitive function of constituting our picture of the world. When it comes to rational behavior, he argues, we should use what he calls *degrees of confidence* to evaluate expectations of utility. Within Bayesian decision theory we can identify coherent degrees of confidence with personal probabilities. He prefers to avoid using such terminology as "degree of belief" or "degree of partial belief" in order to minimize confusion of the practical function of degrees of confidence (which can be evaluated numerically) with the cognitive function of beliefs (which are only qualitative). I have a slight preference for the term *degree of conviction* over *degree of confidence* (because "confidence" has a technical meaning in statistics that I do not want to attach to degrees of conviction), but both terms are acceptable. In addition, I shall sometimes use the more neutral term *degree of credence*, which should serve well enough for those who accept Maher's view and those who do not.[8]

There are two immediate objections to taking mere subjective degrees of conviction as probabilities. The first is that, in general, we cannot expect them to constitute an admissible interpretation of the probability calculus. The second is that we are not prepared to take raw subjective feelings as a basis for *rational* action. So, one standard suggestion is to require that the probabilities constitute *rational* degrees of conviction.

There are, I think, various grades of rationality. At the lowest level, philosophers have, for the most part, taken logical consistency as a basic rationality requirement. If one believes that no mammals hatch their young from eggs, that the spiny anteater (echidna) is a mammal, and that it hatches its young from eggs, standard deductive logic tells us that something is wrong — that some modification should be made — that at least one of these beliefs should be rejected. But logic, by itself, does not tell us what change should be made. Logical consistency can be regained — with regard to this set of beliefs — by rejecting any one of the three. Logical consistency provides one minimal sort of rationality.[9] In traditional logic, the price of adopting an inconsistent set of beliefs is that everything follows from them. Let us say that anyone whose total set of beliefs is logically consistent qualifies for *basic deductive rationality.*

A minimal kind of probabilistic rationality is represented by the subjective Bayesians who require only *coherence*. Coherence is a sort of probabilistic consistency requirement. A set of degrees of conviction that violate the relationships embodied in the standard axioms of probability is incoherent. If, for example, I hold a probability of 2/3 that this coin will land heads up on the next toss, and a probability of

2/3 that it will land tails up on the next toss, and that these two outcomes are mutually exclusive, then I am being incoherent. This shows that there is something fundamentally wrong with the foregoing set of degrees of conviction. As is well known, the price of incoherence is that a so-called *Dutch book* can be made against anyone who holds incoherent degrees of conviction and is prepared to use them as fair betting quotients. A set of bets constitutes a Dutch book if, no matter what the outcome, the bettor loses. It is obvious how this happens in the case in which 2/3 is assigned as the probability of heads and also of tails. Adopting the terminology of L. J. Savage, we usually refer to subjective probabilities that are coherent as *personal probabilities*. It follows immediately from their definition that personal probabilities constitute an admissible interpretation of the probability calculus, and thus overcome the first objection to subjective probabilities.

Violation of the coherence requirement shows that something is wrong with a set of degrees of conviction. Probability theory, however, does not tell us what modification to make in order to repair the difficulty, any more than pure deductive logic tells us which statement or statements of the foregoing inconsistent triad should be rejected. In the probabilistic example, we could achieve coherence by deciding that there is a 1/3 probability of heads, a 1/3 probability of tails, and a 1/3 probability that the coin will come to rest standing on edge, where these three alternatives are mutually exclusive and exhaustive. L. J. Savage held an official view that probabilistic coherence is the only requirement of rationality, but unofficially, so to speak, he adopted others.

Abner Shimony defined a notion of *strict coherence*, and he proved a kind of Dutch book result. Strict coherence demands, in addition to coherence, that no logically contingent molecular statement (statement not involving quantifiers) should have a probability of zero or one.[10] He has shown that anyone who violates strict coherence can be put in a position of making a set of bets such that, no matter what happens, the bettor cannot win, though he or she will break even on some outcomes. Carnap has always considered strict coherence a basic rationality requirement.

A fundamental distinction between the requirements of coherence and strict coherence should be noted. Almost without exception, coherence requirements impose restrictions on relationships among probability values, as found, for instance, in the addition rule, the multiplication rule, and Bayes's theorem. They do not generally yield individual probability values. The two exceptions are that logical truths have probability one and logical contradictions have probability zero. Moreover, the requirement of coherence does not constrain individual probability values beyond the universal constraint that they must lie within the closed unit interval from zero to one. Strict coherence imposes a hefty restriction on a large class of probabilities, where the statements to which they apply (contingent molecular statements) are *not* logical truths or falsehoods. We shall look at the rationale for the strict coherence requirement below.

Strict coherence is a sort of *openmindedness* requirement; various authors have demanded other sorts of *openmindedness* requirements. In characterizing a view he called *tempered personalism*, Shimony[11] argues that, in evaluating and comparing scientific hypotheses, we should allow hypotheses seriously proposed by serious scientists to have nonnegligible prior probabilities. We should also accord a substantial prior probability to the *catchall hypothesis* — the "none of the above" supposition — the notion that we have not yet thought of the correct hypothesis. It is important to note that Shimony offers empirical justifications for these openmindedness conditions. He observes historically that science has progressed remarkably by taking seriously the hypotheses proposed by qualified scientists. He also points out that in many historical situations the set of hypotheses under consideration did not include some hypothesis, proposed only later, that turned out to be successful.

The principles I have mentioned thus far pertain to what might be called the *statics* of degrees of conviction. You look at your total body of knowledge at a given moment, so to speak, and try to discover whether it contains any logical inconsistencies or probabilistic incoherencies (or violations of strict coherence). If such defects are discovered some changes in degrees of conviction are required, but we have not provided any basis for determining what alterations should be made. For the inconsistent triad a modest amount of research reveals that there are several species of egg-laying mammals, and that the echidna is one of them. Hence, we should reject the first statement while retaining the other two. Additional empirical evidence has provided the solution. In the case of the coin, a modification was offered, but it was not a very satisfactory one. The reasoning might have been this. There are three possibilities: the coin could land heads up, or tails up, or on edge. Applying some crude sort of principle of indifference, we assigned equal probabilities to all of them. But silly as this may be, we did eliminate the incoherence.

Clearly we have not said nearly enough about the *kinematics* of degrees of conviction — the ways our personal probabilities should change over time. There are, I believe, three sorts of occasion for modification of personal probabilities. First, there is the kind of situation in which an incoherency is discovered, and some change is made simply to restore coherence. As we have seen, this can be accomplished without getting further information. When an alteration of this type occurs, where modifications are made willy-nilly to restore coherence (or strict coherence), there is change of degrees of conviction, but no kinematics thereof. We can speak properly of a kinematics of degrees of conviction only when we have kinematic principles that somehow direct the changes.

Second, personal probabilities are sometimes modified as a result of a new idea. Suppose we have been considering two rival hypotheses, H_1 and H_2, and the catchall H_c. We have a probability distribution over this set. Someone thinks of a new hypothesis H_3 that has never before been considered. We now have the set H_1, H_2, H_3, and a new catchall $H_{c'}$, where H_c is equivalent to $H_3 \vee H_{c'}$. It is possible, of course, to assign new probabilities in a way that does not involve changing any of the old probabilities, but that is not to be expected in all cases. Indeed, a new idea or a new hypothesis may change the whole picture as far as our plausibility considerations are concerned.[12] From what has been said so far, however, we do not have any principles to guide the way such modifications take place. At this point it looks like no more nor less than a psychological reaction to a new factor. We still have not located any kinematical principle.

Third, personal probabilities must often be modified in the light of new evidence. In many such cases the natural approach is to use Bayes's theorem. This method is known as *Bayesian conditionalization.* It is important to realize that coherence requirements alone do not mandate Bayesian conditionalization. Consider a trivial example. Suppose someone is tossing a coin in the next room. There are only two possibilities, namely, that the coin is two-headed or it is standard. I cannot examine the coin, but I do get the results of the tosses. My prior probability for the hypothesis that a two-headed coin is being tossed is 1/10; that is, I have a fairly strong conviction that it is not two-headed. Now I learn that the coin has been tossed four times, and that each toss resulted in a head. If the coin is two-headed that result is inevitable; if the coin is standard the probability of four heads in a row is 1/16. Using Bayes's theorem you calculate the posterior probability of the two-headed hypothesis and find that is 0.64. You tell me that I must change my mind and consider it more likely than not that the coin is two-headed. I refuse. You show me the calculation and tell me that I am being incoherent. I see that you are right, so I must somehow modify my degrees of conviction. I tell you that I was wrong about the prior probability, and I change it from 1/10 to 1/100. Now I calculate the posterior probability of the two-headed hypothesis using *that* value of the prior, and come out with the result that it is just about 0.14. My system of personal probabilities is now coherent and I still consider it more likely than not that the coin is a standard one. I satisfied the coherence requirement but failed to conform to Bayesian conditionalization.

To many Bayesians, Bayesian conditionalization is a cardinal kinematical principle of rationality. It says something like this. Suppose you are considering hypothesis H. Before you collect the next bit of evidence, announce your prior probabilities and your likelihoods. When the next bit of evidence comes in, calculate the posterior probability of H using those priors and likelihoods. Change your degree of conviction in H from the prior probability to the posterior probability.

Bayesian conditionalization works effectively only if the prior probabilities do not assume the extreme values of zero or one. A cursory inspection of Bayes's theorem reveals the fact that no evidence can change the probability of a hypothesis whose prior probability has one of those extreme values. As L. J. Savage once said about hypotheses being evaluated, one should have toward them a mind that is, if not open, at least slightly ajar. In addition to the requirement of coherence (or strict coherence), many Bayesians accept both Bayesian conditionalization and some sort of openmindedness condition. We shall construe the openmindedness condition as a principle of statics, since it applies to the degrees of conviction that exist at any one time. Nevertheless, we should keep in mind that the main motivation is kinematical — to enable Bayesian conditionalization to work effectively. Bayesian conditionalization is a method for modifying degrees of conviction as a result of considering new evidence.

We are now in a position to distinguish three different grades of rationality in terms of considerations already adduced. Each succeeding grade is to be understood as incorporating the principles involved in the preceding grades:

Basic Deductive Rationality
  Logical Consistency
Static Probabilistic Rationality
  Coherence
  Strict Coherence
  Openmindedness
Kinematic Probabilistic Rationality
  Bayesian Conditionalization

Later in this paper I shall add a fourth grade, to be known as *dynamic rationality*.

The division between the kinematic and static grades of rationality marks a difference between rationality principles that do, and those that do not, involve any relation to the real world. The fact that the static principles do not relate to the real world is not a fault, however, for one sort of rationality consideration is quite independent of objective fact. The basic idea is that one aspect, at least, of rationality involves simply the management of one's body of opinion in terms of its inner structure. It has no concern with the objective correctness of opinions; it deals with the avoidance of blunders of various sorts within the body of opinion — for example, the kind of mistake that occurs when logical inconsistency or probabilistic incoherence is present.

The requirement of strict coherence nicely illustrates the point. Why should we adopt a prohibition against assigning probability zero or one to all molecular statements? Do we really believe that there is a probability greater than zero, for example, of finding a piece of pure copper that is not an electrical conductor? I am inclined to believe that we could search the universe thoroughly from the big bang to the present without finding a nonconducting piece of copper, and that a search from now until the end of time (if such there is) would yield the same result. However, the requirement of strict coherence is not concerned with considerations of this sort. It is designed to prevent a person from having a set of degrees of conviction that would lead to such disavantageous bets as wagering at odds of a million dollars to zero against someone finding a piece of nonconducting copper. It does not matter what the facts of nature are; such a bet would be at best pointless and at worst disasterous. Good housekeeping of one's stock of degrees of conviction recommends against allowing probabilities that, if they are to be used as fair betting quotients, could place one in that kind of situation.[13]

I have no desire to denigrate this type of consideration; it constitutes an important aspect of rationality. But it also seems to fall short of giving us a full-blooded concept of rationality. Surely *rational behavior* demands attention to the objective facts that are relevant to the action one contemplates or undertakes. Examples 1—3 in the first section of this paper were intended to illustrate this point dramatically.

Bayesian conditionalization does, to some extent, provide a connection between opinion and objective fact. It specifies what degrees of conviction should be held constant and which allowed to change under specified circumstances. It tells us how to respond to new items of evidence.

Nevertheless, Bayesian conditionalization cannot provide a sufficient connection between our degrees of conviction and the objective facts. One obvious restriction is that it says nothing about changing our degrees of conviction in the absence of new evidence. It could be construed as prohibiting any change of opinion without new evidence, but that would be unduly restrictive. As we have already seen, there are at least two types of occasions for revising degrees of conviction even if there is no new evidence. The first is required to restore coherence to an incoherent set of degrees of convction. It might be said, in response, that Bayesian conditionalization presupposes a coherent set of degrees of conviction. If we have an incoherent set, any change whatever that will restore coherence (without violating any other rationality condition listed above) will do. Then Bayesian conditionalization should be used.

The second type of situation that calls for revision of degrees of conviction in the absence of new evidence is the occurrence of a new idea. Surely the Bayesian does not want to bar the sort of revision that is based on thought and reflection. However, if that sort of revision is permitted, it can be used to sever connections between degrees of conviction and objective facts. Whenever accumulating evidence — in the form of observed relative frequencies, for example — seems to mandate a certain degree of conviction, a redistribution of personal probabilities can preclude that result. It is because of this weakness of connection between degree of conviction and objective fact, given only Bayesian conditionalization, that I wish to pursue higher grades of rationality.

3. TWO ASPECTS OF PROBABILITY

In *The Emergence of Probability* Ian Hacking maintained that the concept of probability could not appear upon the scene until two notions could be brought together, namely, objective chance and degree of credence.[14] The first of these concepts is to be understood in terms of relative frequencies or propensities; the second is an epistemic notion that has to do with the degree to which one is justified in having confidence in some proposition. In our century Rudolf Carnap codified this general idea in his systems of inductive logic in which there two probability concepts — $probability_1$, inductive probability or degree of confirmation; and $probability_2$, relative frequency. The major part of Carnap's intellectual effort for many years was devoted to his various attempts to characterize degree of confirmation. Because his systems of inductive logic require *a priori* probability measures, which strike me as unavoidably arbitrary, I have never been able to accept his approach.

It is interesting to note, however, that in "The Aim of Inductive Logic," he approaches the problem by beginning with raw subjective degrees of credence and imposing rationality requirements upon them.[15] He adopts coherence, strict coherence, and Bayesian conditionalization. He goes beyond these by imposing additional symmetry conditions. In his "Replies and Systematic Expositions" in the Schlipp volume,[16] he offers a set of 15 axioms, all of which are to be seen as rationality conditions that narrow down the concept of degree of confirmation. These axioms, in other words, are intended to beef up the rather thin notion of rationality characterized wholly by coherence, strict coherence, and Bayesian conditionalization. It is a motley assortment of axioms that require a strange collection of considerations for their justification. The problem of arbitrary apriorism remains in all of his subsequent work.[17] Carnap has stated that the ultimate justification of the axioms is inductive intution. I do not consider this answer an adequate basis for a concept of rationality. Indeed, I think that *every* attempt, including those by Jaakko Hintikka and his students, to ground the concept of rational degree of belief in logical probability suffers from the same unacceptable apriorism.

If the conclusion of the previous paragraph is correct, we are left with subjective probabilities and physical probabilities. By imposing coherence requirements on subjective probabilities we transform them into personal probabilities, and (trivially) these satisfy the standard probability calculus. Moreover, I am convinced — by arguments advanced by F. P. Ramsey, L. J. Savage, and others — that there *are* psychological entities that constitute degrees of conviction. We can find out about them in various ways including the observation of betting behavior. What is required if they are to qualify as *rational* degrees of conviction is the question we are pursuing.

Turning to the objective side, we find propensities and relative frequencies. Although the so-called *propensity interpretation of probability* has enjoyed considerable popularity among philosophers for the last couple of decades, it suffers from a basic defect. As Paul Humphreys pointed out, the probability calculus contains *inverse probabilities*, as in Bayes's theorem, but there are no corresponding inverse propensities. Consider a simple quality control situation. A certain factory manufactures floppy disks. There are several different machines in this factory, and these machines produce disks at different rates. Moreover, each of these machines produces a certain proportion of defective disks; suppose, if you like, that the proportions are different for different machines. If these various frequencies are given it is easy to figure the probability that a randomly selected disk will be defective. It is perfectly sensible to speak of the propensity of a given machine to produce a defective disk, and of the propensity of the factory to produce defective disks. Moreover, if we pick out a disk at random and find that it is defective, it is easy, by applying Bayes's theorem, to calculate the probability that it was produced by a particular machine. It is not sensible, however, to speak of the propensity of this disk to have been produced by a given machine. Consequently, propensities do not even provide an admissible interpretation of the probability calculus. The problem is that the term "propensity" has a causal aspect that is not part of the meaning of "probability."[18]

I have no objection to the concept of propensity as such. I believe there are probabilistic causes in the world, and they are appropriately called "propensities." There is a propensity of an atom to decay, a propensity of a tossed die to come to rest with side 6 uppermost, a propensity of a child to misbehave, a propensity of a plant sprayed with an herbicide to die, etc. Such propensities produce relative frequencies. We can find out about many propensities by observing frequencies.

Indeed, it seems to me that the propensity concept may play quite a useful role in quantum theory. In that context wave equations are often employed, and references to amplitudes are customary. To calculate a probability one squares the absolute value of the amplitude. We can, of course, speak formally about wave equations as mathematical entities, but when they are applied to the description of the physical world it is reasonable to ask what the waves are undulations of. There seems to be no answer that is generally recognized as adequate. In other realms of physics we study sound waves, vibrating strings, light waves, water waves, etc. People sometimes refer to quantum mechanical waves as waves of probability, but that is not satisfactory, for what the wave gives has to be squared to get a probability. In dealing with other kinds of waves we can say what the wave amplitude is an amplitude of: displacement of water molecules, changes in electromagnetic field strength, fluctuations of air density, etc. My suggestion is that the quantum mechanical wave is a wave of propensity — propensity to interact in certain ways given appropriate conditions. The results of such interactions are frequencies, and observed frequencies give evidence as to the correctness of the propensity we have attributed in any given case. If we adopt this terminology we can say that *propensities* exhibit interference behavior; it is no more peculiar to talk about the interference of waves of propensity than it is to speak of the interference of electromagnetic waves. In this way we can avoid the awkward necessity of saying that *probabilities* interfere with one another.

It is my view that — in quantum mechanics and everywhere else — physical probabilities are somehow to be identified with frequencies. One reason for this is that relative frequencies constitute an admissible interpretation of the probability calculus if the axioms require only finite additivity. Hans Reichenbach demonstrated that the axioms of his probability calculus are logical consequences of the definition of probabilities as limiting frequencies.[19] Van Fraassen has pointed out that the limiting frequency interpretation is not an admissible interpretation of a calculus that embodies countable additivity, but I do not consider that an insuperable objection. Complications of this sort arise when we use such mathematical idealizations as infinite sequences and limiting frequencies to describe the finite classes of events and objects with which we deal in the real world. Similar problems arise when we use geometrical descriptions of material objects, or when we use the infinitesimal calculus to deal with finite collections of such discrete entities as atoms or electric charges.

The conclusion I would draw is that there are two kinds of probabilities, personal probabilities and relative frequencies. Perhaps we can legitimately continue thinking of physical probabilities as limits of relative frequencies in infinite sequences; perhaps it will turn out that the concept has to be finitized. This is deep and difficult issue that I shall not pursue farther in this paper. For purposes of this discussion I shall speak vaguely about long-run relative frequencies, and I shall assume that observed relative frequencies in samples of populations provide evidence regarding the relative frequencies in the entire population. The question to which I want now to turn is the relationship between personal probabilities and frequencies.

4. ON RESPECTING THE FREQUENCIES

Section 3 of F. P. Ramsey's essay "Truth and Probability" is entitled "Degrees of Belief,"[20] and in it he develops a logic of partial belief. This essay is rightly regarded as a landmark in the history of the theory of subjective probability. Ramsey mentions two possible approaches to the topic, namely, (1) as a measure of intensity of belief, which could be established introspectively, and (2) from the standpoint of causes of action. He dismisses the first as irrelevant to the topic with which he is concerned, and proceeds to pursue the second. The way he does so is noteworthy.

> I suggest that we introduce as a law of psychology that [the subject's] behaviour is governed by what is called the mathematical expectation; that is to say that, if p is a proposition about which he is doubtful, any goods or bads for whose realization p is in his view a necessary and sufficient condition enter into his calculations multiplied by the same fraction, which is called the 'degree of his belief in p'. We thus define degree of belief in a way which presupposes the use of the mathematical expectation.[21]
>
> We can put this in a different way. Suppose his degree of belief in p is m/n; then his action is such as he would choose it to be if he had to repeat it exactly n times, in m of which p was true, and in the others false. . . .
>
> This can also be taken as a definition of degree of belief, and can easily be seen to be equivalent to the previous definition.[22]

In the time since Ramsey composed the present essay (1926), a good deal of empirical work in psychology has been devoted to the behavior of subjects in gambling situations, and it seems to yield an unequivocal verdict of *false* on Ramsey's proffered law. There is, however, good reason to regard it as a normative principle of rational behavior, and that is entirely appropriate if we regard logic — including the logic of partial belief — as a normative subject.

The situation is relatively straightforward, as Ramsey shows by example. If a given type of act is to be performed n times, and if on m of these occasions it results in a good g, while in the other $n - m$ it results in a bad b (where the sign of b is taken to be negative), then the frequency m/n provides a weighting factor that enables us to calculate the average result of undertaking that action. The total outcome for n occasions is obviously

$$m \times g + (n - m) \times b;$$

the average outcome is

$$[m \times g + (n - m) \times b]/n.$$

If an individual knows the rewards (positive and negative) of the possible outcomes of an act, and also the frequencies of these outcomes, he or she can tell exactly what the total result will be. In that case planning would be altogether unproblematic, for the net result of performing any such act a given number of times could be predicted with certainty. The problem is that we do not know beforehand what the frequency will be.

In view of our advance ignorance of the frequency, we must, as Carnap repeatedly reminded us, use an estimate of the relative frequency, and we should try to obtain the best estimate that is available on the basis of our present knowledge. For Carnap, probability$_1$ fulfils just this function; it is the best estimate of the relative frequency. However, inasmuch as I have rejected Carnap's inductive logic on grounds of intolerable apriorism, his approach is unavailable to me.

Ramsey's approach to this problem is quite different from Carnap's. Ramsey wants to arrive at an understanding of the nature of degree of belief, and he shows that we can make sense of it if (and only if ?) we identify it with the betting quotient of the subject. Using the betting quotient in this way makes sense because of its relation to frequencies. The degree of partial belief, which is to be identified with the subject's betting quotient, can thus be regarded as the subject's best guess or estimate of the relative frequency. If one repeats the same type of act in the same sort of circumstances n times, then, in addition to the actual utilities that accrue to the subject in each type of outcome, the actual frequencies with which the various outcomes occur determine the net gain or loss in utility for the subject. Because of the crucial role played by actual frequencies in this theory, I would say that Ramsey's account of degree of belief is *frequency-driven*. The whole idea is to get a handle on actual frequencies because, given the utilities, the frequency determines what you get.

Ramsey's treatment of the nature of subjective probabilities and their relations to objective probabilities stands in sharp contrast to D. H. Mellor's theory as set forth in *The Matter of Chance*.[23] In that work he explicitly adopts the strategy of basing his "account of objective probability on a concept of partial belief" instead of going at it the other way around. We can say that Mellor offers a *credence-driven* account of objective probability.

On Mellor's view, propensities are not probabilities, and they are not to be identified with chances. A propensity is a disposition of a chance set-up to display a *chance distribution* under specifiable conditions — e.g., upon being flipped in the standard way a fair coin displays the distribution {chance of heads = 1/2; chance of tails = 1/2}. The chance set-up displays this *same distribution* on *every* trial. Obviously the chance distribution, which is always the same for the same chance set-up, is not to be identified with the outcomes, which may vary from one trial to another. In addition, the chance distribution is not to be identified with the relative frequencies of outcomes generated by the chance set-up.

It might seem odd to say that chance distributions are 'displayed' when clearly it is the outcome, not the distribution, that is observable. But obviously, as Mellor is clearly aware, there is nothing in the notion of a disposition that prevents it, when activated, from manifesting something that is not directly observable. A hydrogen atom, for example, has a disposition to emit photons of particular frequencies under certain specifiable circumstances. *To find out what chance distribution a given chance set-up displays*, Mellor maintains, *we must appeal to our warranted partial beliefs*. Chances are probabilities, but they are not to be identified with relative frequencies. Relative frequencies are generated by the operation of chance set-ups having chance distributions. At the foundation we have warranted partial beliefs, which determine chance distributions, which, in turn, yield relative frequencies.

A fuller elaboration of the *credence-driven* point of view, and one that differs from Mellor's in fundamental respects, is provided by David Lewis in "A Subjectivist's Guide to Objective Chance."[24] He begins,

> We subjectivists conceive of probability as the measure of reasonable partial belief. But we need not make war against other conceptions of probability, declaring that where subjective credence leaves off, there nonsense begins. Along with subjective credence we should believe also in objective chance. The practice and the analysis of science require both concepts. Neither can replace the other. Among the propositions that deserve our credence we find, for instance, the proposition that (as a matter of contingent fact in our world) any tritium atom that now exists has a certain chance of decaying within a year. Why should subjectivists be less able than other folk to make sense of that?[25]

Lewis points out that there can be "hybrid probabilities of probabilities," among them credences regarding chances.

> . . . we have some very firm and definite opinions concerning reasonable credence about chance. These opinions seem to me to afford the best grip we have on the concept of chance. Indeed, I am led to wonder whether anyone *but* a subjectivist is in a position to understand objective chance![26]

There is an important sense in which Lewis's guide to objective chance is extraordinarily simple. It consists of one principle that seems to him "to capture all we know about chance." Accordingly, he calls it

> **The Principal Principle**. Let C be any reasonable initial credence function. Let t be any time. Let x be any real number in the unit interval. Let X be the proposition that the chance, at time t, of A's holding equals x. Let E be any proposition compatible with X that is admissible at time t. Then
>
> $$C(A/XE) = x$$

"That," as Lewis says, "will need a good deal of explaining."[27] But what it says roughly is that the degree to which you believe in A should equal the chance of A. I certainly cannot disagree with that principle. The question is, who is in the driver's seat, subjective credence or objective chance? As we have seen, Lewis has stated unequivocally his view that it is subjective credence; I take myself as agreeing with Ramsey that we should leave the driving to some objective feature of the situation. Ramsey opts for relative frequencies, and in a sense I think that is correct. In a more fundamental sense, however, I shall opt for propensities.

Lewis identifies chance with propensity; I take the notion of propensity as the more basic of the two. As I said above, I do not reject that concept, when it is identified with some sort of probabilistic causality (provided it is not taken to be an interpretation of probability). Moreover, I do not have any serious misgivings about attributing causal relations in single cases. On the basis of my theory of causal processes and causal interactions — spelled out most fully in chapters 5—7 of *Scientific Explanation and the Causal Structure of the World* — I believe that individual causal processes transmit probability distributions and individual causal interactions produce probabilistic outcomes (indeterministically, in some kinds of cases, at least).[28] Scientific theories, which have been highly confirmed by massive amounts of frequency data, tell us what the values of these propensities are. Propensities are, as James H. Fetzer has often maintained, entities that are not directly observable, but about which we can and do have strong indirect evidence. Their status is fully objective.[29]

As Lewis formulates his principle, he begins by saying "Let *C* be *any* reasonable initial credence function."[30] This means, in part, that "*C* is a nonnegative, normalized, finitely additive measure defined on all propositions."[31] In addition, it is *regular.* Regularity is closely akin to strict coherence, but regularity demands that no proposition receive the value zero unless it is logically false. In addition, Lewis requires Bayesian conditionalization as *one* (but not necessarily the *only*) way of learning from experience:

> In general, *C* is to be reasonable in the sense that if you started out with it as your initial credence function, and if you always learned from experience by conditionalizing on your total evidence, then no matter what course of experience you might undergo your beliefs would be reasonable for one who had undergone that course of experience. I do not say what distinguishes a reasonable from an unreasonable credence function to arrive at after a given course of experience. We do make the distinction, even if we cannot analyze it; and therefore I may appeal to it in saying what it means to require that *C* be a reasonable initial credence function.[32]

The fact that Lewis does not fully characterize reasonable functions poses problems for the interpretation of his view, but the point that is generally acknowledged by subjectivists or personalists is that reasonable credence functions are not unique. One person may have one personal probability distribution, while another in the same situation may have a radically different one. So it appears that, on Lewis's account, there is no such thing as *unique* objective chance. As I understand the principal principle, it goes something like this. Suppose that I have a credence function C that, in the presence of my total evidence E, assigns the *degree of credence* x to the occurrence (truth) of A. Suppose further that E says nothing to contradict X, the assertion that the *objective chance* of A is x. Then, I can try adding X to my total evidence E to see whether this would change my degree of conviction regarding A. The supposition is that in most circumstances, if my degree of credence in A is x, asserting that it is also the objective chance of A is not going to change my degree of credence to something other than x (provided, of course, that there is nothing in the evidence E that contradicts that statement about the objective chance of A). Under those circumstances I have established the objective chance of A on the basis of my degree of credence in A.

One of the main things that bothers me about this account is that it seems possible that another subject, with a different credence function C' and a different degree of conviction x' in A, will assign a different objective chance to the truth of A. There is, of course, no doubt that different subjects will have different estimates of the chance of A, or different opinions concerning the chance of A. That does not help Lewis's account, inasmuch as he is attempting to characterize *objective chance*, not *estimate of objective chance* or *guess at objective chance*. The personalist may appeal to the well-known swamping of the priors — the fact that two investigators with very different prior probabilities will experience convergence of the posterior probabilities as additional evidence accumulates if they share the same evidence and conditionalize upon it. There are two objections to this response. First, and most obviously, the convergence in question secures intersubjective agreement, without any assurance that it corresponds to objective fact. Second, the convergence cannot be guaranteed if the parties do not agree on the likelihoods, and the likelihoods are just as subjective as the prior probabilities are.

Accordingly, I would want to read the principal principle in the opposite direction, so to speak. To begin, we should use whatever empirical evidence is available — either through direct observation of relative frequencies or through derivation from a well-established theory — to arrive at a statement X that the chance (in Lewis's sense) of A is x. Taking that value of the chance, we combine X with the total evidence E, and calculate the subjective degree of conviction on the basis of the credence function C. If $C(A/XE)$ is not equal to x, then C is not a *rational* credence function. In other words, we should use our knowledge of objective chance to determine what constitutes a reasonable degree of conviction. Our knowledge of objective chance, or propensity, must ultimately be based upon observations of relative frequencies. Epistemically speaking, this amounts to a *frequency-driven* account of rational credence. In the following section, however, I shall suggest that, ontically speaking, we should prefer a *propensity-driven* account — in my sense of the term "propensity."

5. DYNAMIC RATIONALITY

In section 2 of this paper we looked at various grades of rationality, static and kinematic, that can appropriately be applied to our systems of degrees of conviction. We noted emphatically that static rationality comprises only internal 'housekeeping' criteria — ones that do not appeal to external objective facts in any way. In that connection, I made mention of my view that, from the standpoint of *rational action*, it is necessary to refer to objective fact.

Kinematic rationality, which invokes Bayesian conditionalization, makes a step in that direction — it does make *some* contact with the real world. I tried to show, however, that the connection it effects is too tenuous to provide a full-blooded concept of rationality. In an attempt to see how this connection could be strengthened we considered several theories concerning the relationships between subjective and objective probabilities. We found in Ramsey a robust account of the connection between partial beliefs and relative frequencies, which led me to characterize his theory as a *frequency-driven* account of subjective probability. In both Mellor and Lewis we also saw strong connections between subjective probabilities and objective chance. Inasmuch as both of these authors base their accounts of objective chance on subjective probabilities, I characterized them as offering a *credence-driven* account of objective probability. Lewis provides a particularly strong connection in terms of his *principal principle*. I am prepared to endorse something like this principle, except that I think it should run in the direction opposite to that claimed by Lewis. He says, in effect, that we can plug in subjective probabilities and get out objective chance (which he identifies with propensity). I think it should be just the other way around.

I should like to confer the title *dynamic rationality* upon a form of rationality that incorporates some sort of requirement to the effect that the objective chances — whether they be interpreted as frequencies or as propensities — must be respected, as well as other such rationality principles as logical consistency, probabilistic coherence, strict coherence, openmindedness, and Bayesian conditionalization. I believe Ramsey was offering a theory of dynamic rationality because frequencies are its driving force. My version of dynamic rationality will establish one connection between propensities and frequencies as well as another connection between propensities and personal probabilities. Since I regard propensities as probabilistic causes, the term "dynamic rationality" is especially appropriate.

In order to see how this works, let us look at an example. In the Pennsylvania State Lottery, the "daily number" consists of three digits, each of which is drawn from the chamber of a separate machine containing ten ping-pong balls numbered from 0 through 9. It is assumed that all of the balls in any of the machines have equal chances of being drawn and that the draws from the several machines are mutually independent. The winning number must have the digits in the order in which they are drawn. Since there are 1000 numbers between 000 and 999, each lottery ticket has a probability of 1/1000 of winning. The tickets cost $1 each, and the payoff for getting the right number is $500. Thus, the expectation value of each ticket is $0.50.[33] As Ramsey emphasized, if a person played for 1000 days at the rate of one ticket per day, and if the actual frequency of wins matched exactly the probability of winning, the individual would win once for a total gain of $500 and an average gain of $0.50 per play.

Unfortunately, the lottery has not always been honest. In 1980 it was discovered that the April 24 drawing (and perhaps others) had been 'fixed'. On that occasion, white latex paint had been injected into all of the balls except those numbered 4 and 6, thereby making them heavier than the others. Since the balls are mixed by a jet of air through the bottom of the chamber, and are drawn by releasing one ball through the top, the heavier ones were virtually certain not to be drawn. The probabilities of drawing the untampered balls were thereby greatly increased. Those who knew about the crooked arrangement could obviously take advantage of it. If the only balls that could be drawn were 4 and 6, then each of the eight possible combinations of these two digits had a probability of 1/8. In that case, the expectation value would work out to about $62.50, which is not bad for a ticket that can be purchased for $1.[34] The actual result was 666.

One reason for going into this example in some detail is the bring out the familiar fact that, when one is going to use knowledge of frequencies in taking some sort of action on a single outcome, it is important to use the appropriate reference class. The class must be epistemically homogeneous — that is, the agent must not know of any way to partition it in a manner that makes a difference to the probabilities of the outcomes. Someone who had no knowledge of the way the lottery had been fixed would have to assign the value 1/1000 to the probability of any given number being the winning number. Someone who knew of the fix could assign a probability of practically zero to some of the numbers and a probability of 0.125 to the remaining ones. As we have seen, the difference between the two probabilities is sufficient to make a huge difference in the expectations.

Three steps in the decision procedure (the decision whether to purchase a lottery ticket or not) have been taken. First, the event on whose outcome the gamble depends (a particular night's drawing) is referred to a reference class (all of the nightly drawings). Second, the probability that one particular number will win is assessed relative to that reference class. Third, the epistemic homogeneity of that reference class is ascertained. If it turns out to be inhomogeneous, a relevant partition is made, and the process is repeated until an epistemically homogenous reference class is found. This is the probability with respect to the broadest epistemically homogeneous references class (what Reichenbach and I have called "weight") that is taken as the probability value in calculating the expection.

In any given situation, the epistemically homogeneous reference class may or may not be objectively homogeneous.[35] In cases such as the present — where we have a causal or stochastic process generating the outcomes — if the class is objectively homogeneous, I would consider the weight assigned to the outcome as the *propensity* or *objective chance* of that mechanism (chance set-up) to produce that outcome. A mechanism with this propensity generates relative frequencies, some of which are identified as probabilties if we adopt a frequency interpretation.

If the reference class is merely epistemically homogeneous, I would regard the weight as our *best available estimate* of the propensity or objective chance. That weight can easily be identified as our subjective degree of credence — or, if it is not actual, as the degree of credence we should have.

I have no wish to create the impression that any sort of crude counting of frequencies is the only way to determine the propensities of various chance set-ups. In the case of the lottery, we have enough general knowledge about the type of machine employed to conclude that the propensities for all of the different numbers from 000 to 999 are equal if all of the balls are of the same size and weight. We also have enough general background knowledge to realize that injecting some but not all of the balls with latex paint will produce a chance set-up that has different propensities for different numbers. In theoretical calculations in quantum mechanics one routinely computes propensities in order to get values of probabilities. I do think that, at the most fundamental epistemic level, it is the counting of frequencies that yields knowledge of values of propensities.

On my view, as I have said above, propensities are not probabilities; rather, they are probabilistic causes. They generate some basic probabilities directly, as, for example, when the lottery machines produce sequences of digits. From the probabilities generated directly by propensities we can compute other probabilities, such as the (inverse) probability that a given number was produced by a crooked machine.

In cases like the lottery machines, where there are actually many repetitions of the same type of event to provide a large reference class, it may seem that the foregoing account makes good sense. In many similar situations, where we do not have a large number of trials but are familiar with the sort of mechanism involved, we may justifiably feel confident that we know what kinds of frequencies would result if many trials were made. Again, the foregoing account makes sense. But how, if at all, can we extend the account to deal with nonrepetitive cases?

Let us begin by considering situations like the lottery where long sequences of trials exist, but where the subject is going to bet on only one trial in that sequence. Using the mathematical expectation, he or she can ascertain what the average return per trial would be, but what point is there in having that number if we can only apply it in a single instance? The answer is that the *same policy* can be used over and over again in connection with *different sequences.*

Imagine a gambling casino in which there are many (say 100) different games of chance, each of which generates a long sequence of outcomes. In these various games there are many different associated probabilities.[36] Let us assume that our subject knows these probabilities and decides to play each game just once. After deciding the size of the bet in each case, he or she can figure the mathematical expectation for each wager. The situation we have constructed now consists of 100 long sequences of trials, each with an associated expectation value. The subject makes a random selection of one item from each of the sequences; these constitute a new sequence of trials, each with an associated expectation value. The main differences between this new sequence and the original 100 are that each item is produced by a different chance set-up, and both the probabilities and the expectation values differ from item to item. In playing through this new sequence of trials the subject will win some and lose some, and in general the tendency will be to win more of the cases with higher probabilities and to lose more of those with low probabilities. The average gain (positive or negative) per play will tend to be close to the average of the 100 expectation values associated with the 100 different basic sequences.

We realize, of course, that our subject might be extremely unlucky one evening and lose every bet placed, but such an overall outcome will be very rare. If he or she repeats this type of performance night after night, the mathematical expectation of the evening's play will be the average outcome per evening in a sequence of evenings, and the frequency of significant departures from that amount will be small. Reichenbach proved what amounts to the same result to justify his policy for dealing with single cases.[37]

Whether we think of physical probability as propensity or as (finite or infinite) long run relative frequency, there is always the problem of the short run, as characterized elegantly by Charles Sanders Peirce:

> According to what has been said, the idea of probability belongs to a kind of inference which is repeated indefinitely. An individual inference must be either true or false and can show no effect of probability; and therefore, in reference to a single case considered in itself, probability can have no meaning. Yet if a man had to choose between drawing a card from a pack containing twenty-five red cards and a black one, or from a pack containing twenty-five black cards and a red one, and that of a red one were destined to transport him to eternal felicity, and that of a black one to consign him to everlasting woe, it would be folly to deny that he ought to prefer the pack containing the larger proportion of red cards, although from the nature of the risk, it cannot be repeated.[38]

There is no question about ascertaining the probabilities in Peirce's example; they are assumed to be well-known. The problem has to do with the application of this knowledge in a concrete situation. It does not much matter whether the application involves one case or a dozen or a hundred, as long as the number of cases is much smaller than the number of members of the entire population.

In a paper published many years ago, I addressed this problem and suggested a pragmatic vindication of a short run rule to the effect that one should assume that the short run frequency will match, as closely as possible, the long run probability.[39] That vindication did not hold up. Now I would be inclined to give a different argument. Assume that I am right in identifying the probability with the long run frequency, and that the long run frequency of draws resulting in red cards from the predominately red deck is 25/26. Assume that this value has been established on the basis of extensive observation of frequencies. It is not crucial to Peirce's problem that only one draw ever be made from the deck; what is essential is that a person's entire fate hinges on the outcome of one draw.

I shall now say that the drawing of a card from the deck is a chance set-up whose propensity to yield a red card is 25/26. The statement about the propensity is established on the basis of observed frequencies as well as such theoretical considerations as the symmetries of the chance set-up. As I have said above, I think of propensities as probabilistic causes. If it is extremely important to achieve a particular outcome, such as a red card, one looks for a sufficient cause to bring it about. If no sufficient cause is available, one seeks the most potent probable cause. In Peirce's example, a draw from the predominately red deck is obviously a more potent cause than is a draw from the predominately black deck. The fact that the outcome cannot be guaranteed is bad luck, but that is the kind of situation we are in all the time. One might call it *the human predicament*. That is why "*for us*, probability is the very guide of life." Taking the strongest measure possible to bring about the desired result is clearly the sensible thing to do. For the moment, at least, I am inclined to regard this as an adequate answer to the problem of the short run.

Having discussed the easy sort of case in which the frequencies, propensities, mathematical expectations, and reasonable degrees of belief are clear, we must now turn to the hard type of case in which the event whose probability we want to assess is genuinely unique in some of its most important aspects. Here the subjective probabilities may be reasonably tractable — especially in New York City, where it seems that most people will bet on anything — but the objective probabilities are much more elusive. I believe, nevertheless, that the notions of propensity and probabilistic cause can be of significant help in dealing with such cases.

Before turning to them, however, we should briefly consider an intermediate kind of case, which appears at first blush to involve a high degree of uniqueness, but where, upon reflection, we see that pertinent frequencies can be brought to bear. The Challenger disaster provides a useful instance. The life of Christa McAuliffe, the school teacher who joined the crew on that flight, was insured for a million dollars (with her family as beneficiary) by a philanthropic individual. I have no idea what premium was charged.

McAuliffe was the first female who was not a professional astronaut to go on a space shuttle flight. Since she had undergone rigorous screening and training, there was reason to believe that she was not at special risk *vis à vis* the rest of the crew, and that she did not contribute a special risk to that particular mission. As Feynman pointed out, however, there was a crucial piece of frequency information available — namely, the frequency of failure of solid rocket boosters. Taking that information into account, and even making the sort of adjustments Feynman suggested, the minimum premium (omitting profit for the insurance company) should have been $10 000.

Feynman's adjustments are worthy of comment. One adjustment was made on the basis of improved technology; whether that is reasonable or not would depend in large measure on whether the relative frequency of failures had actually decreased as time went by. The second had to do with higher standards of quality control. Perhaps there were frequency data to show that such a factor was relevant, but even in their absence one could take it as a probabilistic cause that would affect the frequency of space shuttle disasters.

Another good example comes from the continuing controversy regarding the so-called *Strategic Defense Initiative*, popularly known as *Star Wars*. Amid all the debate two facts seem incontrovertable. First, enormous technological advances will be required if the project is to be feasible at all.[40] Second, extraordinarily complicated software will be necessary to control the sensors, guidance systems, weapons, etc. This software can never be fully tested before it is used to implement a (counter?) attack. If the computers function successfully, the system will attack only in case an enemy launches an attack against the United States. The computers will have to make the decision to launch a (counter?) attack, for there will not be time for human decision makers to intervene. What degree of confidence should we have regarding the ability of writers to produce software that is almost unimaginably complex and that will function correctly *the first time*? — Without any errors that might launch an attack against a nation that had not initiated hostilities against the United States? — Without any errors that would completely foul up the operation of the system in case of actual war, or render it completely non-operational? We all have enough experience with computers — often in use by government agencies — to make a pretty reasonable guess at *that* probability!

The moral of the consideration of these 'intermediate' cases is simple. It is the point of Feynman's criticism of NASA's risk assessment. Do not ignore the available relevant information about frequencies. *Respect the frequencies.*

Let us now turn to the kind of case in which the presumption of the availability of relevant frequency information is much less plausible. Examples from history are often cited to illustrate this situation. Because of the complexity of actual historical cases, I cannot present a detailed analysis, but I hope to be able to offer some suggestive remarks.

One of the most significant occurrences of the present century was the dropping of the atomic bomb on Hiroshima. The background events leading up to the development of the atomic bomb are familiar: the discovery of nuclear fission and the possibility of a self-sustaining chain reaction, the realization that the Nazis were working on a similar project, the fear that they would succeed and — with its aid — conquer the world, and the urging by important scientists of the development of such a weapon. Before the bomb was completed, however, Germany had surrendered, and many scientists felt that the project should halt, or, if a bomb was created, it should not be put to military use. Given this situation in the spring of 1945, what were the probabilistic causes that led to the dropping of the bomb on a major Japanese city? What probabilistic causes led to rejection of the proposal by many scientists to demonstrate its power to the Japanese, and perhaps other nations, instead of using it on a civilian population?

This is obviously an extremely complicated matter, but a few factors can be mentioned. One event of great significance was the death of Franklin D. Roosevelt, and the elevation of Harry S. Truman to the presidency; this seems clearly to have raised the probability of military use. Another contributing factor was the deep racism in America during World War II, making the lives of oriental civilians seem less valuable than those of caucasians. Still another contributing factor was the failure of scientists to anticipate the devastating effects of radiation, and hence, to view the atomic bomb as just another, more powerful, conventional weapon. An additional complicating feature of the situation was the great failure of communication between scientists and politicians, and between scientists and military people.

Some scientists had suggested inviting Japanese observers to Trinity, the test explosion in Nevada, but others feared the consequences if the first bomb turned out to be a dud. Some had suggested erecting some buildings, or even constructing a model city, to demonstrate the power of the bomb, but various objections were raised. After Trinity, some scientists suggested exploding an atomic bomb over the top of Mount Fujiyama, but some of the military argued that only destruction of a city would provide an adequate demonstration.[41] The personalities of key individuals had crucial import.

As one considers a complex historical situation, such as this, it is natural, I believe, to look at the multitude of relevant items as factors that tend to produce or inhibit the outcome in question. These tendencies are propensities — contributing or counteracting probabilistic causes — and we attempt to assess their strengths. In so doing we are relying on a great deal of experience with scientific, political, economic, social, and scientific endeavors. We bring to bear an enormous amount of experience pertaining to the human interactions and the effects of various traits of personality. My claim is that, in assigning the personal probabilities that would have been appropriate in a given historical situation, we are estimating or guessing at the strengths of and interactions among probabilistic causes.

In writing about scientific explanation during a period of nearly 25 years, Hempel has repeatedly addressed the special problems of historical explanation.[42] He has dealt with a number of concrete examples, in an effort to show how such explanations can be understood as scientific explanations conforming to his well-known models of explanation. I would suggest that they can easily be read in terms of attributions of probabilistic causes to complex historical occurrences.

I certainly do not pretend to have numerical values to assign to the propensities involved in the dropping of the bomb on Hiroshima, given one set of conditions or another — or to any other significant historical event — and I do not think anyone else does either. But it does seem reasonable to regard each of the factors we have cited — and others, perhaps — as probabilistic causes that worked together to bring about a certain result. I would suggest that experts who are familiar with the complexities of the situation can make reasonable qualitative assessments of the strengths of the various factors, classifying them as strong, moderate, weak, or negligible. In addition, they can make assessments of the ways in which the various probabilistic causes interact, reinforcing one another or tending to cancel one another out. It is considerations such as these that can be brought to bear in arriving at personal probabilities with respect to the unique events of human history.

6. CAUSALITY, FREQUENCY, AND DEGREE OF CONVICTION

If one were to think of causality solely in terms of constant conjunction, then it would be natural to identify probabilistic causality with relative frequency in some fashion or other. As I tried to show in *Scientific Explanation and the Causal Structure of the World*, a far more robust account of causality can be provided in terms of *causal processes* and *causal interactions*.[43] This account is intended to apply to probabilistic causal relations as well as to causal relations that can be analyzed in terms of necessary causes, sufficient causes, or any combination of the them. It is also intended to give meaning to the notions of *production* and *propagation*.

My aim in the present paper is to bring these causal considerations to bear on the problem of the relationship between objective and subjective probabilities, and to relate them to rationality. As we noted above, Ramsey stressed the crucial role of mathematical expectations of utility in rational decision-making. For purposes of this discussion, let us make the grossly false assumption that we know the utilities that attach to the various possible outcomes of the actions we contemplate.[44] What we would *really* like to have, given that knowledge, is the ability to predict accurately the outcome of every choice we make, but we realize that this is impossible. Given that fact, what we would *really* like to have is knowledge of the actual frequencies with which the outcomes will occur. That would also be nice, but it, too, is impossible. Given that fact, the next best thing, I suggest, would be to know the strengths of the probabilistic causes that produce the various possible outcomes. They are the agencies that produce the actual frequencies. I think of them as actual physical tendencies, and have called them *propensities*. It is the operations of physical devices having these propensities — chance set-ups, including our own actions — that produce the actual short-run frequencies, on which our fortunes depend, as well as the long-run frequencies which I am calling *probabilities*.[45]

The best estimate of the actual short-run frequency is, I would argue, the possible value closest to the propensity. Recall Peirce's example. If the agent draws from the deck with the preponderance of red cards, that act has a propensity of degree 25/26 to yield a red card. The actual frequency of red in a class containing only one member must be either one or zero; one is obviously closer to the propensity than is zero. In the other deck, the propensity for red is 1/26, and this value is closer to zero than it is to one. Since the agent wants a red card, he or she chooses to draw from the deck whose propensity for red is closer to the desired short run frequency.

The best way to look at an important subset of subjective or personal probabilities is, I think, to consider them as estimates of the strengths of probabilistic causes. In many cases, several probabilistic causes may be present, and the propensity of the event to occur is compounded out of them. Smoking, exercise, diet, body weight, and stress are, for example, contributing or counteracting causes of heart disease. In such cases we must estimate, not only the strengths of the several causes, but also the interactions among them. Obviously, two causes may operate synergistically to enhance the potency of one another, they may tend to cancel each other out, or they may operate independently. Our assessments of the strengths of the probabilistic causes must take into account their interactions as well as the strengths of the component causes.

The obvious problem that must now be faced is how we can justifiably estimate, guess, infer, or posit the strengths of the propensities. When we attempt to assign a propensity to a given chance set-up, we are dealing with a causal hypothesis, for a propensity is a probabilistic cause. Since Bayes's theorem is, in my view, the appropriate schema for the evaluation of scientific (including causal) hypotheses, I should like to offer a brief sketch of a Bayesian account that appeals to propensities.

There is a widely held maxim, which I regard as correct, to the effect that the meaningful collection of scientific data can occur only in the presence of one or more hypotheses upon which the data are supposed to have an evidential bearing. It is therefore important, I believe, to take a brief excursion into the context of invention (discovery) in order to consider the generation of hypotheses about propensities — i.e., about probabilistic causes.[46] It is a rather complex matter. In the first place, we have to identify the chance set-up and decide that it is an entity, or set of entities, worthy of our interest. This is analogous to selecting a reference class as a basis for an objective probability — i.e., a long run relative frequency. We must also identify the outcome (or set of outcomes) with which we are to be concerned. This is analogous to the selection of an attribute class (or sample space) for an objective probability relationship. Without these, we would have no hypothetical probabilistic causal relation to which a value of a propensity could be assigned.

But, as I emphasized above, not all objective probabilities are propensities. Inverse probabilities are not propensities.[47] Many correlations do not stand for propensities. It would be a joke to say that a barometer is a chance set-up that exhibits a high propensity for storms whenever its reading drops sharply. It would not necessarily be a joke to say that the atmosphere in a particular locale is a chance set-up with a strong propensity for storms when it experiences a sharp drop in pressure. Thus, whenever we hypothesize that a propensity exists, we are involved in a rather strong causal commitment.

When we have identified the chance set-up and the outcomes of interest, and we hypothesize a probabilistic causal relation between them, we need to make hypotheses about the strength of the propensity. The idea I want to suggest is, roughly, that an observed frequency fulfills one essential function relating to the formulation of a hypothesis about the propensity. Actually, we need more than just a single hypothesized value of the propensity; we need a prior probability distribution over the full range of possible values of the propensity, i.e., the closed unit interval. My proposal is that, in the absence of additional theoretical knowledge about the chance set-up, the observed frequency determines the value of the propensity for which the prior probability is maximal, namely, the value of the observed frequency itself.[48]

There is an additional assumption that is often made by propensity theorists. A chance set-up is defined as a mechanism that can operate repeatedly, such that, on each trial, it has the same propensity to produce the given outcome. This means that the separate instances of its operation are independent of one another; for example, the outcome in any given case does not depend probabilistically upon the outcome of the preceeding trials. This constitutes a strong factual assumption. Another, closely related, assumption has to do with the randomness of the sequence of outcomes. There is a strong temptation to make these assumptions. If we do, we can use the Bernoulli theorem to calculate the probabilities of various frequency distribution, given various values for the associated propensity.

Imagine that we have a black box with a button and two lights — one red, the other green — on the outside. We notice that, when the button is pressed, either the red or the green comes on; we have never seen both of them light up simultaneously, nor have we seen a case in which neither comes on when the button has been pushed. However, we have no *a priori* reason to believe that either of the latter two cases is impossible. This is a generic chance set-up. Suppose that we have observed 10 cases, in 6 of which the green light has been illuminated. We now have a prior distribution for green, peaking at 0.6; a prior distribution for red, peaking at 0.4; and prior distributions for both lights and for neither light, each peaking at 0. So far, I would suggest, it is all context of invention (discovery). At the same time, and in the same context, we may make hypotheses — of the sort mentioned above — about such matters as the independence of the outcomes or the randomness of the sequence. All such hypotheses can be tested by additional observational evidence. We can check such questions as whether the red light goes on every time the number of the trial is a prime number,[49] whether green goes on whenever (but not only whenever) there have been two reds in a row, or whether red occurs on every seventh trial. Answers to all of these questions depend upon the frequencies we observe.

Not all philosophers agree that there is a viable distinction between the context of invention (discovery) and the context of appraisal (justification). But those of us who do have emphasized the psychological aspects of the former context. Such psychological factors would, I believe, have considerable influence in shaping the distribution of prior probabilities — e.g., how flat it is or how sharply peaked at the value of the observed frequency. Prior probabilities represent plausibility judgments; the prior distribution exhibits the plausibilities we assign to the various possible values of the propensity.

The fact that some aspects of the context of invention (discovery) are psychological does not prevent objective features of the situation from entering as well. In particular, I am suggesting, observed frequencies —

the paradigm of objective fact — play a decisive role. In this connection, I have long maintained a view regarding the confirmation of scientific hypotheses that might suitably be called *objective Bayesianism*.[50] One key feature of this approach to confirmation is that the prior probabilities of hypotheses are to be construed as frequencies. A plausibility judgment should identify a hypothesis as one of a given type, and it should estimate the frequency with which hypotheses of that type have succeeded. I shall not rehearse the arguments here, but we should recall, as noted above, that Shimony supports his treatment of prior probabilities in his *tempered personalism* by an explicit appeal to frequency considerations. This is another way in which, I believe, the frequencies should be respected as an aspect of a fully objective Bayesian account.

7. CONCLUSION

Ramsey's conception of rationality accords a central role to mathematical expectations of utility. The probabilities occurring in the expression for the expectation are, in Ramsey's terms, degrees of partial belief. His rationale for this approach brings into sharp focus the actual short-run frequencies that determine the net outcomes of our choices. It seems clear that Ramsey regarded his subjective probabilities as estimates of (short- or long-run) frequencies. For this reason I regard his view as a *frequency-driven* conception of rationality.

Dynamic rationality, as I would define it, consists in the attempt to use propensities — i.e., probabilistic causes — as the weighting factors that occur in the formula for expected utility. Since we cannot be sure that our choices and decisions will be fully efficacious in bringing about desired results, it is reasonable to rely on the strengths of probabilistic causes. This line of thought treats our voluntary choices, decisions, and actions as probabilistic causes of what happens as a result of our deliberations. Dynamic rationality involves a *propensity-driven* view of objective probabilities and short-run frequencies.

Because the values of propensities are not known with certainty, we have to make do with the best estimates we can obtain. I have been urging that some personal probabilities are, precisely, best estimates of this sort. There are, however, physical probabilities that cannot be identified with propensities; consequency, I believe, there are personal probabilities that cannot straightforwardly be construed as estimates of propensities. These personal probabilities may be taken as estimates of frequencies. Where we do not have access to the pertinent propensities, we do best by using estimates of frequencies. Since, however, I take it that frequencies are generated by propensities, my approach involves a *propensity-driven* account of degree of conviction.

In dealing with universal laws in the sciences we are used to the idea that some laws — e.g., the ideal gas law — provide empirical regularities without furnishing any causal underpinning. We look to a deeper theory — the kinetic-molecular theory — for a causal explanation of that regularity. In the domain of statistical regularities a similar distinction can be made. There are statistical laws that express statistical regularities without providing any kind of causal explanation. Such statistical regularities are physical probabilities (long-run frequencies). In some cases a deeper statistical theory exists that embodies the propensities that constitute the causal underpinning. In both the universal and the statistical cases, however, the unexplained empirical regularities can serve as a basis for prediction and decision-making.

A major thesis of this paper is that observed frequencies often provide the best evidence we have concerning long-run frequencies and the strengths of probabilistic causes. Sometimes our evidence is of a less direct and more theoretical variety. At the most primitive level, I believe, observed frequencies constitute the evidential base.

Induction by enumeration has traditionally been regarded as a rule of primitive induction for *justifying* inferences to values of physical probabilities (long-run frequencies). Whether it may be a suitable rule for this aim is an issue I shall not go into here.[51] Since my main concern in this paper has been with propensities and probabilistic causes, it has *not* been at the level of primitive induction. Causal attributions are more complicated than mere statistical generalization. For purposes of the present discussion, I have suggested that we consider induction by enumeration instead as a method — in the context of invention (discovery) — that makes a crucial contribution to the generation of hypotheses. Used in this way, it provides, I believe, a concrete method for *respecting the frequencies*. It thereby serves as a counterpart to David Lewis's *principal principle* — providing the link between objective and subjective probabilities, but going *from* the objective (observed frequency) *to* the subjective (personal probabilities).

Observed frequencies furnish knowledge of physical probabilities, and propensities. They also constitute the basis for personal probabilities — both those that correspond to propensities and those that do not. These personal probabilities, in turn, provide the weights that should figure in our calculations of mathematical expectations. The mathematical expectations that result from this process then constitute our *very guide of life.*

University of Pittsburgh

NOTES

[1] Peter Wyden, *Day One* (New York: Simon and Schuster, 1984), p. 50.

[2] *Ibid.*, pp. 50—51. Wyden mentions this incident, but without throwing any light on the questions I am raising.

[3] Eliot Marshall, "Feynman Issues His Own Shuttle Report, Attacking NASA's Risk Estimates," *Science,* **232** (27 June 1986), p. 1596. All of the following quotations from this article appear on the same page.

[4] [If the probability of failure of a given rocket is 0.01, the probability of at least one failure in 50 firings is 0.395; the probability of at least one failure in 100 firings is 0.634 W.C.S.].

[5] [Marshall offers the following interpretation of that figure: "It means NASA thinks it could launch the shuttle, as is, every day for the next 280 years and expect not one equipment-based disaster." One could hope that NASA statisticians would not follow Marshall in making such an inference. I presume that the number 280 was chosen because that many years contain a little over 100 000 days — 102 268 to be precise (allowing for leap years). However, it must be recalled, as was pointed out in the article, that each launch involves 2 rockets, so the number of rocket firings in that period of time would be 204 536. If the probability of failure on any given rocket is 1 in 100 000, the probability of at least one failure in that many firings is about 0.870. In one century of daily launches, with that probability of failure, there is about a 50—50 chance of at least one failure. This latter estimate is, of course, absurd, but to show the absurdity of the NASA estimate we should try to get the arithmetic right.]

[6] A recent discussion of this issue can be found in Mark J. Machina, "Decision-Making in the Presence of Risk," *Science* **236** (1 May 1987), pp. 537—543.

[7] Patrick Maher, "The Irrelevance of Belief to Practical Action", *Erkenntnis* **24** (1986), pp. 263—284.

[8] Rudolf Carnap and David Lewis, among others, use the *credence* terminology.

[9] I realize that a great deal of important work has been done on the handling of inconsistent systems, so that, for example, the presence of an inconsistency in a large body of data does not necessarily bring a scientific investigation to a grinding halt. See, for example, Nicholas Rescher and Robert Brandom *The Logic of Inconsistency* (Oxford: Basil Blackwell, 1980). However, inasmuch as our main concern is with probabilistic coherence, it will not be necessary to go into that issue in detail.

[10] A molecular statement is one that involves no quantifiers. The negation of a molecular statement is obviously also molecular. As the term "molecular" is being used

here, basic statements (statements without quantifiers or binary connectives) are considered molecular.

[11] Abner Shimony, "Scientific Inference," in Robert G. Colodny, ed., *The Nature and Function of Scientific Theories* (Pittsburgh: University of Pittsburgh Press, 1970), pp. 79—172.

[12] When the modification is drastic enough, Thomas Kuhn calls it a scientific revolution.

[13] Another way of handling these betting considerations — one that I would prefer — is to remember that two factors bear upon our bets. The first is the degree of probability of the event that we are betting on; the second is our degree of assurance that that probability is objectively correct. In his theory of probability$_1$ Carnap, in effect, eliminated this second consideration by making all such probability statements either logically true or logically false. This meant that the caution factor had to be built into the value of the probability; strict coherence does the job.

[14] Ian Hacking, *The Emergence of Probability* (Cambridge: Cambridge University Press, 1975). Hacking argues that this could not have occurred before the 17th century. I am not at all sure that his historical thesis about the time at which the concept of probability emerged is correct, but his conceptual analysis strikes me as sound.

[15] Rudolf Carnap, "The Aim of Inductive Logic," in Ernest Nagel, Patrick Suppes, and Alfred Tarski, *Logic, Methodology, and Philosophy of Science* (Stanford: Stanford University Press. 1962).

[16] Rudolf Carnap, "Replies and Systematic Expositions," in Paul Arthur Schilpp, ed., *The Philosophy of Rudolf Carnap* (La Salle, Ill.: Open Court Publishing Co., 1963), pp. 859—1018.

[17] Carnap continued work on the project of constructing an adequate inductive logic until his death in 1970. His later work on the subject is published in *Studies in Inductive Logic and Probability*, vols. I—II (Berkeley, Los Angeles, London: University of California Press, 1971, 1980). The first volume is edited by Rudolf Carnap and Richard C. Jeffrey. Jeffrey is the sole editor of volume II.

[18] See Wesley C. Salmon, "Propensities: A Discussion-Review", *Erkenntnis* **14** (1979), pp. 183—216, for a fuller discussion of this topic.

[19] Hans Reichenbach, *The Theory of Probability* (Berkeley & Los Angeles: University of California Press, 1949), §18.

[20] Frank Plumpton Ramsey, *The Foundations of Mathematics*, edited by R. B. Braithwaite (New York: Humanities Press, 1950), pp. 166—184. In order to discuss quotations from Ramsey's article I shall revert to his terminology and refer to *partial beliefs* and *degrees of belief*.

[21] [Ramsey clearly means that the subject behaves so as to maximize his or her expected utility. W.C.S.]

[22] *Ibid.*, p. 174.

[23] D. H. Mellor, *The Matter of Chance* (Cambridge: Cambridge University Press, 1971). In discussing Mellor's views I shall continue to use the traditional "partial belief" and "degree of belief" terminology. I have treated Mellor's book in considerable detail in Wesley C. Salmon, "Propensities: A Discussion Review," *op. cit.* This article also contains a general discussion of propensity theory with special emphasis on the work of Karl Popper, its originator.

[24] David Lewis, "A Subjectivist's Guide to Objective Chance," in Richard C. Jeffrey,

ed., *Studies in Inductive Logic and Probabaility*, vol. II (Berkeley, Los Angeles, London: University of California Press, 1980), pp. 263—293. In discussing Lewis's views, I shall use his terminology of "partial belief" and "credence."

[25] *Ibid.*, p. 263.

[26] *Ibid.*, p. 264, Lewis's italics.

[27] *Ibid.*, p. 266. In explaining this principle Lewis makes excursions into nonstandard analysis and possible worlds semantics. I do not think either of these tools is required for our discussion.

[28] Wesley C. Salmon, *Scientific Explanation and the Causal Structure of the World* (Princeton, N.J.: Princeton University Press, 1984).

[29] James H. Fetzer, "Dispositional Probabilities", in R. Buck and R. Cohen, eds., *PSA 1970* (Dordrecht: D. Reidel Publishing Company, 1971), pp. 473—482; and *Scientific Knowledge* (Dordrecht: D. Reidel Publishing Company, 1981).

[30] Lewis, *op. cit.*, p. 266, my italics.

[31] *Ibid.*, p. 267.

[32] *Ibid.*, p. 268.

[33] Someone who buys a ticket each day for a year pays $365 for a set of tickets whose expectation value is $182.50. This is poor investment indeed — especially for people at or near the poverty level — but regrettably it is not an uncommon occurrence.

[34] Fortunately, this particular incident was discovered, and the perpetrators were brought to trial and punished. Presumably this sort of thing is not going on any more.

[35] I have attempted to characterize this rather complicated concept in *Scientific Explanation and the Causal Structure of the World* (Princeton: Princeton University Press, 1984), chapter 3.

[36] We can think of different roulette wheels as different games. The same goes for slot machines, black jack tables, etc. It is not necessary that every game have a probability distribution different from all others.

[37] Hans Reichenbach, *The Theory of Probability* (Berkeley & Los Angeles: University of California Press, 1949), §56, 72.

[38] Charles Sanders Peirce, *Collected Papers*, edited by Charles Hartshorne and Paul Weiss (Cambridge, MA: Harvard University Press, 1931), vol. II, §2.652.

[39] Wesley C. Salmon, "The Short Run," *Philosophy of Science* **22** (July, 1955), pp. 214—221.

[40] This assertion is based upon a report sponsored by the American Physical Society, reported in *Science News* **131** (May 2, 1987), p. 276. The report refers to the need for "improvements of several orders of magnitude."

[41] My primary source for these remarks is Wyden, *op cit.*

[42] I refer primarily to "The Function of General Laws in History," *Journal of Philosophy* **39** (1942), pp. 35—48, reprinted in Hempel, *Aspects of Scientific Explanation*; "Explanation in Science and in History," in Robert G. Colodny, ed., *Frontiers of Science and Philosophy* (Pittsburgh: University of Pittsburgh Press, 1962), pp. 7—34; "Aspects of Scientific Explanation," §7—10, in *Aspects of Scientific Explanation*, pp. 447—487.

[43] See chapters 5—6.

[44] The unrealistic nature of this assumption is nicely conveyed by the old saying. "When the gods want to punish us they grant us our prayers."

[45] It may be that some probabilities are not generated by propensities as I am

construing them. Ian Hacking discusses this issue in his paper, "Grounding Probabilities from Below," in P. Asquith and R. Giere, eds., *PSA 1980* (East Lansing, Mich.: Philosophy of Science Assn., 1982), pp. 110—116, and offers an interesting actual example in which this appears to be the case.

[46] I am adopting the felicitous terminology, "context of invention" and "context of appraisal," proposed by Robert McLaughlin in "Invention and Appraisal," in Robert McLaughlin, ed., *What? Where? When? Why?* (Dordrecht: D. Reidel Publishing Co., 1982), pp. 69—100, as a substitute for the traditional "context of discovery" and "context of justification."

[47] See p. 14 above.

[48] This approach bears some strong resemblances to a suggestion offered by Ian Hacking in §9 of "One Problem About Induction," in Imre Lakatos, ed., *The Problem of Inductive Logic* (Amsterdam: North-Holland Publishing Co., 1968), pp. 52—54. It has taken me almost twenty years to appreciate the value of this suggestion.

[49] Supposing that red was illuminated on the second, third, fifth, and seventh trials.

[50] See, for example, Wesley C. Salmon, *The Foundations of Scientific Inference* (Pittsburgh: University of Pittsburgh Press, 1967), chap. VII.

[51] The considerations advanced by Hacking — see note 48 — make me suspect that it is not.

PART I

PROBABILITY, CAUSALITY, AND MODALITY

WILLIAM EDWARD MORRIS

HUME'S REFUTATION OF INDUCTIVE PROBABILISM*

I

Wesley Salmon once argued that an important piece of "unfinished business" for contemporary philosophers is the task of responding successfully to the challenge Hume's statement of the problem of induction poses.[1] I agree. But there is a prior piece of business we need to attend to first. We need to figure out what Hume's argument actually is.

This may seem ridiculous. Hume's argument is the most familiar piece of philosophy written in English.[2] Practically everyone who has any exposure to philosophy reads at least part of it.

But despite its currency — perhaps *because* of it — there is no consensus about how Hume's argument should be read. Its structure and content are widely debated. Opinions differ wildly on what Hume is saying. There are nearly as many Humes as there are philosophers who have written on his famous argument.

Hume himself is often blamed for this situation. John Passmore is hardly alone in complaining that Hume is

> one of the most exasperating of philosophers. Each separate sentence in his writings — with very few exceptions — is admirable in its lucidity: the tangled syntax and barbarous locutions which bedevil the reader of Kant and Hegel are completely absent. And yet, although in a different way, Hume is at least as difficult as Hegel. In his editorial introduction to the *Enquiries*, Selby-Bigge summed up the Hume problem thus: "He says so many things in so many different ways and different connections, and with so much indifference to what he has said before, that it is very hard to say positively that he taught or did not teach this or that particular doctrine. . . . This makes it easy to find all philosophies in Hume, or, by setting up one statement against another, none at all."[3]

This charge of "local lucidity and global obscurity"[4] is not only wrong, it is also grossly unfair. Though Hume's argument is more involved than his interpreters generally appreciate, it is neither obscure nor convoluted. And it is certainly not inconsistent. The misreading of Hume is due more to mistaken presuppositions brought to his arguments by

James H. Fetzer (ed.) Probability and Causality. 43—77.

would-be interpreters than to whatever unclarity resides in the arguments themselves.

Hume first states what we now call "the problem of induction" in Book I, Part III, Section vi of the *Treatise*.[5] That argument is summarized in his anonymous *Abstract* of the *Treatise*.[6] He restates it in compressed form, with more attention to structure, in Section IV of the first *Enquiry* — a section appropriately titled "Sceptical Doubts Concerning the Operations of the Understanding."[7]

Hume's conclusion in these passages is entirely negative. It is that our causal expectations are not based on any form of reasoning: "even after we have experience of the operations of cause and effect, our conclusions from that experience are *not* founded on reasoning, or any process of the understanding." (*Enquiry*, p. 32. italics Hume's).

There is a prominent line of argument, however, due to J. L. Mackie[8] and D. C. Stove,[9] which denies that Hume's conclusion really has the force it appears to have. Hume's argument, they maintain, depends entirely on his "deductivism" — the view that "an argument gives no rational support for its conclusion if the inference to that conclusion is not deductively valid."[10] Hume's deductivism led him to ignore the possibility that there could be "reasonable but probabilistic" arguments. So "inductive probabilism," the view that there *are* probable inductive arguments, is unaffected by Hume's argument. Consequently, the common belief that Hume refuted inductive probabilism is, as Stove puts it, "an entirely imaginary episode in the history of thought." (Chappell, 1966, p. 189).

If Mackie and Stove were correct, their claims would take much of the sting out of the "sceptical doubts" Hume raises in these famous passages. But they are wrong. Hume was not a deductivist and his case against inductive probabilism is not an "imaginary episode." This essay sets the record straight. I look carefully and critically at the substance and structure of the Mackie-Stove interpretation, and argue that it fails to capture much of what is important in Hume's argument. I then provide a corrected account of what Hume is really saying in these passages, to show that he not only considers, but refutes inductive probabilism.

II. THE MACKIE-STOVE INTERPRETATION

Though Mackie and Stove developed their readings of Hume independently, there are many significant points of contact between them. The most striking similarity, of course, is their mutual use of "structure diagrams" to represent the dependency relations they find in Hume's argument.[11] More significant philosophically, however, is Mackie's wholesale absorption of Stove's account into his own, where he uses it as an important lemma in his rendition of Hume's argument. Because of this, their separate readings are frequently considered as though they constituted a single, unified account. I treat them this way in my discussion.

I begin with Mackie's rendition of Hume's argument. Where Stove's account becomes relevant for Mackie's, I discuss it in detail, both on its own and in its role as an integral part of Mackie's interpretation.

Mackie develops his interpretation of Hume in Chapter One of *The Cement of the Universe.* His main purpose there is to assess Hume's contribution to our understanding of the concept of causation. Consequently, his account goes beyond the purely negative phase of Hume's argument. Since my sole concern here is with that part of the argument, I consider only those portions of Mackie's reading which bear directly on Hume's statement of his "sceptical doubts."

Mackie's account of the argument is outlined in the diagram given in Figure 1. The diagram reads from left to right, with the arrows indicating lines of support. According to the diagram, then, (A) should be the central conclusion of the argument. But Mackie actually takes "the main conclusion" of Hume's argument to be this claim about necessity:

(B) Necessity is in the mind, not in objects, though we tend to project it onto the objects

Hume's conclusion about causation

(A) Causation in the objects, so far as we know, is only regular succession

is regarded by Mackie as merely "a corollary" to (B).

Strictly speaking, both (A) and (B) go beyond the scope of Hume's negative argument. But it is important to look briefly at Mackie's disucssion of them. What he says about these propositions reveals much about what is structurally amiss in his account of Hume's argument.

In the first place, Hume never asserts (A). The closest he comes is something considerably weaker. Hume never claims anything about what causation in the objects really is. Nor does he make weaker claims

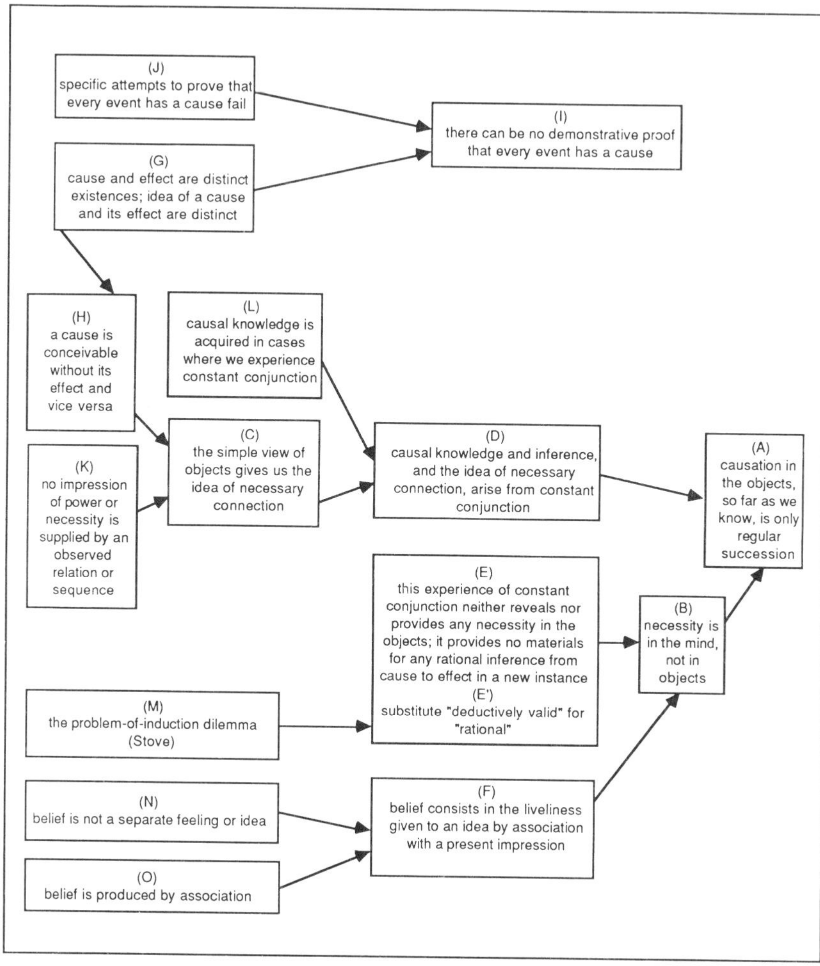

Fig. 1. J. L. Mackie: Structure diagram of Hume's account of causation.

about the objects "as we know them." That is something which we not only do not know *now*, it is something we can *never* know.[12] We do make claims about objects, or at least speak as if we could move from our ideas of objects to claims (however incomplete) about the objects themselves. But Hume is insistent that we should never confuse this tendency of the mind to "spread itself out on objects" with anything we actually know about them. He makes this clear later in the *Treatise*, criticising the "*philosophical* hypothesis" that we can move from our perceptions to the objects themselves. He clearly has Locke[13] in mind here, and probably Descartes[14] as well:

> The only existences, of which we are certain, are perceptions. . . . The only conclusion we can draw from the existence of one thing to that of another, is by means of the relation of cause and effect. . . . The idea of this relation is deriv'd from past experience, by which we find, that two beings are constantly conjoin'd together, and are always present at once to the mind. But as no beings are ever present to the mind but perceptions; it follows that we may observe a conjunction or a relation of cause and effect between different perceptions, but can never observe it between perceptions and objects. 'Tis impossible, therefore, that from the existence or any of the qualities of the former, we can ever form any conclusion concerning the existence of the latter . . . . (*Treatise*, p. 212)

Hume's only conclusion is, appropriately, about *our idea* of causation. Tracing that idea back to the external impressions which give rise to it yields only constant conjunction, it is true, but this point is quite different from the strong claim that is made in (A). Hume is very carefully, and given his epistemology, rightly agnostic about what causation in the objects actually consists in.

In addition, (A) — or a suitably corrected version of it — is not the main conclusion of Hume's argument. Something like (B) is. This, however, does not confirm Mackie's view that (A) is just a corollary to (B). (A) should be support for (B), not the other way round.[15]

Nor does Hume make the claim Mackie attributes to him in (B). While our *idea* of necessity is just that of constant conjunction plus the transition of the mind that custom or habit produces, Hume is careful to remain agnostic about whether or whatever necessity may be in the objects. Not only are the "secret natures" of objects entirely unknown to us, but whatever connections or "secret powers" they have or do not have with other objects is unknown — and unknowable — as well.[16]

Mackie was probably led to these misstatements about Hume's views through his concern with the topic of causation. Since Hume's work is the basis for so much current thinking on this topic, it is tempting — perhaps unconsciously — to recast his claims so that they appear to speak directly to current concerns about that concept. Unfortunately, Hume's *actual* concerns and claims do not always fit comfortably into the mold the modern metamorphosis of "Hume's problem" has constructed for them. This lack of fit is painfully apparent here.

III

Mackie's mistakes in stating Hume's conclusions should make us suspicious that he has also misconstrued Hume's argument in the sections that concern us most. This suspicion is borne out by a close look at the rest of Mackie's interpretation of Hume.

According to Mackie, (B) is supported by "three lines of thought." He treats these lines differently at different times. Sometimes he speaks as though they were three separate strains of argument, each producing its own largely independent support for (B). At other times, however, he characterizes them as converging parts of a unified argument which yields (B).

The conclusions of these three lines of support are represented in Mackie's diagram by (D), (E), and (F).

What Mackie distinguishes as Hume's "third line of thought" consists of propositions (N) and (O), which together support (F). They are concerned with what Mackie calls Hume's "psychological theory of belief." I question whether Mackie accurately reproduces Hume's argument concerning the nature of belief and its relevance for his account of causation. But since this sub-argument concerns the positive aspect of Hume's theory, I do not discuss it further here.

The other two lines of argument converging on (B), however, are directly concerned with Hume's negative argument.

"Hume's second line of thought" is the chain leading from (M) to (E). Mackie calls (M) "the problem-of-induction dilemma." He intends that phrase to summarize the argument he takes over from Stove's work.

As Mackie reads it, the dilemma is based on the claim that reasoning concerning experience must depend upon the principle that the future must resemble the past. Such a principle could either be supported by demonstrative or probabilistic reasoning. Neither works. Demonstrative reasoning cannot support it since its falsity is conceivable. And the attempt to support it by probabilistic reasoning fails due to circularity. (Mackie, 1974, p. 9)

Mackie rather oddly comments that while "the problem-of-induction dilemma has tremendous relevance for other parts of philosophy, it is only marginally relevant to our topic of causation." (Mackie, 1974, p. 15) He may intend this puzzling remark as a partial justification for his failure to discuss this part of Hume's argument. But while one might justify separating these issues in other contexts, it is difficult to see how

Mackie could think it possible to separate *Hume's* discussion of the problem of inductive reasoning from his discussion of causal relations.

In any case, Mackie omits any extended discussion of the dilemma, claiming that he "need not analyze it because it has been very thoroughly analyzed by D. C. Stove whose conclusions I shall simply take over and apply." (Mackie, 1974, p. 9) He does, however, summarize — as well as endorse — Stove's account:

> Hume's premise that 'reason' would have to rely on the principle of uniformity holds only if it is assumed that reason's performances must all be deductively valid. . . .
>
> Reasonable but probabilistic inferences, then, have not been excluded by Hume's argument, for the simple reason that Hume did not consider this possibility. . . . the only reasonings that [the] second horn [of the dilemma] condemns as circular are deductively valid arguments in which the principle of uniformity is both a premise and a conclusion. (Mackie, 1974, pp. 15—16)

The fate of Mackie's argument, then, at least for this crucial "line of thought," rests entirely with Stove. So we need to look very carefully at Stove's rendition of this part of the argument.

IV. STOVE'S ARGUMENT

Mackie's summary of the "problem-of-induction dilemma" effectively isolates the central thesis of Stove's argument — that Hume was a "deductivist." According to Stove, Hume thought that the only reasonable arguments were deductively valid ones. But what Mackie's summary omits is Stove's crucial claim that Hume's alleged deductivism never explicitly surfaces in his statement of the argument. It is supposedly a suppressed premise of that argument instead.

This is graphically apparent in Stove's structure diagram given in Figure 2. The thesis of deductivism does not fit into the flow of Hume's argument, even on Stove's reconstruction. It must be added to the diagram for the conclusion to follow.

Stove's argument turns on the question whether imputing deductivism to Hume is legitimate. Is it really a suppressed premise in his argument? To answer this question, we need to look at Stove's case for this claim in some detail.

A controversial feature of Stove's account focuses on the proposition-pairs (J)—(J′) and (E)—(E′). According to Stove, (J′) and (E′) are "translations," into "modern philosophical terminology," of (J) and (E). But are they?

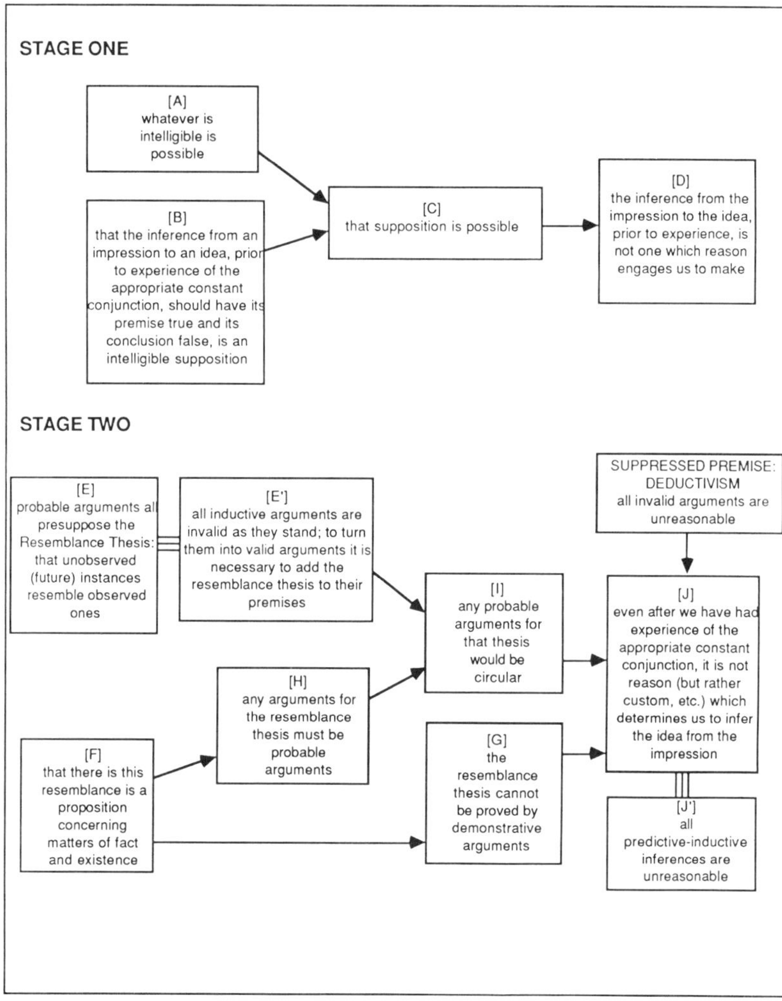

Fig. 2. D. C. Stove: Structure diagram of Hume's account of causation.

We just saw, in looking at Mackie's interpretation, that there are general problems with attempting to "help" a historical figure by trying to recast his arguments in "more precise" modern terminology. Stove seems aware of them. In considering the difficulty of deciding whether Hume's text contains any argument against inductive probabilism, he even says: "What difficulty there is in the case is due to what may be called the 'translation problem.' Hume's philosophical language is not ours, and the danger of taking his words in a sense he did not intend is acute in the present case."[17]

I could not agree more with the good sense of this passage. The problem is that Stove appears to have forgotten his own warning. His "translations" are not equivalent to their originals. More importantly, they are quite remote from anything it is legitimate to regard Hume as intending with his original propositions.

V

Stove's "translation" from (E) to (E′) is certainly suspect. While (E) seems to be a reasonable summary of an important thesis Hume actually held, it is difficult to see how (E′) can be regarded as in any way a *restatement* of that same thesis.[18] For (E′) appears to say much more than (E) does. (E′)contains substantive claims that: (i) inductive arguments which do not employ the resemblance thesis are invalid, and (ii) adding the resemblance thesis is necessary to make these arguments valid. (E) doesn't seem to say anything about either claim.

If we look closely at (E), however, we see that it contains a key term, "presuppose," which Hume himself never uses. Hume says that "all our experimental conclusions proceed upon the supposition that the future will be conformable to the past" (*Enquiry*, p. 35), that "all reasonings from experience are founded on the supposition, that the course of nature will continue uniformly the same" (*Abstract*, p. 651), and that "if reason determin'd us, it wou'd proceed upon that principle, *that instances, of which we have had no experience, must resemble those, of which we have had experience. . . .*" (*Treatise*, p. 89) These claims, and other similar statements of the same point, might be legitimately paraphrased using "presuppose," so long as that term retains its ordinary meaning viz. "to suppose beforehand," or even "to require as an antecedent in logic or fact."[19] So read, the claim that probable arguments "presuppose" the resemblance thesis is uncontroversial. It implies no particular *interpretation* of what Hume actually says.

This is not the way Stove reads "presuppose." He employs the term for a particular philosophical purpose, one not surprisingly tied to his own reading of the argument. He introduces his "silent translation" this way:

> It should be noticed, however, that even the version given above as the original contains one translation of sorts. This is the word 'presuppose', in premise (E) of Stage Two. Hume does not use this word. . . . I have used it because it is the obvious choice of one word which is to stand for a rather remarkable variety of phrases, all equally unclear, which Hume employs when he is stating his premise (E).[20]

This would be fine, if Stove left it at that. But he immediately goes on to consider various *philosophical* uses of "presuppose," in order to decide what *Hume's sense* is! Without irony, he eventually picks this one: "Sometimes when we say of an argument from *p* to *q*, that it presupposes *r*, our meaning is as follows: that, as it stands, the argument from *p* to *q* is not valid, and that, in order to turn it into a valid argument, it would be necessary to add to its premisses the proposition *r*."[21]

It is hardly surprising that Stove is able to regard (E′) as a "translation" of (E) once he reads "presuppose" this way. But this is only because the real problem lies in the tacit translation which produced (E) in the first place. This "translation" has no obvious basis in Hume's text. It depends on Stove's peculiar reading of that text, for which he has yet to argue.

In the passages with which Stove is concerned, Hume is considering what evidential support can be given for "our reasonings concerning cause and effect." Nothing he says arbitrarily limits the forms of support that can be given for "reasonings" of this sort. His intention is to cast his net as widely as he can, so that every possible kind of "reasoning" will be included in his critique.

This is in keeping with his primary objective, which is to raise "sceptical doubts about the operations of the understanding." While these "sceptical doubts" question whether legitimate support can be given for causal inferences, Hume's real purpose in raising them is to show that our causal expectations are not based on reason — or reason*ing* — at all. It does not help his case, given this strategy, simply to insist that our reasoning is *bad*, because deductively invalid, unless the resemblance thesis is added to our reasoning. For this leaves open the possibility — exploited by Stove — that Hume's argument either ignores, or unfairly eliminates, non-deductive forms of reasoning which we might use in such cases. This seems unlikely given that the real point of Hume's "sceptical doubts" is to show that we are not reasoning *at all* in such cases.[22]

VI

As a "translation," (J′) is perhaps worse. Even if were true that (J′) accurately represents something Hume holds (or even something that his view commits him to), it is hard to see why it should not be included separately, instead of being introduced as a "translation" of (J). For (J) does represent something Hume actually asserted, though whether it has just the place Stove assigns it in the argument's structure is another matter.

While it is perhaps true that, for Hume, *if* we reasoned about the future, *and if* in doing so, we used predictive-inductive arguments, then we would be *un*reasonable, as (J) says, in using such (bad) arguments. But Hume does not think that we are unreasonable in this way. That is not because we have some *better* form of argument which we actually employ in place of predictive-inductive arguments. For Hume, there is no such "better" argument. It is rather that we do not reason *at all* — well or ill — when we form causal expectations. And this is exactly what (J) says. Our habits of forming expectations of this kind, given the appropriate experiences, are thus better described as "nonreasonable," if anything.

So Stove's "translations" at best seriously distort Hume's argument. They also misconstrue his intentions and objectives in providing it.

VII

Irrespective of the question whether (J′) is an adequate translation of (J), *if* (J′) *were* a thesis Hume held independently, it does not follow from anything Stove offers as Hume's argument. It requires Stove's own addition of the deductivist premise to the argument. So *Stove* clearly needs deductivism for this argument, whether or not *Hume* did.

Stove argues, as we have seen, that deductivism is a missing or supressed premise of Hume's argument. Hume never stated it explicitly, even in contexts apart from the argument under consideration. Stove claims this despite the fact he goes to some lengths to applaud Hume for his straightforward presentation of this same argument:

If there is anything about this argument of Hume which is more admirable than its content, it is the explicitness of it. Almost always, the main obstacle to the evaluation of an argument, and often an insuperable obstacle, is the difficulty of identifying it — of finding out what the argument actually is. How seldom when men argue, in philosophy or elsewhere, can one confidently draw up the structure diagram . . . of their arguments! In the case of Hume's argument for predictive-inductive scepticism, however, one can do so. One can even, as we have seen, substitute for Hume's phraseology at the places where it could nowadays be misleading, other phrases which one can be confident express his meaning. (Stove, 1973, p. 46)

We have seen the disastrous consequences of Stove's failure to heed his own warning about the perils of "translating" the words of an historical figure when we looked at his seemingly innocent and innocuous attempt to "substitute for Hume's phraseology . . . other phrases which . . . express his meaning." Now we have Stove abandoning his rhetoric about the "explicitness" of Hume's argument to argue that being "admirably explicit" is not enough. For he continues:

Admirably explicit as it is, Hume's argument is yet not entirely explicit. Philosophers almost always . . . intend their own arguments to be valid ones, and in this respect Hume was no exception. . . . Yet it is obvious that his argument is not valid as it stands, either in stage one or stage two. Hume has supressed, as being too obvious to require expression, certain propositions which are nevertheless necessary for the validity of his argument. (Stove, 1973, p. 46)

The important premise that Stove regards as missing from Hume's otherwise "admirably explicit" argument is, of course, deductivism. But Stove provides no independent textual grounds for thinking that deductivism *is* a missing premise in this argument. His case rests on the point that the argument would not be valid unless deductivism were added to it. This would be a legitimate and serious charge, were it clear that *Hume's* argument required that premise for its validity. But as Stove mounts this charge, it is by no means clear that *Hume's* argument is the one under discussion. When we look at the argument he is actually discussing, it is his own "translation" of Hume's argument whose validity is being evaluated. This would also be legitimate *if* Stove had already shown that this reading of the argument is sufficiently warranted by the text. But as we have seen, his case for that interpretation depends on his supplying a convincing independent case for Hume's alleged deductivism. So Stove's argument fails on both counts. It begs the questions it was supposedly introduced to resolve.

VIII

There is one consequence of adding deductivism to the argument which should give one additional pause as to whether Stove has really captured Hume's position. Ian Hinkfuss noticed, in his perceptive and thorough review of Stove's book, that the deductivist premise, plus the first clause of (E′), yields Hume's sceptical conclusion — (J′) as Stove construes it — immediately. If this is correct, then Hume had no real need for introducing or discussing causation and the resemblance thesis in the way *he* seemed to think was essential for his argument (Hinkfuss, 1974, p. 274).

Stove, however, may not regard this as odd. While he never comments directly on the move Hinkfuss pointed out, he does at one point cite this truncated argument as "Hume's argument." He says: "Now, suppose it were suggested that Hume's argument, 'All inductive inferences are invalid, all invalid inferences are unreasonable, so all inductive inferences are unreasonable,' is open to the same objections." (Stove, 1973, p. 108)

What is odd about this is that it makes the rest of Hume's argument simply superfluous, if not irrelevant. How could an account which leaves out so much of apparent importance pretend to capture Hume's real thesis? It is, of course, *possible* that Hume unwittingly included much material in his presentation that is unnecessary for his argument. But then Stove should explain why this is so, if his account of the text is to be convincing.

IX

Even as it stands, Stove's argument is somewhat strange. It consists of two "stages" which are entirely separate; Stove never brings them together as two parts of an integrated argument for some particular conclusion.

Stove does say that the conclusion of the second stage "echoes" or "reiterates" the conclusion of the first stage. He regards this as an "extremely important feature" of his representation of Hume's argument. And it might seem to account for why he never brings the two separate stages together. But when Stove explains what he means by this, it is clear that something has gone wrong in his interpretation. He says:

> . . . the conclusion of the second stage of the argument "reiterates" or "echoes" that of the first stage. For it can already be seen that the conclusion (D), which in stage one

Hume drew concerning the *a priori* inference, is in its content the same as the conclusion (J) which in stage two he drew concerning the predictive-inductive inference. That is, it is not "reason" which "engages" or "determines" us to make inferences of that kind. What Hume means by saying this, will be determined later; but it is important to observe that it is what he concludes about both the *a priori* and the predictive-inductive inference alike. (Stove, 1973, p. 32)

One problem with Stove's remarks is that it is difficult to see why, if Stage Two truly "reiterates" or "echoes" that of Stage One, Hume should have thought it necessary to provide both parts of the argument. Stove seems unaware, as he was with the point discussed in the previous section, of any need to explain this. Without any explanation, however, it counts against the textual adequacy of his reconstruction of the argument.

But there is a more serious problem here. Stove calls attention to the fact that Stage Two "reiterates" that of Stage One because he thinks it makes it "possible to verify independently the necessity of ascribing this thesis [deductivism] to Hume." (Stove, 1973, p. 50) He offers this remarkable analysis of the argument in Stage One in support of his claim:

Here Hume draws, concerning the *a priori* inference, the same conclusion in (d) as he draws later in (j) concerning the predictive-inductive inference, viz. that it is unreasonable. But the only stated premisses of stage 1, viz. (a) and (b), clearly entail no more than this, that all *a priori* inferences are such that it is possible for them to have true premisses and false conclusions; that is, they are all invalid. Yet Hume certainly concluded that they are unreasonable. He must, therefore, have assumed in stage 1 of his argument, as well as in stage 2, though he did not state, that all invalid inferences are unreasonable. (Stove, 1973, p. 50).

This is incredible. Even ignoring Stove's gross overstatement — it is presumably a slip — that Hume's claim concerns *all a priori inferences*, the argument of Stage One cannot possibly provide evidence of Hume's commitment to deductivism in any form required by the argument of Stage Two. In Stage One Hume is exclusively and explicitly concerned with deductive forms of argument. If one tried to construct a demonstrative argument to support a causal belief, that argument could be shown to fail by the conceivability argument. The premisses of such an argument could be true and its conclusion false. Such an argument would indeed be unreasonable. But it is unreasonable because *it was presented as a deductive argument*, and failed the test — validity — that is a necessary condition for such an argument's being a good one. This at most commits Hume to the claim that *all invalid deductive arguments are unreasonable.* It says nothing at all about whether the only reasonable arguments are deductively valid ones. We cannot move from that claim to the stronger claim that *all deductively invalid arguments are unreasonable.* But Stove requires this stronger claim for his reconstruction of Hume's argument. He has, however, provided no reason for believing that Hume held any such thing. So Stove's case for Hume's alleged deductivism fails.

X

The failure of Stove's reconstruction has serious consequences for Mackie's rendering of Hume's argument as well. Mackie's acceptance of Stove's thesis that Hume was a deductivist lead him to modify his proposition (E), which the problem-of-induction dilemma was supposed to support, in a way that follows Stove's lead:

> Proposition (E), then, can be established only if we amend it, reading 'deductively valid' in place of 'rational,' and interpreting 'necessity' in such a way that the denial that constant conjunction reveals necessity is equivalent to the denial that it furnishes the materials for deductively valid inferences about new instances. Let us call the amended version (E′). (Mackie, 1974, p. 16)

This does not help Mackie's argument. The amended structure is only as strong as Stove's case for deductivism. Indeed, there is no real evidential movement here at all. (E′) merely repeats Stove's conclusion — and we just saw that *it* is inadequately supported.

On the other hand, if Mackie sticks with (E), Stove's argument fails to support it, unless "rational inference" is read as "reasonable inference." Even then it is no better supported than (E′). On either version, Mackie's argument suffers severely from its reliance on Stove's weak case for deductivism.

XI

All that is left in Mackie's argument, then, is Hume's "first line of thought" — (D) and the propositions supporting it. One initial problem with this line of argument surfaces in (L) and is transferred to (D) as a result. (L) says that we acquire "causal knowledge" as a result of experiencing constantly conjoined events. This is inappropriate. While we do acquire what *causal beliefs* we have as the result of experiencing constantly conjoined events, we abandoned hope of acquiring *causal knowledge* — on Hume's view, at least — when we passed from the level of relations between ideas to the realm of matters of fact.

There are other problems as well. (H) and (G) are best regarded as independent ways of making the same point. But if there is any dependency between them, it goes from (H) to (G), not the other way round. And even if we accepted the rest of this line, (B) and (D) together do not establish (A).

Mackie's account, then, is vitiated by its own internal difficulties as well as by its excessive reliance on Stove's faulty argument. The problems we have seen with both accounts are such that local repairs seem impracticable, if not downright impossible. Accordingly, we need to look elsewhere for an adequate account of Hume's negative argument.

XII. HUME'S NEGATIVE ARGUMENT

Hume's argument actually begins with a central thesis in his account of the mind. It is that all mental operations are ultimately traceable to either of two faculties: the understanding and the imagination. In Section IV of the *Enquiry*, he raises a series of "sceptical doubts concerning the operations of the understanding." These "doubts" are introduced to question the ability of the understanding to provide us with any reasoning which will take us beyond the immediate deliverances of the senses and memory.

Hume describes this negative project, not without irony, as "an easy task." He is quite clear about what this task is:

> I shall content myself, in this section, with an easy task, and shall pretend only to give a negative answer to the question here proposed. I say then, that, even after we have experience of the operations of cause and effect, our conclusions from that experience are not founded on reasoning, or on any process of the understanding. This answer we must endeavour both to explain and to defend. (*Enquiry*, p. 32)

"The question here proposed" is the question of "what is the nature of that evidence which assures us of any real existence and matter of fact, beyond the present testimony of our senses, or the records of our memory." (*Enquiry*, p. 26)

As I set out the details of Hume's attempt "to explain and defend" his "sceptical doubts," I use material not only from the *Enquiry*, but from his parallel argument in the *Treatise* as well. I also make use of the summary account he offers in the *Abstract.* Hume is clearer structurally about his project in the *Enquiry* than he is elsewhere, though he often explains his points more fully in the *Treatise.* When he does I freely help myself to those passages, for I take him to be developing essentially the same argument in all three texts.

XIII

I begin by backtracking a bit. This will allow us to see how Hume gets to the place where his "sceptical doubts" are appropriate.

Hume's project is to provide an account of human nature, one which – as the subtitle of the *Treatise* makes plain – attempts "to introduce the experimental method of reasoning into moral subjects." This means that while Hume's aim is "to explain the principles of human nature" (*Treatise*, p. xvi), he views this task as one of *describing* the operations and contents of the mind, not one of providing theoretical *explanations* of them. He will *explain* by *describing.* So he characterizes his work as "mental geography," a "delineation of the distinct parts and powers of the mind." (*Enquiry*, p. 13) Only in this way, he thinks, can we accurately determine "the proper province of human reason" (*Enquiry*, p. 12), as opposed to whatever theoretical role we might speculatively assign it.

It is easy enough to begin the project. The mind passively receives impressions, some of which are retained as ideas. This is adequate for recording input, but it fails to account for the ways in which some ideas become connected in the mind with others. This network of connections, which every human mind possesses, is as important to our ability to understand the world and act in it as are the impressions and ideas which are the subject-matter of those connections. These connections are not only pervasive in human thought, they are predictable. As Hume puts it: "It is evident that there is a principle of connexion between the different thoughts or ideas of the mind, and that, in their appearance to the memory or the imagination, they introduce each other with a certain degree of method and regularity." (*Enquiry*, p. 23)

It is tempting to regard these connections as the result of the activity of the understanding. Simply *assuming* this, however, would be to violate Hume's descriptive method. Observing the contents of the mind yields *data* concerning the interconnections of our ideas. We need to observe and describe, not theorize and explain, how these connections come to be.

Hume tries to do just this. He begins by enumerating the kinds of connections he actually observes among ideas. Though he grants that enumerating observed connections is insufficient to ensure that his categorization is exhaustive, he believes nonetheless that he has isolated the only "principles of connection among ideas." They are: resemblance, temporal and spatial contiguity, and causation.[23]

The most important connections between ideas are those which take us beyond "the present testimony of our senses, and the records of our memory." Resemblance and contiguity are able only to compare ideas which are present to the mind; they cannot, therefore, account for the way in which we are able to go beyond sensation and memory.

Yet there must be *some* principle by which the mind does this. When we hear a voice in the dark, and become convinced that there is someone nearby, or when we receive a letter from afar and believe that the sender is in that place, we go beyond anything given to our senses or retained in our memory. And "here it is constantly supposed that there is a connexion between the present fact and that which is inferred from it. Were there nothing to bind them together, the inference would be entirely precarious." (*Enquiry*, p. 27) The certainty with which we hold many such beliefs leads us to think that they are not "entirely precarious."

The only principle left which is capable of connecting ideas is that of cause and effect. Attending to cases reveals that it is indeed this relation which produces the connection: "if we anatomize all the other reasonings of this nature, we shall find that they are founded on the relation of cause and effect, and that this relation is either near or remote, direct or collateral." (*Enquiry*, p. 27)

To find out whether the understanding is operative in producing the connections we find between things present to the senses and those that are not, we must ask whether the understanding is operative when we relate events as causes and effects: "If we would satisfy ourselves, therefore, concerning the nature of that evidence, which assures us of matters of fact, we must enquire how we arrive at the knowledge of cause and effect." (*Enquiry*, p. 27) So the important question concerning the interconnections between our ideas that gives us our coherent picture of the world turns out to be the question of how we are able to go beyond the present testimony of the senses. And *that* question has, in turn, *become* a question about the basis of our beliefs that are based on cause and effect.

The operative theoretical candidate is still reason. So we need to determine whether our causal beliefs are due to the operations of the understanding, to some process of reasoning which supports or justifies the causal beliefs we form.

This is the state of play as Hume begins to raise his "sceptical doubts" about whether reason and the understanding are capable of playing such a role.

XIV

There are quite a few ways in which we form beliefs which go beyond our present experience. We predict events in the distant future on the basis of past and present ones: we predict a comet's return from records of its past appearances. We come to believe that a past event took place in a certain location from historical or archeological evidence: we conclude that a battle occured at a site where we dig up bullets and arrowheads. We become convinced that someone unobserved is present because of what we experience: I come to believe that Kirk is in the room with me when I hear his voice in the dark. We also reason from cause to effect, from effect to cause, from effect to collateral effect, and from cause to collateral event.

These examples are all significantly different, but Hume's interest at this point is in stressing what they have in common. All are cases where we connect something which is present to the senses with something which is not; where we connect the unobserved with the observed. As such, they are all examples of what Hume describes as "our reasonings concerning cause and effect."

It will be useful to have a theory-neutral term to describe what happens to us that is common to all these cases. Hume sometimes uses "reasoning" and "inference" in this way to refer to any operation of the mind. But his usage can be seriously misleading, and can verge on the paradoxical, especially when he is arguing that "our reasonings concerning cause and effect" do not depend on "reason."

What happens in all these cases is that, based on what I observe at present, I *expect* that something I do not now observe will happen, or did happen, or is now happening. Since what I expect is based on my belief that there is a causal connection between what I observe and what I don't, we can preserve a helpful neutrality by describing my state as that of forming *causal expectations.*

In a typical case where I form a causal expectation, the *data* are something like this: I have just done several things I frequently do. I arranged a number of charcoal brickettes into a rough pyramid-like shape, and then soaked them with charcoal lighter. Now I put a lighted match to the charcoal. I *expect* the charcoal to light.

Hume asks whether my expectation that the charcoal will light is based on reasoning. If so, on what reasoning is it based? I must, it seems, answer this question before I can determine whether my expectation is good or bad, well- or ill-founded.

XV

In order to decide whether my expectations about the charcoal are based upon reasoning, we must first determine what reasoning of this sort would have to concern. Accordingly, Hume begins Section IV of the *Enquiry* with his infamous division of "all the objects of human reason or enquiry" into *relations of ideas* and *matters of fact.*[24] His argument moves from this classification through an exhaustive examination of whether it is plausible to regard causal expectations as based on reasoning from either category. It ends with the negative conclusion that our causal expectations are not based on reasoning at all.

I diagram my structural account of Hume's negative argument in Figure 3. The argument begins at proposition (8) with Hume's classification of the possible categories reasoning can concern. The argument then divides into two main parts, beginning with propositions (4) and (9). Each part corresponds to one category of reasoning.

The first part considers whether reasoning which results in causal expectations could concern relations between ideas. It is rather straightforward. Such reasoning must either be intuitive or demonstrative. Hume argues that neither will suffice to account for our causal expectations.

The second part asks whether the reasoning could concern matters of fact. This complex argument divides first into the possible ways in which matters of fact can be established: by experience alone, or by reasoning. Experience alone is insufficient. The reasoning involved must be either demonstrative or probabilistic. Hume quickly shows that the reasoning cannot be demonstrative.

He then turns to the final, but involved, leg of the argument.

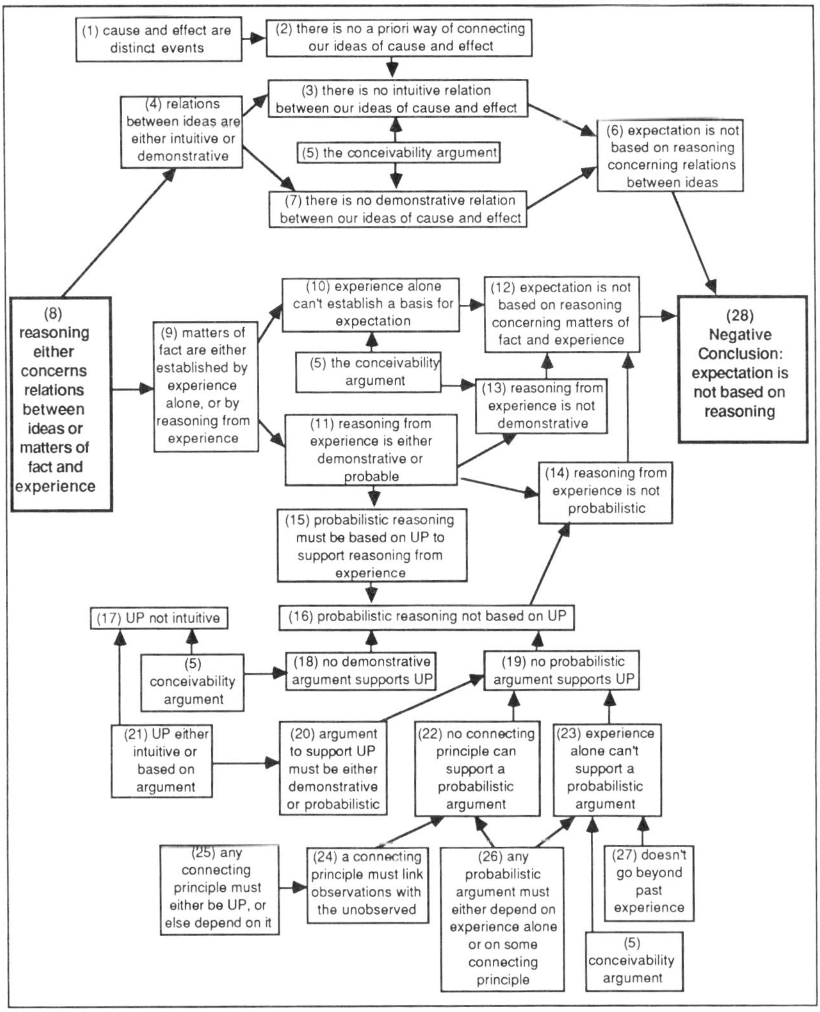

Fig. 3. Hume's refutation of inductive probabilism.

Probabilistic reasoning depends on the principle that nature is uniform (UP). So our question *becomes* a question about what might establish or support UP. Experience alone is again insufficient, as is demonstrative reasoning. Probabilistic support for UP ultimately fails, too. It either fails to provide a connection between the observed and the unobserved, or does so only by invoking, in a way that is viciously circular, either UP itself or some equivalent principle.

So Hume's answer is negative in both cases. The two parts of the argument then converge to establish Hume's negative conclusion (28). To appreciate it, we need to look closely at the detailed structure that supports each part.

XVI. RELATIONS BETWEEN IDEAS

Relations between ideas are either *intuitive* or *demonstrative* (4). Hume gives no precise definition of either sub-category. He does say that an *intuitive* connection is one that is "discoverable at first sight" (*Treatise*, p. 70) without appeal to experience.

This is enough to show that the events we relate as cause and effect cannot be connected intuitively. Such events are "distinct" (1). Neither idea is "contained" in the other. My idea of applying the match to the charcoal does not "include" the idea of the charcoal's catching fire, in the way that my idea of a triangle "includes" the idea of its being three-sided. So there seems to be no *a priori* way of establishing a connection between the two ideas. (2) Experience seems to be required.

It is tempting, however, to think that there *must* be some intuitive connection in cases like this, if only because the expectation is so immediate and familiar. Thus

> we are apt to imagine that we could discover these effects by the mere operation of our reason, without experience. We fancy, that were we brought of a sudden into this world, we could have at first inferred that one Billiard-ball would communicate motion to another upon impulse; and that we needed not to have waited for the event, in order to pronounce with certainty concerning it. (*Enquiry*, p. 28)

Hume is aware of this temptation, but he also has a diagnosis of it. His diagnosis telegraphs something of the positive account he will offer later: "Such is the influence of custom, that, where it is strongest, it covers not only our natural ignorance, but even conceals itself, and seems not to take place, merely because it is found to the highest degree." (*Enquiry*, pp. 28—29) But he is quite aware that this alternative account is not by itself a refutation. So he offers an argument to supplement his diagnosis. It is the first appearance of an argument which will recur throughout his discussion: the conceivability argument (5).

The conceivability argument exploits the fact that, regardless of how familiar an event may be to me, no matter how strongly held is my conviction that it will occur, I must admit that I can *conceive* of the cause event occurring without the effect event also occurring. While I am convinced that the charcoal will light, I can easily *imagine* its not lighting. Whatever causal connection there is between applying the lighted match and the charcoal's catching fire must be *external*, not *internal.*

So our causal expectations, however familiar, are not intuitive (3). If these expectations are the result of reasoning concerning relations between ideas, it must be in virtue of some demonstrative argument which connects the occurrence of the cause event with my expectation that the effect event will occur.

Hume provides no example of what such a demonstrative argument might be. In fact, he never explains what a demonstrative argument really is. He may have assumed that doing so was unnecessary. Perhaps he took it for granted that his readers were familiar with this form of argument. But it may also be the case that he had good reason for not committing himself to any particular account of what "demonstration" involves.[25] At any rate, it is clear in this context that a demonstrative argument must be one that is at least deductively valid. It must be impossible for its premises to be true and its conclusion false.

Because of this, it matters little for Hume's argument what *else* constitutes a demonstrative argument, or what version(s) might be introduced in an attempt to connect cause and effect. Such an argument would have to mention the cause event in its premises and the effect event in its conclusion. That allows Hume to apply the conceivability argument (5) again, this time to the alleged demonstration. For any such argument, we can imagine the premises being true and the conclusion false. So no demonstrative argument — no purely *a priori* argument concerning only the relations between ideas — could ever establish that cause and effect are connected in such a tight way that the conceivability argument could not show that they were actually independent (7).

This means that causal expectation can not be based on any reasoning concerning relations between ideas (6). The first part of Hume's argument is now complete. If our causal expectations are based on any reasoning, it must be reasoning which concerns matters of fact.

XVII. MATTERS OF FACT

Matters of fact can also be established in two ways: by experience alone, or by reasoning from experience (9).

Experience alone can't support our causal expectations. For while experience is capable of establishing particlar matters of fact, it cannot *connect* these experiences in the way the causal relation requires. Nor can it *project*, or go beyond, the data recorded in actual experience. Causal expectation involves both.

Experience may record that my application of the lighted match to the prepared charcoal was followed almost immediately by the charcoal's catching fire. These observations establish that both events occurred, and were related spatially and temporally. But this conjunction of events, in Hume's view, does not constitute causation.

With the aid of memory, I may collect much more data of a similar character. My recording of these experiences may take place over a wide variety of times, places, and conditions. But all this experience, taken by itself, provides is a record of the *constant conjunction* of similar events: the application of matches under certain conditions, and the subsequent lighting of charcoal under these same conditions.

Nor can experience alone account for the projective aspect of my causal expectations. My prediction, just before I apply the match, that the charcoal will catch fire, goes well beyond my present experience (observing the match, preparing the charcoal, and the like) as well as beyond the data my memories provide of similar past occasions when I applied similar matches to similarly prepared charcoal.

Experience itself can only record and report. It cannot take me beyond what it records and reports. If I consider the matter carefully, I can see that, for all experience alone provides, I can conceive of applying the lighted match to the charcoal with no result at all, or with the charcoal exploding like fireworks, or with some even more bizarre result (5).

My experiential record alone, accurate and thorough though it may be, gives me no reason to connect the experiences I have recorded with those I expect to have. For Hume, this is not at all surprising. When we reflect on what establishing such a connection would be, we see that it would involve an additional idea over and above my recorded and retained experiences. How could such an idea be generated? Not from these experiences alone:

> As our senses show us in one instance two bodies, or motions, or qualities in certain relations of succession and contiguity; so our memory presents us only with a multitude of instances, wherein we always find like bodies, motions, or qualities in like relations. From the mere repetition of any past impression, even to infinity, there never will arise any new original idea . . . and the number of impressions has in this case no more effect than if we confin'd ourselves to one only. (*Treatise*, p. 88)

But without such an idea to provide the required connection between what I have experienced and what I expect, I have no reason for thinking that what I expect to happen will indeed come to pass.

Experience by itself, then, provides neither the *connection* nor the *projection* that the project of underpinning my expectations requires (10). That support must come from somewhere else.

XVIII. REASONING FROM EXPERIENCE

The only remaining way to establish a conclusion concerning matters of fact is through reasoning from experience. The reasoning must be either *demonstrative* or *probabilistic* (11).

That the reasoning involved might be *demonstrative* is frequently ignored by interpreters of this argument. Their reason for doing so is that demonstration traditionally involves deducing conclusions from necessarily true premises. They assume that Hume accepted and employed this traditional conception of demonstrative argument. Were this the case, demonstrative arguments would concern only relations between ideas, and would have been dealt with already.

It is not clear, however, that this is *Hume's* conception of demonstrative reasoning. There is good reason to think that he conceived of demonstration in a significantly broader way, so that while it includes all the traditional demonstrative arguments, it also encompasses any conclusion deduced from an experiential premise.

Some "stretched" version of the concept of demonstration seems necessary if we are to understand the distinction Hume has in mind when he contrasts reasoning in the more abstract realms of mathematics with what he calls "mixed mathematics." The former "reasonings" clearly involve demonstration in the traditional sense. But the conclusions of "mixed mathematics" — the phrase generally refers to applications of mathematical theories to empirical subject-matter — are derived by deduction from experiential premises. He clearly regards these forms of reasoning as legitimate.[26]

This may not be an innovation of Hume's part. Barbara Shapiro's work on probability and certainty in England strongly suggests that the break with the traditional conception of demonstration began nearly a century before. (Shapiro, 1983, Chapter 2) This "stretching" of the concept of demonstration is very likely another aspect of the response to contemporaneous shifts occurring in the related concepts of *scientia* and *opinio*. As Ian Hacking has shown, the modern concept of probability emerged as a result of these shifts, and with it, the modern problem of induction — "Hume's problem." (Hacking, 1975) Perhaps the fully stretched notion of demonstration first emerges fully in Hume as well.

Hume's admission of stretched demonstrations — deductions from empirical premises — does not mean that legitimate support for our causal expectations can be given by using this form of reasoning with premises which record past experience of constantly conjoined events. The conceivability argument comes into play again to show that we can easily conceive of a change in the course of nature which would thwart even the best-founded argument of this kind. But the interest of this category of possible arguments is not in what it proves. It is rather in what it says about the remaining category of probabilistic arguments. Hume obviously has a different form of reasoning in mind with this final category than the one just considered. Given that, it is extremely unlikely that he conceives of it along deductivist lines. Otherwise, it could easily be accommodated in the present category.

XIX. PROBABILISTIC REASONING

The only possibility left is what Hume calls "moral, or probable reasoning." This form also fails to provide the *connection* the justification of our causal expectations requires, unless we add an important premise to our reasoning (15). The premise concerns the connection between past experiences and expectations. Indeed, it mandates the connection. It says that my causal expectations are justified because nature is uniform. The future will be like the past. Hume's version of the principle is, arguably, even stronger. It says that "*instances, of which we have had no experience, must resemble those, of which we have had experience, and that the course of nature continues always uniformly the same.*" (*Treatise*, p. 89)

Stove and Mackie call this "the resemblance thesis." It is also frequently called "the similarity thesis." If prefer to call it "the uniformity-of-nature principle" (UP).

Hume states the principle loosely, almost casually. He seems indifferent between the version just quoted and several others he offers, even though they seem to vary significantly in strength. It is tempting to object that we need a more precise statement of what the principle is if we are to give his argument an accurate evaluation.

But Hume ignores these natural questions, to move directly to the question of what support or justification can be given for UP. Once again, the form of the question we are asking has changed significantly.

It is not necessary that I actually invoke, or consciously consider, some such principle when I reason from experience. It is sufficient that UP can be regarded as a background principle in all my formations of causal expectations. But we do need to know what the justification for it is. If it is not adequately supported, then any reasoning which relies on it will also provide inadequate support for whatever conclusions that reasoning is supposed to justify.

UP might be intuitive. If it is not, then it must be justified by argument (21). That argument might either be demonstrative or probabilistic (20).

UP is certainly not intuitive (17). There is nothing incoherent about imagining the future as not being like the past. Hume once again invokes the conceivability argument (5) to show that even though UP, or some similar principle, might seem obviously true, this is due to the familiarity and experienced regularity of our experience. It is not because UP is intuitively obvious: "I shall allow, if you please, that the one proposition may justly be inferred from the other: I know, in fact, that it always is inferred. But if you insist that the inference is made by a chain of reasoning, I desire you to produce that reasoning. The connexion between these propositions is not intuitive." (*Enquiry*, p. 34) If UP has support, it must be from an argument. It is time to produce that "chain of reasoning."

By this time, it should not be surprising that the argument can't be a demonstrative one (18). The reason is due, equally unsurprisingly, to the conceivability argument (5), applied this time to UP. There is no reason that the future must be like the past. We can certainly conceive of nature not being — or not continuing to be — uniform: "We can at least conceive of a change in the course of nature; which sufficiently proves, that such a change is not absolutely impossible. To form a clear idea of any thing, is an undeniable argument for its possibility, and is alone a refutation of any pretended demonstration against it." (*Treatise*, p. 89)

So if there is an argument supporting UP, it must be a probabilistic argument. Hume has not discussed probabilistic reasoning before now. He has told us only that the degree of certainty a probabilistic argument provides is less than that provided by a demonstrative argument. In the *Treatise* account, he pauses briefly at this point to give a short "conceptual analysis" of probabilistic reasoning.

Probabilistic reasoning involves at least three necessary conditions: (a) impressions, either of present sensation or of memory, (b) inference, to (c) an idea of something which is present neither in sight nor memory. (*Treatise*, p. 89) This mixture of impressions and ideas is required because

> Probability, as it discovers not the relations of ideas, consider'd as such, but only those of objects, must in some respects be founded in the impressions of our memory and senses, and in some respects on our ideas. Were there no mixture of any impression in our probable reasonings, the conclusion wou'd be entirely chimerical: And were there no mixture of ideas, the action of the mind, in observing the relation, wou'd, properly speaking, be sensation, not reasoning. (*Treatise*, p. 89)

But the causal relation, as we have already seen, is the only relation that is capable of taking us beyond our immediate impressions and memories to that which is unobserved. So the causal relation must be the foundation of any inference involved in probabilistic reasoning. The game is really over at this point, but Hume goes on to spell it out.

Any probabilistic argument which might be thought capable of justifying UP must either depend on experience alone, or else on some connecting principle.

Experience alone obviously cannot drive an argument, even a probabilistic one, which could support UP (23). Experienced uniformities, however extensively and carefully recorded, are only a small number of all the instances where uniformity, and thus ultimately, UP, is in question. How could such a small evidence base possibly provide adequate support for such a sweeping principle?

Even if we did not discount the adequacy of the evidence provided by experienced uniformities, the conceivability argument (5) clearly shows that past uniformities can provide no guarantee that things will continue in the future in the same way as they did in the past. We can easily imagine a change in whatever uniformities we record.

But the real objection to using experience alone to support a probabilistic argument for UP is that UP provides the connection it does between the observed and the unobserved by going far beyond anything given in actual experience. It covers unobserved past and present events as well as future ones. We saw earlier that experience was by itself incapable of providing a link or bridge from observed events to the unobserved. Why should we now think that it is suddenly capable of supporting an even stronger connection? (27)

So the work in a probabilistic argument which could support UP would have to be based on a connecting principle, though of course such a principle may work with observations experience provides.

What might such a connecting principle be like? To support UP, it clearly must link observations with the unobserved. Such a principle, however, must be UP itself (though perhaps in disguise!), or else must depend on it.

Hume originally invoked UP because it seemed the only way in which past experiences might reasonably be connected with our causal expectations. He stated it in an extremely loose and general form. This seemed inexcusably sloppy, if not actually objectionable, at the time. Now we can see that Hume's choice of this rather unspecific formulation of UP was deliberate. He wanted to introduce it in this form, so that it could easily be seen to encompass more specific versions. In this general and unspecific form, UP mandates that future regularities will conform to past ones. Any other connecting principle must be virtually equivalent to it.

The only possible way out involves invoking an apparently different type of connecting principle. Such a principle would use recorded past regularities, in some such form as "Past futures resembled past pasts." A principle of this kind — call it CP — is of course vulnerable to the charge of being underdetermined by its evidence base.

But there is a more serious objection to CP. It is a principle which connects past observations with other past observations. For CP to provide support for UP, we need to establish a connection between CP and the future.

How might such a connection be made? It seems to require another general premise, one which connects past past-future connections with past-future connections. Such a principle would assure us of the uniformity of nature across time. But any such principle would just be another version of UP, brought in this time to support CP. It would not be the independent connecting principle the argument requires.

It is difficult to see how this might be avoided. Either we lack the connection we require to justify reasoning via UP, or else we attempt to provide that connection. If we lack the connection, then we have insufficient justification for reasoning based on UP. If we try to provide this connection, then, we can see upon analysis that this putative justification is either viciously circular or viciously regressive. It is either question-begging because it invokes the very principle it was introduced to support, or else the chain of justification continues to invoke principles which themselves require support, while giving us no reason to believe that the chain will ever terminate.

Even if we think of the chain of support in causal terms, we are no better off. As Hume says in the *Treatise*, summarizing his argument at just this juncture, "the same principle cannot be both cause and effect of another" (*Treatise*, p. 89). If we think of the support chain as not being traditionally justificatory, but as causal, then UP serves both as effect and as cause in the same linear chain. The causal roles it is assigned on this interpretation are incompatible. They make the hypothetical causal support chain ultimately incoherent.

Probabilistic reasoning, then, cannot provide the support UP requires (19). But without a basis for UP, probabilistic reasoning cannot establish or support our causal expectations (14). This completes the second part of Hume's argument: our causal expectations are not based on any reasoning concerning matters of fact and experience (12). It only remains for Hume to draw the two strains of his argument together. Reasoning must concern either relations between ideas or matters of fact (8). Since our causal expectations are not based on reasoning concerning relations between ideas (6), that reasoning must concern matters of fact, if reasoning is involved at all. But since we have just shown that our causal expectations have no basis in reasoning concerning matters of fact, we must conclude that no reasoning is actually involved. Hume appears to have established his negative conclusion (28), and vindicated his "sceptical doubts concerning the operation of the understanding." Our causal expectations have no basis in reasoning *at all.*

XX. CONCLUSION

Hume's negative argument is powerful and compelling. Within its parameters, it is a good one. Does this mean that Hume's vindication of his "sceptical doubts" forces *us* to be sceptics? Our choice of responses to this question may appear to be cut out for us. Either we deny that there is a problem, or we solve it.

One obvious tack is to attempt to reject the problem by rejecting the parameters in which it is stated. Hume's argument shows that UP is not reducible to impressions. But this means either that UP is unjustified or that Hume is wrong. His empiricism dictates the justification conditions for principles such as UP. Perhaps we should reject Hume's empiricism, not UP.

This is tempting because Hume relies so heavily on "the theory of ideas" and so much of that theory is so obviously flawed. It is grounded in theses in epistemology and the philosophy of mind which most of us regard as false, if not incoherent. We should reject these aspects of Hume's empiricism. But this will not make the problem go away. As Stroud (1978) has argued, much of what is interesting in Hume's account survives when the theory of ideas is abandoned. "Hume's problem" does not require Hume's flawed framework.

It is also tempting to meet Hume's problem head-on, and offer a solution to it. This may also require one to reject some of Hume's background empiricism, especially if the solution employs non-logical necessary connections, synthetic *a priori* principles, or sophisticated uses of modal notions. Attempts at solution along these lines walk a fine line between falling prey to a descendent of Hume's original arguments, and changing the subject altogether.[27]

Theories of this sort are a motley lot. But they all characteristically offer what Saul Kripke (1982) calls a "straight solution" to Hume's sceptical doubts. Straight solutions claim to *answer* Hume. If one of them were correct, Hume's sceptical worries would be proven unwarranted. Theories of this kind proliferate. But it is significant that no one of them holds sway. It is even unclear what a satisfactory straight solution to Hume's problem would *look like*.

Hume proposed his own solution. But it is not a "straight solution." He calls it a "sceptical solution," and it breaks the dilemma of responses with which we seem to be faced.

The radical nature of this "sceptical solution" is not yet fully understood. Much recent work on Hume's views about causal expectation concerns whether he really was a sceptic. One line of thought emphasizes his causal scepticism.[28] Another denies it.[29] Both lines largely ignore the precise character of his sceptical doubt, his rejection of straight solutions, and what he meant by offering a sceptical solution to the doubts he raises. But the nature and extent — indeed, the *meaning* — of Hume's "scepticism" will not be adequately appreciated until we are clear about *all* these aspects of his view.

Hume's statement of his positive view — his "sceptical solution" — begins with his acknowledgement of the consequences of his negative argument. The explanatory model of human behavior which makes reason prominent and dominant in thought and action is indefensible. Scepticism about it is well-founded. The model must go.

Hume replaces it with a *description* of how we *do* form our causal expectations. This naturalistic account does assign a role to reason, though it is a truncated and subordinate one compared to the role it played in the explanatory model.[30] In this positive context, Hume has interesting and surprisingly modern things to say about variant and invariant sequences, proofs and probabilities, and rules by which to judge of causes and effects. These aspects of Hume's thought have not been totally overlooked.[31] But they need to be understood as essential components of Hume's "sceptical solution," not as isolated sets of remarks. Working out the details of this positive picture of human nature is perhaps the central challenge faced by students of Hume. It is a project that is of more than historical interest, however. For we may find that when the picture is complete, we will have finished Salmon's "unfinished business."

University of Cincinnati

NOTES

* My research was supported by a grant from the Taft Faculty Committee, University of Cincinnati, for which I am most grateful.

I have discussed these topics, always with profit, with John Biro, James Cargile, and John McEvoy. My views have improved as a result of the criticisms participants made in seminars I gave on Hume and Causation at the University of Cincinnati and the University of Virginia. Especially helpful were Robert Friedman, Jack Fuchs, Sally Haslanger, Eric Melvin, Lonnie Plecha, Charles Stephan, James L. White, and everyone who urged that, if I persisted in criticizing Mackie's and Stove's structure diagrams, I should at least be willing to provide one of my own. Christopher Gauker, Donald Gustafson, Larry Jost, and Miriam Solomon provided helpful comments on a final draft.

I am especially grateful to Robert Richardson, James H. Fetzer, and Linda Weiner for their extensive comments, helpful suggestions, and encouragement.

1 Salmon, 1978.

[2] It is tempting to declare Hume's argument the most familiar piece of Western philosophy, period. But Descartes' *Meditations* — or at least, *Meditation I* — provide stiff competition. It is interesting to note, however, that opinions differ *equally wildly* on the interpretation of *Meditation I*.

[3] Passmore, 1952, p. 1.

[4] Robinson, 1962, in Chappell, 1966, p. 129. Robinson cites, with approval, Passmore's remarks just quoted.

[5] Hume, 1978. References cited parenthetically hereafter as "*Treatise*." The 'first' is a double one: *Hume's* first statement of the problem is *the* first published statement of the problem. See Hacking, 1975, p. 176.

[6] Hume, 1938. References cited parenthetically hereafter as "*Abstract*."

[7] Hume, 1975. References cited parenthetically hereafter as "*Enquiry*."

[8] Mackie, 1974.

[9] Stove first published his views in his article, "Hume, Probability, and Induction" (Stove, 1965). He revised and expanded them in his book, *Probability and Hume's Inductive Scepticism* (Stove, 1973).

[10] Penelhum, 1975, p. 50.

[11] Mackie invented the "structure diagrams" he and Stove both use. Though they claim to be using the arrows to indicate dependency relations in the same way, it is far from clear that they in fact do so. Stove uses converging arrows to represent elements of an argument for a conclusion (the proposition on which the arrows converge). Mackie sometimes uses converging arrows to indicate separate lines of support which independently provide "some support" for a conclusion. At other times, however, he uses them much in the way Stove does.

My representations of Mackie's and Stove's structure diagrams differ in one important respect from their originals: they use schematic letters in their diagrams, and provide separate "dictionaries" explaining what propositions the letters stand for. My versions combine dictionary and structure diagram into one figure.

[12] See Sievert (1974).

[13] Locke is notoriously inconsistent on this issue. Despite the passages which make him vulnerable to Hume's attack, he also says things which are very much in the spirit of Hume. See Locke (1975), Book IV, Chapter III, Section 24, for example.

[14] Descartes' arguments in *Meditation III* and perhaps in *Meditation VI* are clear representatives of the sort of "philosophical hypotheses" Hume has in mind here.

[15] Robinson (1980) sees that Mackie has reversed the dependency relation here. But his account is flawed by his willingness to accept Mackie's mistaken and misleading formulations of Hume's statements.

[16] Sievert (1974) is quite perceptive and helpful here.

[17] Stove, in Chappell (1966), p. 192. Stove makes substantially the same claim in (1973), p. 32.

[18] This would, of course, be true *even if* Hume also held something like (E′).

[19] These are slight paraphrases of the entry in *Webster's New Collegiate Dictionary*.

[20] Stove, 1973, p. 32. See also Stove's remarks in Chappell (1966), p. 200.

[21] Stove, 1973, p. 43. Stove discusses other senses of "presuppose" from pp. 42—44.

[22] This point is also overlooked by Beauchamp and Mappes (1975) as well as by Beauchamp and Rosenberg (1981), who base their view that Hume was not really a "sceptic" about induction on a very limited picture of who his targets were.

[23] *Enquiry*, Section III, pp. 23—24. Hume's account of relations in the *Treatise* (Book I, Sections IV and V) is more complex, but it does not really affect the discussion here.
[24] *Enquiry*, p. 25. The *Treatise* account does not make this distinction explicit, but effectively presupposes it.
[25] This is certainly the case if I am correct in thinking that Hume was willing to "stretch" his use of "demonstrative argument" and "demonstration" beyond the bounds of the traditional definition. See below, Section XVIII.
[26] "Mixed mathematics" traditionally includes applications of mathematical theories to astronomical, mechanical, and physical phenomena. It was occasionally extended to any attempt to mathematically describe or quantify empirical data. Hume may be tacitly proposing to extend the range of mixed mathematics even further.
[27] Or the alleged solution might just beg the question against Hume's arguments. An excellent example is Russell's naturalistic response to inductive scepticism in *Human Knowledge: Its Scope and Limits*. (Russell, 1948). Russell's "postulates" either fall prey to Hume's original arguments or beg the question against them, as is nicely pointed out by Salmon (1967), pp. 43—48. I show how epistemically impoverished Russell's approach is in Morris (1979).
[28] Recent examples include Robison (1977) and Fogelin (1983) and (1985).
[29] Prominent examples include Beauchamp and Mappes (1975) and Beauchamp and Rosenberg (1981).
[30] Hume's account *is* "naturalistic," but as with many of the features of his positive account, we are by no means clear about what these terms mean when applied to his view. Many recent writers on Hume have followed Kemp Smith's (1905; 1941) lead in stressing Hume's "naturalism," though all too often the naturalism is invoked to show that Hume was not *really* a sceptic. These "straight solution" co-optings of Hume's naturalism do little to help us get clear about what his view really is.
[31] Helpful discussions of these topics can be found, for instance, in Hacking (1978), Wilson (1983), and Ferreira (1986).

BIBLIOGRAPHY

Beauchamp, Tom L. and Thomas Mappes (1975) "Is Hume Really a Sceptic About Induction?" *American Philosophical Quarterly* **12**, 119—129.

Beauchamp, Tom L. and Alexander Rosenberg (1981) *Hume and the Problem of Causation*, Oxford: Oxford University Press.

Chappell, Vere C. (1966) *Hume: A Collection of Critical Essays*, New York: Doubleday Anchor Books.

Ferreira, M. Jaime (1986) *Scepticism and Reasonable Doubt*, Oxford: Oxford University Press.

Fogelin, Robert J. (1983) "The Tendency of Hume's Scepticism" in *The Sceptical Tradition*, edited by Myles Burnyeat, Berkeley: University of California Press, 397—412.

Fogelin, Robert J. (1985) *Hume's Scepticism in the Treatise of Human Nature*, London: Routledge and Kegan Paul.

Hacking, Ian (1975) *The Emergence of Probability*, Cambridge: Cambridge University Press.

Hacking, Ian (1978) "Hume's Species of Probability" *Philosophical Studies* **33**, 21–37.
Hinkfuss, Ian (1974) "Review of Stove (1973)" *Australasian Journal of Philosophy* **52**, 269–276.
Hume, David (1975) *Enquires Concerning Human Understanding and Concerning the Principles of Morals*, Third Edition, with text revised and notes by P. H. Nidditch, Oxford: Oxford University Press.
Hume, David (1978) *A Treatise of Human Nature*, Second Edition, with text revised by P. H. Nidditch, Oxford: Oxford University Press.
Hume, David (1938) *An Abstract of a Treatise of Human Nature*, edited by J. M. Keynes and P. Sraffa, Cambridge: Cambridge University Press.
Locke, John (1975) *An Essay Concerning Human Understanding*, edited by P. H. Nidditch, Oxford: Oxford University Press.
Mackie, J. L. (1974) *The Cement of the Universe*, Oxford: Oxford University Press.
Morris, William Edward (1978) "Moore and Russell on Philosophy and Science" *Metaphilosophy* **10**. 111–38.
Passmore, John (1952) *Hume's Intentions*, Cambridge: Cambridge University Press.
Penelhum, Terence (1975) *Hume*, New York: St. Martin's Press.
Robinson, H. M. (1980) "Mackie's Interpretation of Hume" *Analysis* **40**. 19–24.
Robinson, J. A. (1962) "Hume's Two Definitions of 'Cause'" as reprinted in Chappell (1966), 129–147.
Robison, Wade L. (1977) "Hume's Causal Scepticism" in *David Hume*, edited by W. D. Todd, Edinburgh: Edinburgh University Press.
Russell, Bertrand (1948) *Human Knowledge: Its Scope and Limits*, New York: Simon and Schuster.
Salmon, Wesley C. (1967) *The Foundations of Scientific Inference*, Pittsburgh: University of Pittsburgh Press.
Salmon, Wesley C. (1978) "Unfinished Business: The Problem of Induction" *Philosophical Studies* **33**, 1–19.
Shapiro, Barbara J. (1983) *Probability and Certainty in Seventeenth-Century England*, Princeton: Princeton University Press.
Sievert, Donald (1974) "Hume, Secret Powers, and Induction" *Philosophical Studies* **24**, 247–260.
Smith, Norman Kemp (1905) "The Naturalism of Hume" *Mind* **11**, 148–173; 335–347.
Smith, Norman Kemp (1941) *The Philosophy of David Hume*, London: Macmillan.
Stove, David C. (1965) "Hume, Probability, and Induction" as reprinted in Chappell (1966), 187–212.
Stove, David C. (1973) *Probability and Hume's Inductive Scepticism*, Oxford: Oxford University Press.
Stroud, Barry (1978) *Hume*, London: Routledge and Kegan Paul.
Wilson, Fred (1983) "Hume's Defense of Causal Inference" *Dialogue* **22**, 661–694.

ABNER SHIMONY

AN ADAMITE DERIVATION OF THE PRINCIPLES OF THE CALCULUS OF PROBABILITY*

1. A RECONSTRUCTION OF ADAM'S REASONING

If Adam was a rational man even before he had garnered much experience of the world, and if the ability to reason probabilistically is an essential part of rationality (as Bishop Butler maintained when he wrote that "But, to us, probability is the very guide of life,"[1]), then Adam must at least tacitly have known the Principles of the Calculus of Probability. Specifically, Adam must have known that the epistemic concept of probability — probability in the sense of "reasonable degree of belief" — satisfies these Principles, for it is the epistemic concept, rather than the frequency concept or the propensity concept, which enters into rational assessments about uncertain outcomes.[2] But what warrant did Adam have for either an explicit or a tacit assertion that epistemic probability satisfies the Principles of the Calculus of Probability?

Today, the best known and most widely accepted justification of this assertion is the "Dutch Book" Theorem, originally proved independently by F. P. Ramsey[3] and B. de Finetti[4] for a subjectivist or personalist version of epistemic probability, but applicable also to non-subjectivist or not-entirely-subjectivist versions.[5] It is most implausible, however, to attribute awareness of this theorem to Adam, partly because it requires some mathematics that is not completely trivial, partly because of the rarity of gambling in the Garden of Eden, and partly because in the one case in which Adam did indulge in a gamble he exhibited little skill at decision theory. A less well known but quite ingenious proof that epistemic probability satisfies the standard Principles was given independently by R. T. Cox[6] and I. J. Good[7] and refined by J. Aczél[8]. But mathematically, this type of proof is even more complicated than the Dutch Book Theorem, and furthermore the proof is actually incomplete without some recourse to considerations of betting.[9] It is my contention that Adam must have had a much simpler warrant for the Principles of the Calculus of Probability than either of

James H. Fetzer (ed.) Probability and Causality. 79—89.

these, and the purpose of this paper is to reconstruct his simple and sturdy reasoning.

The crux of the reconstructed reasoning is the rough idea that epistemic probability is somehow an estimate of relative frequency. In R. Carnap's list[10] of the informal concepts which can be identified with epistemic probability, or probability$_1$, as he calls it, we find first "a measure of evidential support," second "a fair betting quotient," and third "an estimate of relative frequency," but I am supposing that the third is the most primitive. This idea is particularly attractive and appropriate if the hypothesis h asserts that an individual drawn from a particular collection has the property M, while the evidence e gives some information about the collection and about the randomness of the mode of drawing. Then the epistemic probability statement $P(h/e) = r$ can reasonably be construed as stating an estimate of the relative frequency of individuals with the property M in the collection. To say this is not tantamount to identifying the epistemic concept of probability with the frequency concept, because the latter refers to the *actual frequency* of M in the collection, whereas the former refers to an *estimate of the actual frequency*.

Estimation thus plays a central role in the reconstruction of Adam's probabilistic reasoning. It will not do to explicate estimation in the standard manner of sophisticated probability theory, viz., the estimate upon evidence e of a quantity A which may have one of the values $a_1, \ldots, a_n$ is

$$E(A/e) = \sum_{i=1}^{n} a_i P(h_i/e),$$

where h_i is the hypothesis that A has the value a_i. This standard explication is out of place in our reconstruction, because it would generate a circularity. I suggest, therefore, that estimation be taken as a primitive operation, essentially that which is conveyed by "most reasonable guess of a quantity." (In Section 2 we shall be more cautious and represent the estimate of relevant quantities by an interval of the real line rather than by a single real number.)

The explication of epistemic probability as an estimate of relative frequency now permits an extremely simple demonstration of the four basic Principles of the Calculus of Probability, which we state as follows:[11]

(i) If $P(h/e)$ is a well defined real number, it lies in the interval [0, 1],

(ii) if e is not the impossible proposition 0 and e logically implies h, then $P(h/e) = 1$.

(iii) if $P(h/e)$, $P(h'/e)$ and $P(h \vee h'/e)$ are well defined real numbers, then $P(h \vee h'/e) = P(h/e) + P(h'/e)$.

(iv) if $P(h/e)$, $P(h'/h \,\&\, e)$, $P(h/e \,\&\, h')$, and $P(h \,\&\, h'/e)$ are well defined real numbers, then $P(h \,\&\, h'/e) = P(h/e)P(h'/h \,\&\, e) = P(h'/e)P(h/e \,\&\, h')$.

If the epistemic probability expression $P(h/e)$ is replaced by an appropriate expression for probability in the frequency sense, then Principles (i)—(iv) would hold trivially, since all would be simple statements about ratios of the cardinalities of specified sets. But with little modification, a similar justification of Principles (i)—(iv) holds when $P(h/e)$ is interpreted, as we have done, as epistemic probability. A useful notation in presenting the argument is to let C designate the collection referred to in the evidential proposition e and let C_M designate the subset of C with the property M, while $n(C)$ and $n(C_M)$ respectively are the cardinalities of C and C_M. The Principles are now justified as follows:

(i) Since $n(C_M)/n(C)$ is necessarily a real number in the interval [0, 1] if it is defined at all (i.e., if $n(C) \neq 0$), the estimate of this ratio also lies in this interval.

(ii) If e logically implies h then $n(C_M) = n(C)$, and hence if $n(C) \neq 0$ the ratio $n(C_M)/n(C)$ equals 1.

(iii) If e implies the falsity of $h \,\&\, h'$, and h and h' attribute the properties M and M' respectively to the randomly chosen individual, then C_M and $C_{M'}$ are disjoint sets, and hence $n(C_M \cup C_{M'} = n(C_M) + N(C_{M'})$. But $P(h \vee h'/e)$ is the estimate of the relative frequency $n(C_M \cup C_{M'})/n(C)$, and if the denominator $n(C) \neq 0$ then this ratio is well defined and equals the sum of $n(C_M)/n(C)$ and $n(C_{M'})/n(C)$, which are estimated by $P(h/e)$ and $P(h'/e)$ respectively.

(iv) If h and h' attribute the properties M and M' respectively to the randomly selected individual, then $P(h \,\&\, h'/e)$ is the estimate of $n(C_M \cap C_{M'})/n(C)$, $P(h'/h \,\&\, e)$ is the estimate of $n(C_M \cap C_{M'})/n(C_M)$, and $P(h/e \,\&\, h')$ is the estimate of $n(C_M \cap C_{M'})/n(C_{M'})$. But if $n(C_M) \neq 0$ (and *a fortiori*

> $n(C) \neq 0$), then $n(C_M \cap C_{M'})/n(C) = [n(C_M \cap C_{M'})/n(C_M)]\,[n(C_M)/n(C)]$. It is reasonable then that the estimate of the lhs equals the product of the estimates of the two factors of the rhs, establishing the first half of the Principle. The second half follows in the same way.

One more step is necessary for the reconstruction of Adam's reasoning. In many cases the pair (h, e) in $P(h/e)$ neither specifies a definite collection nor attributes a definite property to a randomly selected member of the collection; for h need not refer to an individual at all, or it may refer to an individual by a proper name without indicating a reference class, and there is a great variety of forms of evidential propositions. The foregoing justification of Principles (i)—(iv) for epistemic probability, which clearly was parasitic upon the standard and trivial justification of the corresponding Principles for the frequency concept of probability, can be extended by the appropriate construction of a reference class associated with the pair (h, e). Adam can consider a very large class of pairs (h_i, e_i), each referring to a different and independent situation, but such that $P(h_i/e_i) = P(h/e)$ for each i. The ground for this equation may be a subjective judgment of indifference, if Adam was a personalist, or it may be something additional, if he was a logical probabilist or a tempered personalist;[12] no commitment need be made on this point for the purposes of the argument to come. Now the statement $P(h/e) = r$ can be explicated straightforwardly as an estimate of an appropriate relative frequency as follows. Let C be the set of the pairs (h_i, e_i) in which e_i is true, and C_M the subset of C in which h_i is also true. Then $P(h/e) = r$ means that the estimate of the quotient $n(C_M)/n(C)$ is r. (Even if Adam was cautious and used an interval estimate when $n(C)$ is small, the independence of the situations eliminates the grounds for suspicion of correlations among the outcomes of the h_i, and hence the estimate of $n(C_m)/n(C)$ should diminish practically to a point as $n(C)$ becomes very large.[13]) The arguments above for Principles (i)—(iv) can now be paralleled with little modification.

I submit that this line of reasoning, which is informal and unrigorous, but simple and sturdy, can be considered to be a reconstruction of Adam's explicit or (more likely) tacit justification that epistemic probability satisfies the four Principles of the Calculus of Probability.

2. A REFINEMENT OF THE ADAMITE DERIVATION

It is interesting to give a rigorous version of the Adamite derivation, in order to allay the suspicion that illegitimate steps may lurk in the informal reasoning, and to make explicit the mild but nontrivial assumptions upon which the reasoning depends.

We shall take as primitive the concept of *equiprobability* or *indifference*, but in order to avoid the appearance of an antecedent commitment to the existence of a quantitative probability $P(h/e)$ for an arbitrary pair of propositions (h, e), we shall use the notation $I(h, e; h', e')$ to mean "h is as probable on evidence e as h' is on evidence e', and conversely."

In our argument we use the term "situation," and for our purposes it will suffice to take situations to be finite Boolean algebras of propositions. A situation is thus not specified by a factual assertion but rather by a finite set of possible facts, the set being closed under negation and disjunction and consequently under conjunction. A situation is thus a mode of discriminating possibilities.

The concept of "independent situations" will be taken as primitive. We have good intuitive judgments that certain situations are independent of each other or irrelevant to each other, e.g., B_1 consisting of all Boolean combinations of a finite set of propositions concerning the outcome of the state lottery, and B_2 consisting of all Boolean combinations of a finite set of propositions concerning the decay of a nucleus in a distant nebula. It would be a major enterprise to explicate clearly and judiciously the concept of "independent situations," but for our purposes there is no need to attempt even a crude explication. The reason is that we shall never actually have to decide whether two situations are independent, and hence we do not have to be equipped with a criterion for such a decision. Instead, we shall be engaged in thought experiments in which an arbitrarily large class of independent situations is available for the purpose of making estimates of truth frequencies, each of which is equivalent — in a sense to be specified — to a given situation of interest, but the detailed content of each is irrelevant.

By the *equivalence of situations* B_1 and B_2 we mean the following: B_1 and B_2 are isomorphic as Boolean algebras, and if h_1 and e_1 are the counterparts of h_2 and e_2 respectively under the isomorphism, then $I(h_1, e_1; h_2, e_2)$.

If B is a situation and $\{B_i\}$ is a finite set of independent situations equivalent to B, then we shall take as primitive the concept

$$E(h/e; \{B_i\}),$$

which is an interval estimate of the relative truth frequency of counterparts of h & e to counterparts of e in the set $\{B_i\}$. In other words, if h_i and e_i are the counterparts in B_i of h and e, and if $n(e; \{B_i\})$ and $n(h \,\&\, e; \{B_i\})$ are respectively the numbers of true e_i and true h_i & e_i in the $\{B_i\}$, then

$$E(h/e; \{B_i\}) = [r_1, r_2]$$

means "r_1 and r_2 are the glb and lub respectively for reasonable values of the quotient $n(h \,\&\, e; \{B_i\})/n(e; \{B_i\})$."

We now make two mild assumptions which connect the concept of estimation to mathematical properties of relative truth frequencies.

E_1: If e, h belong to situation B, e is distinct from the impossible proposition 0, $\{B_i\}$ is a finite set of independent situations equivalent to B, and h_i and e_i are respectively the counterparts in B_i of h and e, then if it is necessary that the quotient $n(h \,\&\, e; \{B_i\})/n(e; \{B_i\})$ belongs to the interval $I \subseteq \mathbb{R}$ whenever the denominator $n(e; \{B_i\})$ is non-zero, it follows that $E(h/e; \{B_i\}) \subseteq I$.

E_2: Let $\{B_i\}$ be a finite set of independent situations equivalent to B, h^α and e^α be propositions in B ($\alpha = 0, 1, \ldots, k$), and h_i^α and e_i^α be the counterparts of h^α and e^α in B_i. If q^α is defined as $n(h^\alpha \,\&\, e^\alpha; \{B_i\})/n(e^\alpha; \{B_i\})$ and if the functional relation $q^0 = Q(q^i, \ldots, q^k)$ holds whenever all the q^α are well-defined (i.e., when each $n(e^\alpha; \{B_i\})$ is non-zero), then

$$E(h^0/e^0, \{B_i\}) \subseteq [R_1, R_2],$$

where

$$R_1 = \text{glb}\,\{Q(x_1, \ldots, x_k) \mid x_1 \,\varepsilon\, I_1, \ldots, x_k \,\varepsilon\, I_k\}$$
$$R_2 = \text{lub}\,\{Q(x_1, \ldots, x_k) \mid x_1 \,\varepsilon\, I_1, \ldots, x_k \,\varepsilon\, I_k\},$$

and

$$I_\alpha = E(h^\alpha/e^\alpha, \{B_i\}).$$

We also make the following assumption about the existence of independent situations equivalent to a given situation.

S_1: If B is a situation and $e^{\alpha}(\alpha = 1, \ldots, k)$ is a set of propositions in B, each different from the impossible proposition 0, then there exists a countable sequence $\{B_i\}$ of independent situations each equivalent to B in which there are infinitely many true propositions among the counterparts e_i^{α} of the e^{α}.

(*Note:* Assumption S_1 guarantees arbitrarily large reference classes for each of the e^{α}, but this does not mean that the relative frequency of true e_i among all the e_i of the $\{B_i\}^N$ must approach a non-zero value as N approaches ∞ ($\{B_i\}^N$ being the initial segment of length N in the infinite sequence $\{B_i\}$).

We now have the materials at hand to define epistemic probability and to demonstrate Principles (i)—(iv).

Definition: If h is an arbitrary proposition and e is a proposition distinct from the impossible proposition 0, then $P(h/e) = r$ (r a real number) holds if and only if for every situation B to which h and e belong and every countable sequence $\{B_i\}$ of independent situations equivalent to B and containing infinitely many true counterparts of e the following two limiting relations hold:

$$\lim_{N \to \infty} \text{lub}[E(h/e; \{B_i\}^N)] = r,$$

$$\lim_{N \to \infty} \text{glb}[E(h/e; \{B_i\}^N)] = r.$$

Comment: Because of assumption S_1 there exists a sequence of independent situations equivalent to every situation B containing h and e, with infinitely many true counterparts of e, and therefore the class of countable sequences mentioned in the *Definition* is not empty. As a result, the warrant of $P(h/e) = r$ cannot be the trivial one of the emptiness of the class of sequences; and consequently it is impossible that $P(h/e) = r$ and $P(h/e) = r'$ if $r \neq r'$. Clearly, however, the *Definition* and assumptions do not guarantee that there exists a real number r such $P(h/e) = r$ for each e distinct from the impossible proposition 0.

Proof of Principle (i): Suppose $P(h/e) = r$, and let $\{B_i\}$ be one of the sequences mentioned in the *Definition*. Then for every N large enough that $\{B_i\}^N$ contains at least one true counterpart of e the quotient $n(h \,\&\, e; \{B_i\}^N)/n(e; \{B_i\}^N)$ is well defined and lies in the interval [0, 1]. Hence, by assumption E_1 $E(h/e; \{B_i\}^N) \subseteq [0, 1]$, and therefore the limit r, which is asserted to exist by the *Definition*, must also lie in [0, 1].

Proof of Principle (ii): If e logically implies h, then in any situation (i.e., finite Boolean algebra) containing h and e the conjunction $h \,\&\, e$ is identical with e. Hence, in any situation B_i equivalent to B the counterpart of $h \,\&\, e$ is identical with the counterpart of e. Hence, for any N large enough that the denominator of the quotient $n(e \,\&\, h; \{B_i\}^N)/n(e; \{B_i\}^N)$ is non-zero this quotient is 1, and hence by assumption E_1 both the lub and the glb of the interval $E(h/e; \{B_i\}^N)$ are 1. By the *Definition* it follows that $P(h/e) = 1$.

Proof of Principle (iii): Let e be distinct from the impossible proposition 0 and let it logically imply $\sim(h \,\&\, h')$. Generate a Boolean algebra B from e, h, and h'. Suppose $P(h/e) = r$, $P(h'/e) = r'$, and $P(h \vee h'/e) = r''$. Then given any countable sequence $\{B_i\}$ of independent situations equivalent to B with infinitely many true counterparts of e, and given $\varepsilon > 0$, there exists an integer N_ε such that all of the following relations hold:

$$\begin{aligned}
&|\text{lub}\,[E(h/e; \{B_i\}^{N_\varepsilon})] - r| < \varepsilon,\\
&|\text{glb}\,[E(h/e; \{B_i\}^{N_\varepsilon})] - r| < \varepsilon,\\
&|\text{lub}\; E(h'/e; \{B_i\}^{N_\varepsilon}) - r'| < \varepsilon,\\
&|\text{glb}\; E(h'/e; \{B_i\}^{N_\varepsilon}) - r'| < \varepsilon,\\
&|\text{lub}\; E(h \vee h'/e; \{B_i\}^{N_\varepsilon}) - r''| < \varepsilon,\\
&|\text{glb}\; E(h \vee h'/e; \{B_i\}^{N_\varepsilon}) - r''| < \varepsilon.
\end{aligned}$$

By assumption S_1 there exists such a sequence $\{B_i\}$, and for $\varepsilon > 0$ let N_ε be the appropriate integer for this sequence with the properties displayed above. Then if we write

$$\begin{aligned}
&E(h/e; \{B_i\}^{N_\varepsilon}) = [s_1, s_2],\\
&E(h'/e; \{B_i\}^{N_\varepsilon}) = [t_1, t_2],
\end{aligned}$$

and

$$E(h \vee h'/e; \{B_i\}^{N_\varepsilon}) = [u_1, u_2],$$

we have

$$r - \varepsilon < s_1 \leqslant s_2 < r + \varepsilon,$$
$$r' - \varepsilon < t_1 \leqslant t_2 < r' + \varepsilon,$$
$$r'' - \varepsilon < u_1 \leqslant u_2 < r'' + \varepsilon.$$

Since e implies $\sim(h \;\&\; h')$, it follows that $e \;\&\; h$ is the same proposition as $e \;\&\; h \;\&\; \sim h'$, and $e \;\&\; h'$ is the same proposition as $e \;\&\; h' \;\&\; \sim h$. Hence, $e_i \;\&\; h_i$ and $e_i \;\&\; h_i \;\&\; \sim h_i'$, which are the counterparts in B_i of $e \;\&\; h$ and $e \;\&\; h \;\&\; \sim h'$ respectively, are identical propositions; and likewise the counterparts in B_i of $e \;\&\; h'$ and $e \;\&\; h' \;\&\; \sim h$ are identical propositions. It follows that the sets of true counterparts of $e \;\&\; h$ and of $e \;\&\; h'$ in $\{B_i\}^{N_\varepsilon}$ are mutually exclusive, and since the set of counterparts of $e \;\&\; (h \vee h')$ is the union of these two sets, we obtain

$$n(h/e; \{B_i\}^{N_\varepsilon})/n(e; \{B_i\}^{N_\varepsilon}) + n(h'/e; \{B_i\}^{N_\varepsilon})/n(e; \{B_i\}^{N_\varepsilon})$$
$$= n(h \vee h'/e; \{B_i\}^{N_\varepsilon})/n(e\; \{B_i\}^{N_\varepsilon}),$$

provided that N_ε is taken large enough that the denominators in all these quotients is non-zero. By assumption 2 we infer

$$[u_1, u_2] \subseteq [r + r' - 2\varepsilon, r + r' + 2\varepsilon].$$

Hence,

$$r + r' - 3\varepsilon \leqslant r'' \leqslant r + r' + 3\varepsilon,$$

and letting $\varepsilon \to 0$ we obtain

$$r'' = r + r'.$$

Proof of Principle (iv): Suppose that $P(h/e) = r$, $P(h'/e \;\&\; h) = r'$, $P(h' \;\&\; h'/e) = r''$. Let B be the Boolean algebra generated from e, h, and h' and $\{B_i\}$ a countable sequence of independent situations equivalent to B with infinitely many true counterparts of e and $e \;\&\; h$. Then by the *Definition* if $\varepsilon > 0$ there exists an integer N_ε such that the following relations hold:

$$|\operatorname{lub}[E(h/e); \{B_i\}^{N_\varepsilon})] - r| < \varepsilon,$$
$$|\operatorname{glb}[E(h/e); \{B_i\}^{N_\varepsilon})] - r| < \varepsilon,$$
$$|\operatorname{lub}[E(h'/h \;\&\; e); \{B_i\}^{N_\varepsilon})] - r'| < \varepsilon,$$
$$|\operatorname{glb}[E(h'/h \;\&\; e); \{B_i\}^{N_\varepsilon})] - r'| < \varepsilon,$$
$$|\operatorname{lub}[E(h \;\&\; h'/e; \}B_i\}^{N_\varepsilon})] - r''| < \varepsilon,$$
$$|\operatorname{glb}[E(h \;\&\; h'/e; \{B_i\}^{N_\varepsilon})] - r''| < \varepsilon.$$

If

$$E(h/e; \{B_i\}^{N_\varepsilon}) = [s_1, s_2],$$
$$E(h'/h \,\&\, e; \{B_i\}^{N_\varepsilon}) = [t_1, t_2],$$
$$E(h \,\&\, h'/e; \{B_i\}^{N_\varepsilon}) = [u_1, u_2],$$

then

$$r - \varepsilon < s_1 \leqslant s_2 < r + \varepsilon,$$
$$r' - \varepsilon < t_1 \leqslant t_2 < r' + \varepsilon,$$
$$r'' - \varepsilon < u_1 \leqslant u_2 < r'' + \varepsilon.$$

By assumption E_2

$$[u_1, u_2] \subseteq [rr' - \varepsilon r' - \varepsilon r - \varepsilon^2, rr' - \varepsilon r' - \varepsilon r + \varepsilon^2]$$

(where the lower bound on the rhs is specified conservatively). Hence

$$rr' - \varepsilon r' - \varepsilon r - \varepsilon^2 - \varepsilon \leqslant r'' \leqslant rr' - \varepsilon r' - \varepsilon r + \varepsilon^2 + \varepsilon.$$

Letting $\varepsilon \to 0$ we obtain $r'' = rr'$. Parallel reasoning yields the second part of Principle (iv).

Our rigorous derivation of Principles (i)—(iv) is now complete. This derivation follows the general strategy of the Adamite proof given in Section 1, but in view of the unavoidable epsilontics, perhaps it should be called a "Solomonic" derivation.

The most disagreeable part of the derivation is the existence assertion in assumption S_1. This assumption may appear palatable if one considers the possibility of taking each B_i to be a finite Boolean algebra of propositions concerning subsets of a region R_i of space. Neither the infinity of different situations (if space is infinite) nor the independence of the various B_i is troublesome; and the equivalence of each B_i to B may be achievable by an appropriate partitioning of R_i. An alternative to this strategy would be to rephrase the *Definition* of $P(h/e)$ counterfactually, asserting the limiting relations conditionally upon the existence of the infinite sequence $\{B_i\}$ and, of course, construing the counterfactual not to be automatically true if the condition is not factually fulfilled. We should not wish to attribute such counterfactual reasoning to Adam, but it would not be beyond the powers of Solomon.

Boston University

REFERENCES AND NOTES

* This paper is based upon a lecture given at the University of Pittsburgh in January 1982. A similar thesis is developed independently by Bas van Fraassen in "Calibration: A Frequency Justification for Personal Probability," in *Physics, Philosophy, and Psychoanalysis: Essays in Honor of Adolf Grünbaum*, ed. by R. S. Cohen and L. Laudan (D. Reidel: Dordrecht and Boston, 1983). I dedicate the paper to Wesley C. Salmon, because he explored with penetration and devotion the epistemic applicability of the frequency concept of probability.

[1] Butler, Bishop Joseph, *The Analogy of Religion, Natural and Revealed, to the Constitution and Course of Nature*, ed. by. G. R. Crooks (Harper: New York, 1868), (originally published in 1736), third paragraph of the Introduction.

[2] Some philosophers have maintained, however, that the frequency concept of probability can be applied epistemically, for example, Hans Reichenbach, *The Theory of Probability* (U. of California: Berkeley, 1949), and Wesley C. Salmon, *The Foundations of Scientific Inference* (U. of Pittsburgh: Pittsburgh, 1966), pp. 83—96, and Wesley C. Salmon, "Statistical Explanation," in *The Nature and Function of Scientific Theories*, ed. by R. G. Colodny (U. of Pittsburgh: Pittsburgh, 1970), pp. 173—231.

[3] Ramsey, Frank P., "Truth and Probability," in *The Foundations of Mathematics and Other Logical Essays* (Routledge and Kegan Paul: London, 1931). Reprinted in *Studies in Subjective Probability*, ed. by H. Kyburg and H. Smokler (Wiley: New York, 1964).

[4] DeFinetti, Bruno, "La prévision: ses lois logiques, ses sources subjectives," *Annales de l'Institut Henri Poincaré* **7**, 1—68 (1937). Reprinted in English translation in *Studies in Subjective Probability*, ed. by H. Kyburg and H. Smokler (Wiley: New York, 1964).

[5] Shimony, Abner, "Coherence and the Axioms of Confirmation," *Journal of Symbolic Logic* **20**, 1—28 (1955).

[6] Richard T. Cox, "Probability, Frequency, and Reasonable Expectation," *American Journal of Physics* **14**, 1—13 (1946).

[7] Good, I. J., *Probability and the Weighing of Evidence* (C. Griffin: London, 1950).

[8] Aczél, J, *Lectures on Functional Equations and their Applications* (Academic Press: New York, 1966).

[9] Shimony, Abner, "Scientific Inference," in *The Nature and Function of Scientific Theories*, ed. by R. G. Colodny (U. of Pittsburgh: Pittsburgh, 1970), 79—172, especially pp. 108—110.

[10] Carnap, Rudolf, *Logical Foundations of Probability* (U. of Chicago: Chicago, 1950), pp. 168—175.

[11] Only in Principle (ii) has it been explicitly stated that e is not the impossible proposition, but this restriction is a necessary condition for clauses like "$P(h/e)$ is a well defined real number" which occur as antecedents in Principles (i), (iii), and (iv).

[12] See Ref. 9, especially Section III.

[13] For a discussion of the effect of correlations on the character of the estimate see Arthur Hobson, "The Interpretation of Inductive Probabilities," *Journal of Statistical Physics* **6**, 189—193 (1972) and Abner Shimony, "Comment on the interpretation of inductive probabilities," *Journal of Statistical Physics* **9**, 187—191 (1973).

ILKKA NIINILUOTO

PROBABILITY, POSSIBILITY, AND PLENITUDE

One recurrent theme of Western metaphysics is the thesis Arthur Lovejoy called the Principle of Plenitude: no genuine possibility can remain forever unrealized. Another traditional idea is the claim that probability is degree of possibility. This equation is by no means unproblematic, since probability and possibility can be interpreted in a variety of ways. Still, it suggests that it is interesting to study what consequences alternative theories of probability may have concerning the validity of the Principle of Plenitude. The aim of this paper is to make such historical and systematic remarks on this question that serve to highlight the subtle differences between the frequency and propensity interpretations of probability.

1. MODALITY AND PROBABILITY AS RESEARCH PROGRAMMES

The logical theory of modality was started by Aristotle with his distinction of the modal notions (necessity, possibility, contingency, impossibility), discussion of the problem of future contingents, and treatment of the basic forms of modal syllogisms.[1] From the Stoics (Diodorus) and the Scholastics (Ockham, Duns Scotus) to the eighteenth century (Leibniz, Wolff, Kant) modality continued to be an important part of logic, but as a research programme it was degenerating — failing to give an adequate analysis of its basic concepts or to lead to important new discoveries.[2] In his 1774 article on Aristotelian logic, Thomas Reid complained that modal syllogistics had, with justice, fallen into "neglect, if not contempt".[3]

Even the birth of modern formal logic in the middle of the nineteenth century did not for some time cure modal theory of its regress. The first syntactic treatment of logical modalities was published by C. I. Lewis as late as in 1912, and the first general accounts of the semantics of intensional languages in terms of possible worlds appeared in the 1950s.

The career of probability theory has been more impressive. The mathematical calculus of probabilities, created by Pascal, Fermat, and

James H. Fetzer (ed.) Probability and Causality. 91—108.

Huygens in the 1650s for games of chance, found soon important applications in many different fields — and the classical formalism received its perfection in the hands of Laplace in 1812. This victorious path was continued in the nineteenth century with the creation of probabilistic theories of scientific inference and with the discovery of the significant role of probability in scientific theory formation, and in our century with the abstract axiomatic treatment of probability measures.[4]

The concept of probability that emerged around 1660 had a dual nature: it was concerned with aleatory chance phenomena, i.e., devices (like dice, cards, roulette) which have powers or abilities to produce series of events with stable long run frequencies, but also with doxastic and epistemic states of opinion and knowledge, i.e., degrees of belief, proof, and certainty.[5] These two faces of probability have been expressed in the *physical* and *epistemic* interpretations: probability as an objective quantitative feature of repeatable events or set-ups in the world, or probability as a rational degree of belief in the truth of a proposition relative to a knowledge situation.[6]

A similar distinction between physical and epistemic interpretations has played a central role in the theory of modality. Philosophers from Scotus and Leibniz to Hegel, Bolzano, and Peirce have argued for the existence of "real" or "physical" possibilities inherent in things. In lieu of this *de re* interpretation, modality has been regarded as a *de dicto* qualifier of our cognitive relation to a judgment. Following Kant's division of modalities into problematic, assertoric, and apodeictic, this epistemic characterization was the official view of the German school logic in the nineteenth century — and it also gave the grounds for Frege and Russell to dismiss modality from their extensional systems of modern logic.[7] New interest in physical modality and probability has been arising in our century, due to the invention of quantum mechanics and the even earlier process Hacking calls the "erosion of determinism".

2. PROBABILITY AS DEGREE OF POSSIBILITY

The feasibility of parallel interpretations, with common standard examples of application, suggests an important link between modality and probability: perhaps the latter is a refined quantitative counterpart to the classificatory notion of modality. In this view, modality and probability would be variants of the same concept — or at least reducible to each other. Leibniz expressed this idea in 1678 with the thesis *probabilitas est gradus possibilitas*.[8]

The equation of probability with degree of possibility, if taken seriously, does not yet tell which of these notions is more basic. Thus, the way is open to two attempted reductions.

First, *probability can be reduced to modality*. This strategy was followed in the classical definition of probability as "the ratio of the number of favourable cases to that of all the cases possible", where the various cases are "equally possible" in the epistemic sense.[9] German textbooks of school logic discussed in this spirit Probability as a subsection in their chapter on Modality. As present-day supporters of this programme, we may count those philosophers who employ the possible worlds semantics in their analysis of physical probability.[10]

Secondly, one may try to *reduce modality to probability*. If a quantitative measure of probability has been defined without invoking other modal concepts, then necessity could be defined by probability one, impossibility by probability zero, and possibility by non-zero probability.[11] A variant of this view is defended by Popper: probabilistic propensities are introduced as theoretical entities (like forces) by scientific theories, and these "degrees of physical possibility" then give us whatever insight we have on modality.[12]

This second strategy is especially appealing to those nominalist empiricists, who wish to give an extensionalist statistical treatment of modal statements. John Venn's evaluation in 1876 is characteristic of this view:

> . . . the logicians, after having had a long and fair trial, have failed to make anything satisfactory out of this subject of the modals by their methods of enquiry and treatment; and . . . it ought, therefore, to be banished entirely from that science, and relegated to Probability.[13]

Venn's *The Logic of Chance* (1866) was the first systematic attempt to develop the frequency interpretation of probability as a "proportion in the long run" — expressed in 1843 by J. S. Mill and R. L. Ellis with the definition of equipossibility by "equally frequent occurrence" of two events.[14]

Perhaps the most prominent of Venn's present successors[15] is Wesley Salmon, who wishes to avoid degrees of possibility in his theory of probability and explanation[16] and regards the possible-worlds approach as "inadvisable" to a "wide variety of philosophical problems".[17]

3. EPISTEMIC POSSIBILITY

The standard definition of epistemic possibility can be formulated in the following way. Let b be a proposition or a theory which expresses our knowledge situation at a given time t. Then the statement h is *epistemically possible* relative to b at time t if and only if b does not entail the negation of h, i.e.,

(1) $b \nvdash \sim h$.[18]

Condition (1) is equivalent to

(2) $\nvdash \sim (b \mathbin{\&} h)$.

Hence, h is epistemically possible at time t if and only if h is logically compatible with our knowledge at t. Further, h is epistemically possible *simpliciter*, i.e., relative to a tautology b, if and only if h is logically consistent.

Let P be an epistemic probability measure over the relevant language, so that $P(h/b)$ is, at time t, our rational degree of belief in the truth of h. Then, by the probability axioms, $P(h/b) = 0$ if h is not epistemically possible relative to b. The converse holds, if P is a so-called *regular* probability measure. Hence.

(3) Assuming that P is a regular probability measure, h is epistemically possible relative to b if and only if $P(h/b) > 0$.

In this sense, *epistemic probability is degree of epistemic possibility*. This conclusion can be strengthened by noting that a regular probability measure for a language **L** has to assign a non-zero weight for each possible state of affairs expressible in **L**.[19]

As epistemic possibility (or consistency) is not a guarantee of truth, our observations so far do not yet have anything to do with the Principle of Plenitude.[20] Suppose now that A is a generic event or event-type which may or may not occur in repeated independent trials at times $1, 2, \ldots, t, \ldots$. Let A_t be the proposition that A occurs at time t. Then we may ask whether the epistemic possibility of A guarantees the occurrence of A at some time t:

(4) If $P(A_t/b) > 0$ for all $t = 1, 2, \ldots$, then A_t for some t.

Principle (4) is trivially valid in the case, where evidence b asserts that A has already occurred at least once.

Another special case is more interesting: b asserts that event A has a constant physical probability r for all t, and the epistemic probability measure P satisfies $P(A_t/b) = r > 0$ for all t. This case reduces to the question about the relation of physical possibility and plenitude (see Sections 4—5 below).

But in general there is no guarantee for the truth of (4): $P(A_t/b) > 0$ only means that A_t is logically compatible with b. Thus, epistemic possibilities, even if repeated infinitely often, need not become actual.

4. STATISTICAL POSSIBILITY

According to Bertrand Russell, propositions can only be true or false, so that "the whole doctrine of modality only applies to propositional functions, not to propositions". A propositional function is necessary, if it is true for all values of its argument; possible, if true for some value; and impossible, if false for all values.[21]

Russell's proposal covers two main non-epistemic *extensional* treatments of modality. First, if the arguments of propositional functions range over individuals, necessity reduces to universal quantification and possibility to existential quantification.[22] For example, 'if x is a man, x is mortal' is necessary, and 'x is a man' is possible. Secondly, if the argument ranges over points of time (e.g., 'It rains in Helsinki at time t'), then necessity equals always true, possibility sometimes true, and impossibility never true.[23]

The Principle of Plenitude is an immediate consequence of the statistical concept of modality: if 'x is a man' is possible, then some actual thing is a man; if rain in Helsinki is possible, then it rains sometimes in Helsinki. More generally,

(5) If 'x is an F' is possible, then there exists an F.

(6) If event A is possible, then A_t for some point of time t.

Thus, no statistical possibility can remain for ever unactualized.

The frequency theory of probability is a direct generalization of the statistical concept of modality. For example, Venn argues that a modal statement of the form 'All poisonings by arsenic are probably mortal' does not apply the predicate 'probably mortal' to each poisoning individually, but rather the predicate 'mortal' is "to be applied to a *portion* (more than half) of the members denoted by the subject".[24] Similarly, the probability that a toss with a given coin gives heads refers to the proportion of tosses with heads in the larger class of all tosses. Thus, according to the finite frequency interpretation, a statement of the form 'An F is a G with probability r', or $P(G/F) = r$, means that the relative frequency of attribute G in the finite reference class F is r:

(7) $rf(G/F) = |G \cap F| / |F| = r.$

If the reference class F is an infinite "series", and F_n is the initial segment of the first n members of F, then the probability $P(G/F)$ is identified with the limit

(8) $\lim_{n \to \infty} rf(G/F_n) = \lim_{n \to \infty} |G \cap F_n| / n.$[25]

Venn's treatment is extensional, since the truth conditions for probability statements refer to the history of *one* world only. Elsewhere he gave a vivid picture of his conception of the external world of past and future facts:

> We contemplate the world of phenomena as if it resembled some vast scroll, unrolled to a certain extent before our eyes, but written upon in the same sort of characters from beginning to end; or rather, since we do not recognize either beginning or end, inscribed with writing which may be traced from the midst indefinitely in both directions. Of the unopened parts we guess at the contents from what we have read of the rest, though even of this opened part we can at present decipher only a fragment.[26]

It follows that the frequency theory of probability is committed to the Principle of Plenitude. For finite reference classes,

(9) When F is finite, $rf(G/F) > 0$ if and only if at least one F is a G.

Relative frequency is here a *degree of statistical possibility* in a straightforward sense. For limits of relative frequencies, we have the result: if $P(G/F) = \lim_{n \to \infty} rf(G/F_n)/n > 0$, then, for some n, $rf(G/F_n) > 0$, so that G occurs in the finite series. Hence, for a frequentist probability P,

(10) If $P(G/F) > 0$, then some F is a G.

For repeatable events A, this result can be formulated as follows:

(11) If $P(A) > 0$, then A_t for some t.

The converses of (10) and (11) do not hold generally, since the limit (8) can be 0 even if G occurs sometimes (but not too often) in the series F. Thus, $P(A) = 0$ does not mean that A is impossible, or $P(A) = 1$ that A is necessary. However, Per Martin-Löf has proposed a frequency theory with a strong condition of randomness for the reference class where the converses of (10) and (11) hold.[27] For Martin-Löf, probability is degree of statistical possibility in the strong sense that

(12) $P(A) > 0$ if and only if A_t for some t.

A. A. Cournot presented in 1843 a frequentist theory where mathematical probability is "the limit of physical possibility":

> . . . it is natural to regard each event as having a stronger tendency to occur, or as being more possible in fact or physically, in proportion as it is produced more often in a large number of cases.[28]

But Cournot also added that some events are known *a priori* to be "physically impossible". He refers here to examples of events which, in modern terms, have measure zero: the center of a disk lands precisely on the intersection point of the diagonals. Cournot's claim that such events never in fact happen is problematic, however. If an experiment is represented by an idealized mathematical description, where the space of outcomes is infinite (e.g., the set of real numbers, the set of infinite binary sequences), then we have to take seriously the idea that each trial realizes an outcome which is "physically impossible" in the sense of having measure zero in the model.[29]

What has been said above applies, with some qualifications, to the *hypothetical frequency interpretation* as well. Some possibilities may fail to become actual for the reason that the chance device is destroyed. For example, it is possible to get heads in a throw with this coin; but this possibility never realizes if I break or melt the coin before any tosses have actually been made. Instead of employing a reference class which belongs to the history of the actual world, the hypothetical frequency interpretation defines the probability $P(G/F)$ as the limit of the relative frequency that G *would* have in an infinite series of F's. Similarly, $P(A)$ is the limit of the relative frequency that event A *would* have in an infinite number of trials.

According to this view, $P(G/F) = r$ is true if and only if the limit of $rf(G/F)$ cquals r in cvcry physically possible world where the extension of F is an infinite series.[30] Hence, the Principle of Plenitude is valid in the conditional form:

(13) If $P(G/F) > 0$ and F is infinite, then at least some F is a G.

Similarly,

(14) If $P(A) > 0$ and A is subjected to an infinite series of trials, then A_t for some t.

Some formulations of the frequency interpretation emphasize the "hypothetical" nature of infinite reference classes F — suggesting somewhat carelessly that such a class or "ensemble" consists not only of the finite number of actual F's but also the potentially infinite number of "possible" F's. Taken seriously, this idea could be formulated by taking F^{ph} to be the union of the extensions F^{w} of F in all physically possible worlds w, and defining $P(G/F)$ as the limit of $rf(G^{ph}/F^{ph})$. It is doubtful that this idea has any meaning, since there seems to be no non-arbitrary way of ordering the infinite class F^{ph}.[31] If F^{ph} were finite, this definition would make sense. However, unlike other formulations of the frequency interpretation, it would not generally satisfy the Principle of Plenitude: $rf(G^{ph}/F^{ph}) > 0$ need not imply that $rf(G/F) > 0$ in the actual world.

5. SINGLE-CASE PROPENSITIES

The classical definition applied epistemic probability to singular events, such as a single toss with a given coin. Jacques Bernoulli's *Ars conjectandi* (1713) contained a celebrated theorem which linked such single-case probabilities to observable relative frequencies.

Let $rf_n(A)$ be the relative frequency of event A in the first n trials. If the trials are independent with a constant probability $P(A_t) + p$ $(t = 1, 2, \ldots)$, then Bernoulli's Theorem says that, for any $\varepsilon > 0$,

(15) $P(|rf_n(A) - p| > \varepsilon) \to 0$, when $n \to \infty$.

E. Borel strengthened this result in 1909 to the form which says that the relative frequency converges *with probability one* to the value p:

(16) $P(rf_n(A) \xrightarrow[n \to \infty]{} p) = 1.$

Bernoulli's and Borel's Theorems have a great significance for those theorists who are ready to assign physical probabilities to single cases, since they indicate that such objective probability statements can be tested by repeated observations. This is the basic idea of the *single-case propensity interpretation*: a singular probability statement of the form $P_x(G/F) = r$ assigns to a physical set-up x a numerical disposition of strength r to produce an outcome of type G in each trial of type F.[32] Such statements are modal or non-extensional, but they can be tested by subjecting x (or other set-ups similar to x) to repeated trials of type F. It is no surprise that most supporters of this view regard propensities as *degrees of physical possibility*.[33]

Venn was a merciless critic of the assumption that one could associate with some chance device (in relation to the agencies that influence it) an "objective probability" which then develops into a sequence exhibiting uniformity. For Venn, this idea is "one of the last remaining relics of Realism, which after being banished elsewhere still manages to linger in the remote province of Probability".[34] He goes so far as to claim that the basis on which the mathematics behind Bernoulli's Theorem rests is "faulty".[35]

Borel's Theorem (16) is even more perplexing for a frequentist. As a special case of the Strong Law of Large Numbers, it is the strongest mathematical result about the connection between probability and relative frequency. Still, it does not guarantee that probability equals limiting frequency — this holds only "almost surely" in the sense that the probability measure of those infinite sequences, where the limit differs from probability, has measure zero.[36] It does not follow that such sequences are physically impossible or never happen.[37]

Thus, the frequency theory which *defines* probability as the limit of relative frequency is much stronger than Borel's Theorem. In my view, this means that the step from condition (16) to definition (8) is illegitimate: a necessary condition for stipulating a definitional connection between two quantities is that they are at least extensionally equivalent in the actual world, but there is no guarantee for this minimum assumption here. Therefore, the frequency theory of probability should be rejected.[38]

If the single-case propensity interpretation is accepted, then the Principle of Plenitude, even in its conditional form, is false. If $P(A) = p > 0$, then Borel's Theorem guarantees that *with probability one* $\lim rff_n(A) = p > 0$, but it does not exclude the possibility that A never occurs. Principles (13) and (14) have to be replaced by

(17) If $P_x(G/F) > 0$ and F is infinite, then with probability one set-up x produces outcome G for some trial F.

(18) If $P(A_t) = p > 0$ for all $t = 1, 2, \ldots$ and A is subjected to an infinite number of trials, then with probability one A_t for some t.

In brief, the Principle of Plenitude is "almost true" — but false nevertheless. This is the decisive difference between the single-case propensity and (actualist or hypothetical) frequency theories of physical probability.[39]

This difference can still be illustrated by a couple of examples.

Bunge points out correctly that "what is merely possible in the short run . . . may become necessary or nearly so in the long run". His most careful formulation is indeed a consequence of Bernoulli's and Borel's Theorems: "all possible repetitive chance events . . . are likely to occur in the long run".[40] However, Bunge also mentions careless slogans which are clearly variants of the Principle of Plenitude: "Give chance a chance and it will become necessity", "Anything that is not forbidden [in microphysics] is compulsory", "Any given ecological niche, where life is possible, ends up by being inhabited".[41]

When Reichenbach's axiomatic treatment of probability in *The Theory of Probability* (2nd ed., 1949) did not yet give him the frequency definition for "normal sequences", he added an Axiom of Interpretation: "If an event C is to be expected in a sequence with a probability converging toward 1, it will occur at least once in the sequence". Salmon quotes Reichenbach approvingly: "This is a rather modest assumption".[42] But we have seen that this is not "modest": it entails the Principle of Plenitude which distinguishes the frequentists from the propensitists.

6. LONG-RUN PROPENSITIES

The existence of single-case propensities (at least those other than 1 and 0) means that objective chance is real — the world is indeterministic. Similarly, Salmon argues that his frequentist probabilities relative to objectively homogeneous reference classes presuppose indeterminism. Such accounts of physical probability can, therefore, be plausibly applied at least to microphysical phenomena, such as radioactive decay.

But what happens with the paradigmatic examples of probabilistic phenomena, the classical games of chance involving macroscopic devices? There is no direct evidence that their behaviour would be influenced by quantum mechanical chance. Moreover, there are perfectly good *deterministic* descriptions of such games within the formalism of classical mechanics: the result of a toss of a coin (heads H or tails T) is uniquely determined by mechanical laws and the initial condition of the coin (its weight, shape, density, initial velocity, angular velocity). Further, these deterministic models explain the characteristic stability of relative frequencies by means of the structure of the initial state space.[43]

The idea of assigning single-case propensities to such deterministic systems seem inappropriate. But also the frequency interpretation (actual or hypothetical) fails for these examples. If a symmetric coin is tossed repeatedly, it *is* physically possible that the result is always H — and the limiting relative frequency of H is thereby 1. This possibility, which could be actual, should not be excluded by a definition.[44] An epistemic interpretation is likewise out of the question, since it would not explain the fact that relative frequencies are observed (rather than merely believed) to be stable.

Several reactions to this argument are possible. One could argue with Popper that it is not helpful to make statistical assumptions about the initial conditions, since this only "shifts the issue one step back".[45] However, the main point in the deterministic explanation of coin tossing is that *any* continuous distribution on the space of initial conditions yields sequences with the same limiting relative frequency (say, 1/2). This explains the fact that our efforts to control the initial conditions have no effect on the long run frequencies. We can never intentionally produce a series of tosses such that all the initial conditions belong to the preimage of H. At best we can guarantee that the initial state belongs to a "macrostate" — and the proportion of microstates leading to H in any such macrostate is 1/2.

Should we then make the probabilistic assumption that the coin with the tossing method has a propensity 1/2 to produce heads?[46] Not if single-case propensities are meant, since this would entail indeterminism.[47] Instead, we could say that the procedure of tossing a symmetric coin has a *long-run propensity* 1/2, i.e., a disposition to produce such infinite sequences where the limit of the relative frequency of H is 1/2.[48]

Alternatively one could claim that the deterministic description of macroscopic devices is idealized, so that the microstates should not be assumed to be physically real. On the level of realistic physical description, the sequences of tosses would then be random in the Kolmogorov sense of maximal complexity[49] or objectively homogeneous in Salmon's sense.[50]

This account is compatible with the long-run propensity interpretation: the tossing device can be said to have a certain physical dispositional property, viz. the ability to produce random sequences with a characteristic long-run frequency. However, when the underlying deterministic description is rejected, this ability is no longer explained by the physical structure of the coin and the nature of the tossing method.

For this reason, I am inclined to prefer the deterministic long-run propensity account for games of chance. This approach has the virtue (as I see it) that it is not committed to the Principle of Plenitude. In the space of all possible infinite sequences of initial states, the class of those sequences with the limiting frequency different from 1/2 has measure zero. Hence, in tossing a symmetric coin, heads H is "destined" to happen sooner or later in Peirce's sense — not according to Plenitude but with probability one.

University of Helsinki

NOTES

1 See Kneale and Kneale (1962), Hintikka (1973), Hintikka *et al.* (1977).

2 For historical essays on modality, see Knuuttila (1981). Cf. also the interesting remarks by Peirce (1901). Of course, I don't want to deny that authors like Scotus, Leibniz, Bolzano, and Peirce had ingenious insights about modality. However, it was only after the efforts of C. I. Lewis, Lukasiewicz, Becker, Prior, von Wright, and others that progressive systematic work in this field could start in the 1950s (Kanger, Hintikka, Kripke).

3 See Reid (1846), p. 703.

4 For the history of probability, see Maistrov (1974) and Hacking (1975).

5 See Hacking (1975). The association of the concept of probability both with frequencies and opinions goes back to ancient and medieval discussions.

6 For discussions of the interpretations of probability, see Kyburg and Smokler (1964), Salmon (1967), Fetzer (1981), Suppes (1984).

7 Cf. Niiniluoto (1987b) and other articles of Knuuttila (1987).

8 See Hacking (1975), p. 125. Not all explications for degrees of possibility are probabilistic. For a recent proposal in the framework of fuzzy logic, see Zadeh (1978).

[9] This is Laplace's formulation (Laplace, 1952, p. 11).
[10] See Suppes (1974, 1984), Gärdenfors (1975), Bigelow (1976), Giere (1976), Smokler (1979a), Fetzer (1981), Niiniluoto (1982). See also Kyburg (1974) and the "modal frequency interpretation" of van Fraassen (1980). For a discussion of the modal nature of epistemic probability, see Smokler (1979b).
[11] Cf. Bunge's (1977) definition of modality in terms of a stochastic theory relative to a discrete probability space. However, Bunge also uses a Bolzano-type definition of real possibility as a "lawful fact", i.e., a state not contradicted by a law of nature (*ibid.*, p. 173). He claims that his definition does not involve any "modal logic". Bunge's aversion to possible worlds semantics seems to be directed against the radical type of modal realism, defended by David Lewis (1986). However, his treatment of law statements as equations relative to the state space can be claimed to presuppose — or to be equivalent with — moderate modal realism, where worlds are represented by some linguistic or set-theoretical structures (cf. Niiniluoto, 1987a). (Already Suppes, 1974, proposes that the elements of a sample space of a random variable correspond to "possible worlds".)
[12] See Popper (1982).
[13] See Venn (1888), p. 296.
[14] See Salmon (1981). Venn's remarks on modality are discussed in Niiniluoto (1987b).
[15] Salmon reports that Hans Reichenbach, who supervised his doctoral dissertation on Venn in 1950, did not have any "first-hand knowledge of Venn's work on probability" (Salmon, 1981, p. 126). This may sound surprising, if one does not note that Reichenbach's early inspiration for a frequentist theory of probability came — instead of from Venn — from von Kries and Poincaré. While Reichenbach *defined* equipossibility in terms of long run frequencies, he *justified* the hypothesis of equipossibility by physical symmetry considerations (see Reichenbach, 1949, §69; 1978). He also emphasized that the existence of such frequentist probabilities in games of chance "involves no contradiction to the principle of causality", i.e., to determinism (Reichenbach, 1978, p. 316) — unlike Salmon, who takes the assumption of "objectively homogeneous reference classes" to entail indeterminism (Salmon, 1984). For a revival of the physical symmetry approach, see von Plato (1983, 1987), Kamlah (1983), Keller (1986), and Suppes (1987a).
[16] See Salmon (1984), pp. 112—113. Salmon's position here is not entirely clear. He promises an account of probabilistic causation, where "we can measure the degree of strength of a tendency without getting involved in degrees of possibility" (*ibid.*, p. 113), but he ends up with the concession that propensities or "probabilistic dispositions" seem to "lie at the foundation of probabilistic causality" (*ibid.*, p. 204). Salmon adds an argument (due to Paul Humphreys) for "refusing to speak of propensity interpretation of probability". Suppose that a box contains 125 defective can openers, 25 produced by machine A and 100 by machine B. The probability that a can opener b randomly picked from the box is produced by B is then 4/5. But, Salmon adds, it does not make sense "to say of the can opener that it has a certain propensity to have been produced by that machine" (*ibid.*, p. 205). Of course not: rather the mechanism used for sampling may have a propensity 0.8 to pick out an opener produced by B. Further, the probability for the hypothesis that b is produced by B, give that b was picked out from the box, is an *epistemic* one. Hence, Salmon's argument should not be construed against

the propensity interpretation, as he does. At best it may show that some probabilities are not propensities.

[17] See Salmon (1984), p. 192n. It should be noted that Salmon's concept of an objectively homogeneous reference class essentially involves something like quantification over physical properties (cf. *ibid.*, p. 62). However, if one does not wish to admit universals in one's ontology, then the most plausible way of characterizing properties is to define them as functions from possible worlds to extensions. This suggests that the three rival views about explanation — epistemic, modal, and ontic (Salmon, 1984) — are perhaps not so exclusive of each other as Salmon thinks. (Cf. the treatment in Niiniluoto, 1982.)

[18] See, e.g., Frege (1967), p. 13, Hacking (1967), Levi (1980), Fetzer (1981). The definition of C. I. Lewis replaces logical entailment $\vdash$ by strict implication.

[19] Thus, the inductive probability $P(h/b)$ depends on the weight of those states of affairs admitted by h among those admitted by b (cf. Niiniluoto, 1983). See also Smokler (1979a).

[20] For the history of this principle, see Knuuttila (1981, 1987).

[21] See Russell (1956), pp. 230—231. Russell's claim can be expressed also by saying that modality is applicable only to generic events — instead of singular events.

[22] Cf. also Frege (1967), p. 13.

[23] For the history of this interpretation, see Hintikka (1973), Knuuttila (1981, 1987). As Russell does not apply this statistical concept of modality to temporally definite propositions, he avoids the problem that teased the Aristotelians: all true propositions turn out to be necessary.

[24] Venn (1888), p. 299.

[25] Cf. Venn (1888), pp. 163—164. For difficulties in this definition, see van Fraassen (1980).

[26] Venn (1907), p. 11.

[27] See Martin-Löf (1966).

[28] See Cournot (1956), pp. 39—48.

[29] von Plato (1977) suggests a "quasi-plenum hypothesis" for the case, where the set of outcomes is an infinite metric space (e.g., the phase space of statistical mechanics). If a continuous probability distribution is defined on this space, and if we consider arbitrarily dense finite partitions of the space, then the following holds: with probability one, there are states as close as desired to any given state which will be realized.

[30] Cf. Kyburg (1974), Giere (1976), Fetzer (1981).

[31] Cf. Niiniluoto (1982). On the other hand, there might be non-arbitrary orderings of each of the classes F^w.

[32] See Popper (1982), Mellor (1971), Giere (1976, 1979), Smokler (1979a), Fetzer (1981), Niiniluoto (1982).

[33] Fetzer (1981) makes a distinction between propensity one and a disposition of "universal strength u". Such a distinction is needed if the outcome space is continuous or if infinite sequences are considered (cf. the remarks on Cournot). Giere (1979) discusses these difficulties for the equation of necessity with propensity one.

[34] See Venn (1888), p. 92. The young Peirce greeted in 1867 Venn's "nominalistic" definition with enthusiasm, but the old Peirce returned to realism in 1910.

[35] Venn (1888), p. 91.

[36] Ergodic theorems may be regarded as generalizations of Borel's Theorem. They state that an ergodic system "visits" a state A with a limiting time frequency which almost everywhere equals the measure $P(A)$ of set A. But again such an equality holds only with probability one. The Boltzmann-Einstein conception of physical probability as a limit of time average is thus a variant of the frequency interpretation — and equally problematic for the same reasons. See von Plato (1987) for a discussion of this problem area.
[37] For example, in a series of n independent tosses with a symmetric coin, all sequences of heads and tails have the same probability $1/2^n$. Thus, all these sequences are equipossible. The same conclusion holds in the limiting case: the infinite sequence consisting of tails only is as equally possible as any other particular infinite sequence of results.
[38] Here I agree with the evaluation of Stegmüller (1973). If the frequency theorist tries to take the second-order probability P in (16) seriously, instead of ignoring its existence, he runs again into troubles: if P is a frequency probability, his definition becomes circular; if P expresses "practical certainty" (see Cramér, 1946, p. 148), his definition reduces physical probabilities to epistemic ones.
[39] In the next section, I argue that this holds also for long-run propensities.
[40] Bunge (1977), p. 207.
[41] These examples could be easily multiplied: every planet, where the conditions make the birth of life possible, will be inhabited by living organisms.
[42] Reichenbach (1949), p. 345; Salmon (1977), p. 78.
[43] For the history of such deterministic accounts, see von Plato (1983). Poincaré's treatment of roulette is discussed by Reichenbach (1949, 1978). A recent clear exposition of coin tossing is given by Keller (1986). He shows that, for a circular balanced coin with initial vertical velocity u and angular velocity ω, the preimages of H and T in the u-ω-plane are alternating strips between hyperbolas, and the vertical width of the strips approaches zero when u increases. See also Suppes (1987a).
[44] This is a reformulation of our earlier remark about the frequency theory and Borel's Theorem.
[45] See Popper (1982), p. 99.
[46] This is Popper's (1982) conclusion. Also Keller (1986) emphasizes that randomness enters his deterministic model of coin tossing through the initial distribution. Suppes (1987a), who does not discuss the interpretation of the initial distribution, sees "no general reason" for not using Keller's model to "compute single-case probabilities" (p. 336). In his reply to M. C. Galavotti, Suppes (1987b) says that the issue between single-case and long-run propensities is "a red herring" (p. 371). But he seems to mean only that, as any propensitist could agree, the propensity theory can be used for making both single-case and long-run frequency *predictions* (p. 372).
[47] Giere (1976) suggests that single-case propensities could be applied to coins *instrumentalistically*, i.e., without a realistic interpretation. Suppes (1987a) presents an example of a purely deterministic three-body system which generates random sequences without any initial distribution. He concludes that "the wedge that has sometimes been driven between propensity and determinism is illusory" (Suppes, 1987b, p. 372). However, if this example involves a propensity, it clearly is a *long-run* propensity with universal strength — and there is no wedge between such propensities and determinism.

[48] See Niiniluoto (1982), p. 447. I now think that this disposition does not have a universal strength.

[49] See Ford (1983).

[50] Salmon's most recent definition refers to physical properties about which a computer may receive (and process) information from a physical detector (Salmon, 1984, p. 68). A microstate represented by a point in continous space requires an infinite amount of information, however.

BIBLIOGRAPHY

Bigelow, J. (1976), 'Possible Worlds Foundation for Probability', *Journal of Philosophical Logic* **5** (1976), 299—320.

Bunge, M. (1977), *The Furniture of the World* (Treatise on Basic Philosophy 3), D. Reidel, Dordrecht, 1977.

Cournot, A. A. (1956), *An Essay on the Foundations of Our Knowledge*, The Liberal Arts Press, New York, 1956.

Cramér, H. (1946), *Mathematical Methods of Statistics*, Princeton University Press, Princeton, 1946.

Fetzer, J. H. (1981), *Scientific Knowledge: Causation, Explanation, and Corroboration*, D. Reidel, Dordrecht, 1981.

Ford, J. (1983), 'How Random is a Coin Toss?', *Physics Today* (1983), 40—47.

Fraassen, B. van (1980), *The Scientific Image*, Oxford University Press, Oxford, 1980.

Frege, G. (1967), *Begriffsschrift* (1879). English translation in J. van Heijenoort (ed.), *From Frege to Gödel: A Source Book in Mathematical Logic, 1879—1931*, Harvard University Press, Cambridge, Mass., 1967, pp. 1—82.

Gärdenfors, P. (1975), 'Qualitative Probability as an Intensional Logic', *Journal of Philosophical Logic* **4** (1975), 171—185.

Giere, R. (1976), 'A Laplacean Formal Semantics for Single-Case Propensities', *Journal of Philosophical Logic* **5** (1976), 321—353.

Giere, R. (1979), 'Propensity and Necessity', *Synthese* **40** (1979), 439—451.

Hacking, I. (1967), 'Possibility', *The Philosophical Review* **76** (1967), 143—168.

Hacking, I. (1975), *The Emergence of Probability*, Cambridge University Press, Cambridge, 1975.

Hintikka, J. (1973), *Time and Necessity*, Oxford University Press, Oxford, 1973.

Hintikka, J. (with U. Remes and S. Knuuttila) (1977), *Aristotle on Modality and Determinism*, Acta Philosophica Fennica **29**: 1, North-Holland, Amsterdam, 1977.

Kamlah, A. (1983), 'Probability as a Quasi-Theoretical Concept — J. V. Kries' Sophisticated Account after a Century', *Erkenntnis* **19** (1983), 239—251.

Keller, J. B. (1986), 'The Probability of Heads', *American Mathematical Monthly* **93** (1986), 191—197.

Kneale, W. and Kneale, M. (1962), *The Development of Logic*, Oxford University Press, Oxford, 1962.

Knuuttila, S. (ed.) (1981), *Reforging the Great Chain of Being*, D. Reidel, Dordrecht, 1981.

Knuuttila, S. (ed.) (1987), *Modern Modalities*, D. Reidel, Dordrecht, 1987.

Kyburg, H. E. Jr. (1974), 'Propensities and Probabilities', *British Journal for the Philosophy of Science* **25** (1974), 358–375.
Kyburg, H. and Smokler, H. (eds.) (1964), *Studies in Subjective Probability*, J. Wiley and Sons, New York, 1964.
Laplace, P. S. (1952), *A Philosophical Essay on Probabilities*, Dover, New York, 1952.
Levi, I. (1980), *The Enterprise of Knowledge*, The MIT Press, Cambridge, Mass., 1980.
Lewis, D. (1986), *On the Plurality of Worlds*, Blackwell, Oxford, 1986.
Maistrov, L. E. (1974), *Probability Theory: A Historical Sketch*, Academic Press, New York, 1974.
Martin-Löf, P. (1966), 'On the Definition of Random Sequences', *Information and Control* **9** (1966), 602–619.
Mellor, D. H. (1971), *The Matter of Chance*, Cambridge University Press, Cambridge, 1971.
Niiniluoto, I. (1982), 'Statistical Explanation Reconsidered', *Synthese* **48** (1982), 437–472.
Niiniluoto, I. (1983), 'Inductive Logic as a Methodological Research Programme', *Scientia: Logic in the 20th Century*, Milano, 1983, pp. 77–100.
Niiniluoto, I. (1987a), *Truthlikeness*, D. Reidel, Dordrecht, 1987.
Niiniluoto, I. (1987b), 'From Possibility to Probability: British Discussions on Modality in the Nineteenth Century', in Knuuttila (1987).
Peirce, C. S. (1901), Articles 'Modality', 'Necessary', 'Possibility', 'Probability', in J. M. Baldwin (ed.), *Dictionary of Philosophy and Psychology*, P. Smith, Cloucester, Mass., 1901.
Plato, J. von (1977), 'The Realization of Possibilities according to Probability Theory, and Statistical Theories of Physics', in I. Niiniluoto, J. von Plato, and E. Saarinen (eds.), *Studia Excellentia*, Reports from the Department of Philosophy, University of Helsinki, N:o 3/1977, pp. 54–62.
Plato, J. von (1983), 'The Method of Arbitrary Functions', *The British Journal for the Philosophy of Science* **34** (1983), 37–47.
Plato, J. von (1987), 'Probabilistic Physics the Classical Way', in L. Krüger, G. Gigerenzer, and M. Morgan (eds.), *The Probabilistic Revolution*, vol. 2, The MIT Press, Cambridge, Mass., 1987, pp. 379–407.
Popper, K. R. (1982), *The Open Universe: An Argument for Indeterminism*, Rowman and Littlefield, Totowa, New Jersey, 1982.
Reichenbach, H. (1949), *The Theory of Probability*, 2nd ed., University of California Press, Berkeley and Los Angeles, 1949.
Reichenbach, H. (1978), *Selected Writings, 1909–1953*, vol. 2, D. Reidel, Dordrecht, 1978.
Reid, T. (1846), *The Works* (ed. by W. Hamilton), vols. I–II, Maclachlan and Stewart, Edinburgh, 1846. (8th ed. 1880.)
Russell, B. (1956), 'The Philosophy of Logical Atomism', in *Logic and Knowledge*, Allen & Unwin, London, 1956.
Salmon, W. (1967), *The Foundations of Scientific Inference*, University of Pittsburgh Press, Pittsburgh, 1967.
Salmon, W. (1977), 'The Philosophy of Hans Reichenbach', *Synthese* **34** (1977), 5–88.
Salmon, W. (1981), 'John Venn's *Logic of Chance*', in J. Hintikka, D. Gruender, and E.

Agazzi (eds.), *Proceedings of the 1978 Pisa Conference on the History and Philosophy of Science*, vol. 2, D. Reidel, Dordrecht, 1981, pp. 125—138.

Salmon, W. (1984), *Scientific Explanation and the Causal Structure of the World*, Princeton University Press, Princeton, 1984.

Smokler, H. (1979a), 'Single-Case Propensities, Modality, and Confirmation', *Synthese* **40** (1979), 497—506.

Smokler, H. (1979b), 'The Collapse of Modal Distinctions in Probabilistic Contexts', *Theoria* **45** (1979), 1—7.

Stegmüller, W. (1973), *Personelle und statistische Wahrscheinlichkeit*, Springer-Verlag, Berlin, 1973.

Suppes, P. (1974), 'The Essential but Implicit Role of Modal Concepts in Science', in K. F. Schaffner and R. S. Cohen (eds.), *PSA 1972*, D. Reidel, Dordrecht, 1974, pp. 305—314.

Suppes, P. (1984), *Probabilistic Metaphysics*, Blackwell, Oxford, 1984.

Suppes, P. (1987a), 'Propensity Representations of Probability', *Erkenntnis* **26** (1987), 335—358.

Suppes, P. (1987b), 'Some Further Remarks on Propensity: Reply to Maria Carla Galavotti', *Erkenntnis* **26** (1987), 369—376.

Venn, J. (1866), *The Logic of Chance*, Macmillan, London, 1866. (3rd ed. 1888.)

Venn, J. (1907), *The Principles of Empirical, or Inductive Logic*, London, 1889. (Reprint of the 2nd ed. in 1907: Chelsea Publ. Co., New York, 1973.)

Zadeh, L. (1978), 'Fuzzy Sets as a Basis for a Theory of Possibility', *Fuzzy Sets and Systems* **1** (1978).

JAMES H. FETZER

PROBABILISTIC METAPHYSICS

The demise of deterministic theories and the rise of indeterministic theories clearly qualifies as the most striking feature of the history of science since Newton, just as the demise of teleological explanations and the rise of mechanistic explanations dominates the history of science before Newton's time. In spite of the increasing prominence of probabilistic conceptions in physics, in chemistry and in biology, for example, the comprehensive reconciliation of mechanistic explanations with indeterministic theories has not gone smoothly, especially by virtue of a traditional tendency to associate "causation" with determinism and "indeterminism" with non-causation. From this point of view, the very idea of *indeterministic causation* seems to be conceptually anomalous if not semantically inconsistent. Indeterminism, however, should not be viewed as the absence of causation but as the presence of causal processes of non-deterministic kinds, where an absence of causation can be called "non-causation".

The underlying difference between causal processes of these two kinds may be drawn relative to events of one type (radioactive decay, genic transmission, coin tossing, and so on) as "causes" C of events of other types (beta-particle emission, male-sex inheritance, coming-up heads, and so forth) as "effects" E. Then if deterministic causal processes are present whenever "the same cause" C invariably brings about "the same effect" E, then indeterministic causal processes are similarly present whenever "the same cause" C variably brings about "different effects" E^1, E^2, . . . belonging to some specific class of possible results. So long as beta-particle emission (male-sex inheritance, coming-up heads, . . .) are invariable "effects" of radioactive decay (of genic transmission, of coin tossing, . . .), respectively, then these are "deterministic" causal processes; but if alpha-particle emission (female-sex inheritance, coming-up tails, . . .) are also possible, under the very same conditions, respectively, the corresponding causal processes are "indeterministic", instead.

A deterministic theory (or a deterministic law), therefore, characterizes every physical system of a specified kind K as an instance of a

James H. Fetzer (ed.) Probability and Causality. 109–132.

deterministic causal process for which "the same cause" C invariably brings about "the same effect" E, where indeterministic theories (and indeterministic laws) are open to parallel definition. Thus, the world itself **W** is an indeterministic system if at least one indeterministic theory describing the world is true, which will be the case if at least one of its causal processes is indeterministic in kind. The conceptual problem that remains, of course, is understanding how it can be possible, in principle, for "different effects" E^1, E^2, . . . to be brought about by "the same cause" C, i.e., under the very same conditions. Recent work within this area, however, suggests the plausibility of at least two different approaches toward understanding "indeterministic causation", namely: various attempts to analyse "causation" in terms of probability, on the one hand, and various attempts to analyse "probability" in terms of causation, on the other.

Since attempts to analyse "causation" in terms of probability tend to be based upon interpretations of probabilities as actual or as hypothetical limiting frequencies, moreover, while attempts to analyse "probability" in terms of causation tend to be founded upon interpretations of probabilities as long-run or as single-case propensities instead, let us refer to frequency-based accounts as theories of "statistical causality" and to propensity-based accounts as theories of "probabilistic causation". Even though there thus appear to be two different types of each of these approaches toward this critical conceptual problem, it is not obvious that any of these theories can satisfy the relevant desiderata. My purpose here is to review why three out of four hold no promise in the reconciliation of mechanistic explanations with indeterministic theories. If these reflections are well-founded, then the single-case propensity approach alone provides an appropriate conception of the causal structure of the world.

1. RELEVANT DESIDERATA: CONDITIONS OF ADEQUACY

The first task confronting this investigation, therefore, is the specification of suitable desiderata for an adequate construction of indeterministic causation. Some of these criteria are required of "probabilistic" interpretations, while others are imposed by the element of "causality". We may assume, as probabilistic conditions, the requirements of admissibility, of ascertainability, and of applicability advanced by Salmon (1967), given the following understanding: (1) that the admissibility criterion requires that relations characteristic of mathematical probabilities (such as those of addition, of summation, and of multiplication) must be satisfied rather than some specific formulation of the calculus of probability; (2) that the ascertainability criterion requires that methods appropriate to statistical inquiries (such as statistical hypothesis-testing techniques) are capable of subjecting probability hypotheses to empirical tests but not that they should be verifiable; and, (3) that the applicability criterion requires not only that probabilities have to be predictively significant for "the long run", but also for "the short run" and for "the single case", i.e., for infinite, finite and singular sequences.

These conditions are important for a variety of reasons. In the first place, if condition (1) required that an acceptable interpretation must satisfy some specific formulation of the calculus of probability rather than relations characteristic of mathematical probabilities, propensity conceptions — which cannot satisfy Bayes' theorem, for example — would be excluded *a priori*. In the second place, if condition (2) required that an acceptable interpretation must be verifiable rather than merely empirically testable, only finite relative frequency constructions — but no limiting frequency or any propensity conceptions — could possibly qualify as acceptable. Thus, since condition (3) appears to impose a requirement that only single-case conceptions could be expected to fulfill, further consideration should be given to its justification and to the possibility that, in principle, it ought to be weakened or revised.

In order to fulfill the function of reconciling mechanistic explanations with indeterministic theories, furthermore, an acceptable account of indeterministic causation ought to generate "mechanistic explanations" that explain the present and the future in terms of the past, rather than conversely. Indeed, this condition accords with several features that seem to be characteristic if not invariable aspects of deterministic causation, such as (a) that "causes" precede their "effects", (b) that "effects" do not bring about "causes", and (c) that "causes" differ from their "effects". Let us also assume that "teleological explanations" either (i) explain the past and the present in terms of the future (the *temporal* criterion) or (ii) explain "effects" by citing motives, purposes, or goals (the *intentional* criterion). Thus, some "mechanistic explanations", none of which are "teleological$_1$" in the strong temporal sense, might still be "teleological$_2$" in the weak intentional sense, which we exemplify when we explain our behavior by citing motives and beliefs.

While "intentional explanations" need not explain the past and the present in terms of the future, therefore, an acceptable account of indeterministic causation should not generate "teleological explanations" of either kind, except with respect to appropriate classes of instances involving the proper ascription of *propositional attitudes* as they occur, for example, within psychology, sociology, and anthropology. Ascribing motives, purposes, or goals to inanimate objects, such as gambling devices and subatomic particles within physics or chemistry, appears to entail the adoption of animistic hypotheses, which certainly should not be a pre-condition for "indeterministic causation". And, insofar as causation assumes its fundamental, albeit not exclusive, role with respect to explanation, we may also suppose, as conditions of causation, that an adequate conception of this kind ought to possess explanatory significance not only for "the long run" but also for "the short run" and for "the single case". Though these conditions jointly entail that an acceptable conception of indeterministic causation should support probabilistic hypotheses that are both predictively and explanatorily significant for infinite, finite and singular sequences, they do not dictate any particular variety of "causal explanations", provided they are not "teleological explanations", necessarily.

Strictly speaking, of course, it is suggestive but inaccurate to say that teleological explanations explain the past and the present in terms of the future, since the temporal criterion hinges upon relations of "earlier than" and of "later than". Indeed, four more-or-less distinct species of causation may be identified by introducing "proximate" and "distant" differentia as follows:

(A) *distant mechanistic causation*:

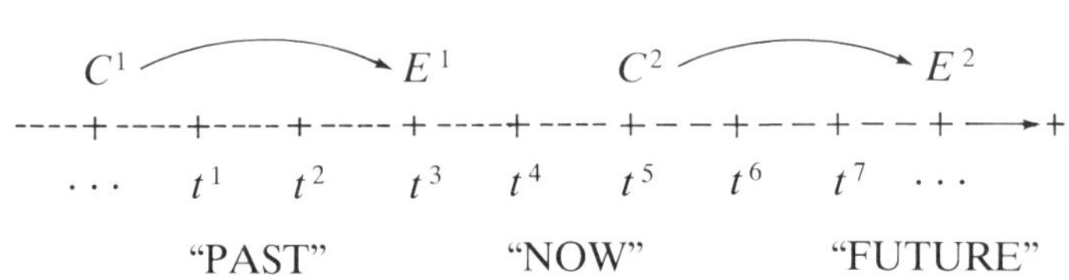

in which spatio-temporally separated "causes" C^1, C^2, . . . bring about their "effects" E^1, E^2, . . . (roughly) by forward "causal leaps" within space-time — whether these causal leaps belong to the "PAST", the "NOW", or the "FUTURE";

(B) *proximate mechanistic causation*:

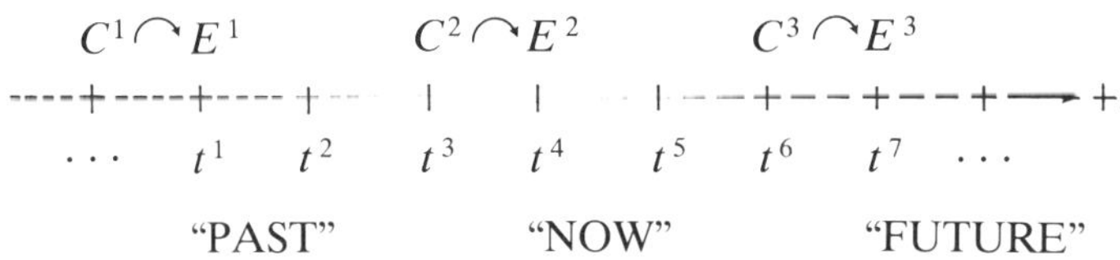

in which spatio-temporally contiguous "causes" C^1, C^2, . . . bring about their "effects" E^1, E^2, . . . by (minimal) forward "causal steps" within space-time — whether these causal steps belong to the "PAST", the "NOW", or the "FUTURE";

(C) *distant teleological$_1$ causation*:

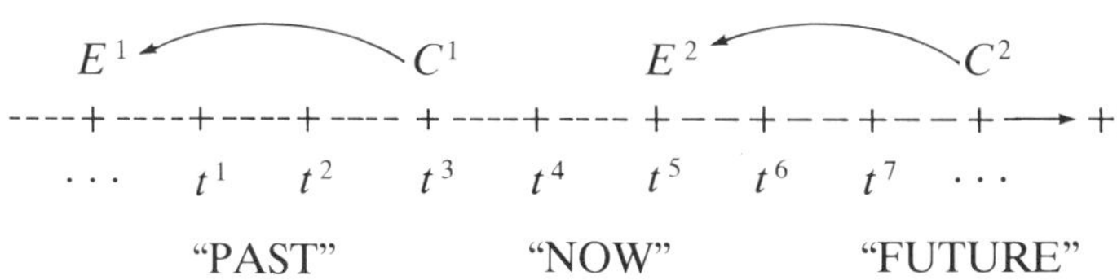

in which spatio-temporally separated "causes" C^1, C^2, . . . bring about their "effects" E^1, E^2, . . . by backward "causal leaps" within space-time — whether these causal leaps belong to the "PAST", the "NOW", or the "FUTURE"; and, finally,

(D) *proximate teleological$_1$ causation*:

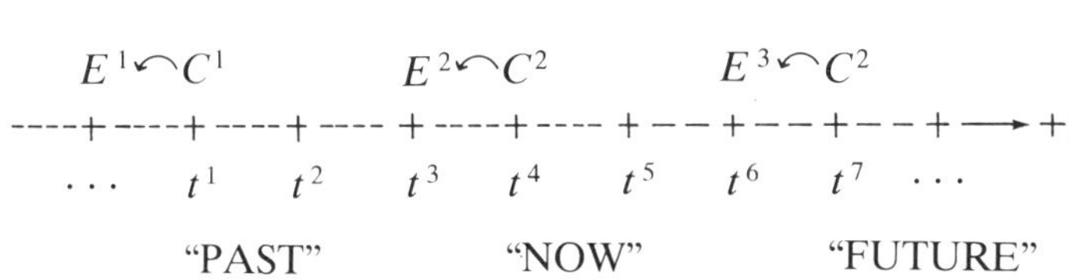

in which spatio-temporally contiguous "causes" C^1, C^2, . . . bring about their "effects" E^1, E^2, . . . by (minimal) backward "causal steps" within space-time — whether these causal steps belong to the "PAST", the "NOW", or the "FUTURE".

While "teleological causation" (in the sense of teleology$_1$) characteristically requires backward (or "retro-") causation, "distant causation" typically entails causation-at-a-distance (within space-time), contrary to the principle of locality of special relativity, which postulates the following requirement:

(E) *the principle of locality*: causal influences are transmitted by means of causal processes as continuous sequences of events with a finite upper bound equal to the speed of light ($\approx$ 186 000 mps);

hence, unless special relativity is to be rejected, there are spatio-temporal restrictions upon causal processes that render both varieties of "distant causation" unacceptable, in principle, since "causal leaps" of either kind would entail violations of the principle of locality. Thus, classical behaviorism, with its tendency to assume that an individual's past history directly determines his present behavior (as an example of distant mechanistic causation), is no more acceptable with respect to its explanatory type within psychology than are appeals to "manifest destiny", with their tendency to presume that present actions are directly determined by future events (as illustrations of distant teleological$_1$ causation), in their explanatory roles within history.

The principle of locality, of course, may be reformulated to reflect the (discrete or continuous) character of space-time by substituting "contiguous" for "continuous" in sentence (E), just as diagrams (A) through (D) may be employed to represent (deterministic or indeterministic) causal processes with the utilization of continuous and discontinuous "causal arrows", respectively. Hence, indeterministic as well as deterministic varieties of distant and proximate (teleological and mechanistic) causation are logical possibilities that do not necessarily also qualify as physical possibilities. Considerations of these kinds, moreover, have very significant consequences for alternative interpretations of indeterministic causation. For, with respect to "the single case", "the short run" and "the long run", conceptions of statistical causality based upon actual and hypothetical limiting frequencies result in probabilistic hypotheses that either are purely descriptive and non-explanatory or else exemplify distant teleological$_1$ causation, while conceptions of probabilistic causation based upon long-run propensities, by comparison, result in probabilistic hypotheses that exemplify both distant mechanistic and proximate teleological$_2$ species of causation, as the arguments that now follow are intended to display.

2. STATISTICAL CAUSALITY: ACTUAL LIMITING FREQUENCIES

Let us begin with theories of statistical causality based upon the actual limiting frequency construction of objective probabilities, which, in turn, is built upon (a) the mathematical definition of the limit of a relative frequency within an infinite sequence, (b) the empirical interpretation of such limits as features of the actual world's own history, and (c) the logical formulation of these probabilistic properties as conditional probabilities (cf. Salmon 1967). Thus, under this construction, the probability of an event of a specific type, *Y*, such as beta-particle emission, male-sex inheritance, etc., relative to the occurrence of an event of another specific type, *X*, such as radioactive decay, genic transmission, etc., can be characterized, in general, along these lines:

(F) $P(Y/X) = p =_{\mathrm{df}} \lim n \rightarrow \infty f^{n}(Ym/Xn) = (p = m/n)$;

that is, the probability for (an event of kind) *Y*, given (an event of kind) *X*, has the value *p* if and only if (events of kind) *Y* occur relative to (events of kind) *X* with a limiting frequency *m*/*n* equal to *p* during the course of the actual history of the world. Insofar as limiting frequencies for events of particular kinds may vary relative to different reference classes *X*, *X′*, *X″* and so forth, various relations of statistical relevance can be defined, such as, for example,

(G) If $P(Y/X \,\&\, Z) \neq P(Y/X \,\&\, -Z)$, then the occurrence of (an event of kind) *Z* is statistically relevant to the occurrence of (an event of kind) *Y*, relative to the occurrence of (events of kind) *X*;

which Salmon (1970) utilized as the foundation for a new account of statistical explanation intended to improve upon Hempel's inductive-statistical theory (Hempel 1962, 1965, 1968). While the mathematical definition of limiting frequency is straightforward when applied within abstract domains (such as the theory of numbers), however, its introduction within causal contexts has raised a variety of problems, most of which are now familiar but remain important, nevertheless.

One such difficulty arises because limiting frequencies, as properties of infinite sequences, are well-defined as features of the physical world only if the number n of occurrences of reference events (of kind X, say) increases without bound. This means that if there were no — or only a few, or many but finite — instances of events of kind X, then neither the probability for (an event of kind) Y given (an event of kind) X nor the relevance relations between (events of kinds) X, Y and Z (for events of any other kind) would be well-defined. So unless there is no end to the number of instances of genic transmission (radioactive decay, etc.) during the history of the actual world **W**, there can be no corresponding limiting frequencies to satisfy the conditions specified by (F) and by (G). Moreover, as long as these events themselves are non-vanishing in their duration, such a conception also entails that either the world's history must go on forever or else the corresponding probabilities cannot exist. So if the existence of limiting frequencies is required for "statistical causality", then neither (probabilistic) "causes" nor "effects" can exist if the world's history is finite.

This problem has invited a number of attempted solutions, the majority of which, following von Mises (1964), treat the infinite sequence condition as an "idealization" concerning what the limiting frequencies would be (or would have been) if the world's history were (or had been) infinite. This defense itself, however, appears to raise more questions than it answers, especially regarding the theoretical justification for these subjunctive and counterfactual claims. An ontological warrant might be generated from the ascription of dispositional properties to the physical world (as an inference to the best explanation, for example), where these properties could be invoked as a foundation for predicting and explaining the occurrence of relative frequencies; but this, as later discussion of the hypothetical frequency and of alternative propensity conceptions should make evident, would be to abandon rather than to repair this position.

Alternatively, a psychological warrant for ascribing hypothetical properties could be discerned in our (inevitable) "habits of mind" within the framework pioneered by Hume; yet, even if our anticipatory tendencies were thereby accurately described, that in itself would provide no foundation for their justification. After all, that we have such expectations does not establish that we ought to have them (although an evolutionary argument could be made to that effect, which would presumably bear the burden not only of separating adaptive from maladaptive expectations but also of deriving their truth from their benefits to survival and reproduction). The evolutionary benefits of expectations almost certainly derive from those "habits of mind" that obtain with respect to the single case and the short run, however, whose precise relations to the long run are not entirely clear. Moreover, while it may be plausible to reason from the truth of an expectation to its (potential) evolutionary benefits, to reason from its (potential) evolutionary benefits to its truth is to beg the question.

Other difficulties arise insofar as limiting frequencies, as symmetrical properties, violate our assumptions about causal relations. Consider, for example, that if 'X' and 'Y' are interpreted as "cause" and "effect", then since

$$\text{(H)} \qquad P(X \,\&\, Y)/P(X) = P(X/Y)P(Y)/P(X) = P(Y \,\&\, X)/P(X) \\ = P(Y/X)P(X)/P(X);$$

— where common reference classes have been suppressed for convenience — whenever "causes" X bring about "effects" Y, those "effects" Y must likewise bring about those "causes" X, in violation of the presumption that "effects" do not (simultaneously) bring about their "causes". Moreover, if 'X' and 'Y' are interpreted as singular events 'Xat' and 'Yat^*', where $t^* > t$, rather than as event types,

$$\text{(I)} \qquad P(Yat^*/Xat)P(Xat)/P(Xat) = P(Xat/Yat^*)P(Yat^*)/P(Xat);$$

then, whenever Xat "causes" Yat^*, Yat^* must also "cause" Xat, in violation of the presumption that "causes" precede their "effects". One avenue of escape from criticisms of this kind, however, might be afforded by emphasizing the temporal aspect of causal relations, where only conditioning events, such as Xat, that are earlier than conditioned events, such as Yat^*, can possibly qualify as "causes" in relation to "effects". Such a maneuver entails the result that, with respect to formulae like (H) and (I), left- and right-hand equalities in numerical value need not reflect similarities in semantical meaning.

Difficulties such as these, of course, have not inhibited Good (1961/62), Suppes (1970), and especially Salmon (1975, 1978, 1980) from elaborating frequency-based accounts of "statistical causality" [although, to be sure, more recently Salmon has endorsed a propensity approach; cf. esp. Salmon (1984)]. Not the least of the problems confronting this program, however, are (i) that statistical relevance relations — positive, negative or otherwise — do not need to reflect causal relevance relations (even when temporal conditions are introduced); and, (ii) that the systematic application of frequency-based criteria of causal relevance (such as those of Suppes and of Salmon) yields the consequence that particular attributes within non-epistemic homogeneous partitions of reference classes invariably occur only with degenerate probabilities equal to zero or equal to one. If statistical relevance relations and temporal conditions were enough to define causal relevance relations, it would turn out to be logically impossible, in principle, for causation to be indeterministic.

As an illustration of this situation (which does not require elaborating these arguments in great detail), let us consider the analysis of statistical causality advanced by Suppes (1970), where he suggests that the basic concept is that of "*prima facie* causation" where X is a *prima facie* "cause" of Y when and only when (a) the event X occurs earlier than the event Y, i.e., if Yat^*, then Xat, where $t^* > t$; and, (b) the conditional probability of Y given X is greater than the unconditional probability of Y, i.e., $P(Y/X) > P(Y)$. Notice that positive statistical relevance is employed here as a measure of positive causal relevance, i.e., Suppes assumes that conditions in relation to which Y occurs more frequently are causally relevant conditions, so that if lung cancer occurs with a higher frequency in smoking populations than in non-smoking populations, then smoking is a *prima facie* cause of lung cancer. To complete his analysis, Suppes further defines a "genuine" cause as a *prima facie* cause that is not a "spurious" cause, which requires that there be no earlier event Z such that the conditional probability of Y given Z & X equals that of Y given X, i.e., $P(Y/Z \& X) \neq P(Y/X)$. Accordingly, the "genuine" cause of an event Y is the event X earliest in time relative to which the probability of Y is highest.

In other to appreciate the magnitude of the obstacles confronting Suppes' constructions, keep in mind that any instance of lung cancer Y will involve a specific person i who possesses many properties in addition to being a smoker or not, as the case happens to be; for any such person will also be married or single, drink gin or abstain, eat caviar or ignore it. Indeed, the frequency for lung cancer Y (or for its absence $-Y$) will vary from class to class as additional properties F^1, F^2, . . . are taken into account until the upper bound of one has been reached; for if i is a gin-drinking, caviar-eating, heavy-smoking married man, there must exist some reference class, X^*, to which i belongs and in relation to which Y (or $-Y$) occurs more frequently than it occurs in relation to any other reference class to which i belongs, namely: a homogenous class X^* for which every member of X^* has lung cancer Y

(or has no lung cancer $-Y$), as it turns out; otherwise, Suppes' conditions cannot have been satisfied. For such a reference class, however, Y (or $-Y$) occurs with probability of one; hence, on Suppes' account, "statistical causality" is a logical impossibility.

These difficulties, of course, are not really hard to understand, insofar as (i) limiting frequencies, as a species of statistical correlation, are only (at best) necessary but not sufficient conditions for causal relations; while, (ii) as properties of the members of reference classes collectively, limiting frequencies are not properties of each member of these classes individually. Consequently, analyses of "statistical causality" that are based upon actual limiting frequencies, as "long run" distributions during the world's history, are purely descriptive and non-explanatory. Even in conjunction with temporal conditions, they are incapable of satisfying the requirements for causal relations. Moreover, as properties of infinite event sequences, they cannot even qualify as properties of singular events themselves. Indeed, although Salmon (1975) thought that causal relevance relations might be analysed as (complex) statistical relevance relations, his more recent efforts (Salmon 1978, 1980, and 1984) suggest, regarding statistical causality, that an analysis in terms of statistical relevance relations and *independent* causal concepts invites a regress, while an analysis in terms of statistical relevance relations and *dependent* causal concepts cannot be correct. Indeed, from premises confined to correlations, only conclusions concerning correlations can be validly derived, and they exemplify neither mechanistic nor teleological species of causation.

3. STATISTICAL CAUSALITY: HYPOTHETICAL LIMITING FREQUENCIES

If these considerations are well-founded, then an analysis of statistical causality based upon the actual limiting frequency construction offers no hope for understanding indeterministic causation; thus, perhaps an account based upon the hypothetical limiting frequency construction might fare a little better. Now the principal difference between the actual and the hypothetical frequency conceptions is not a matter of mathematical frameworks but rather of empirical interpretation, since these limits are features, not of the actual history of the world, $\mathbf{H(W)}$, but of hypothetical extensions of that history, $H_i(\mathbf{W})$. As a consequence, these objective probabilities may be identified with the limiting frequency, if any, with which *Y* would occur, were the number of events of kind *X* that occur during the history of the world to increase without bound [as an illustration, see van Fraassen (1980)]. The hypothetical frequency approach thus promises to improve upon the actual frequency approach in at least three respects, since (a) hypothetical sequences are available, even when actual sequences are not, ensuring that relevant reference classes have enough members; (b) systematic comparisons of statistical correlations then may be refined for improved analyses of causal relations; and, (c) specific attributes within non-epistemic homogeneous partitions of these reference classes perhaps might occur with non-degenerate probabilities. As appealing as this account may initially appear, however, it confronts several rather imposing difficulties of its own.

The major difficulty arises from the hypothetical character of these "long run" frequencies, a linguistic manifestation of which is the use of subjunctive conditionals (concerning what limits would obtain if sequences were infinite) in lieu of indicative conditionals (concerning what limits do obtain when sequences are infinite). Thus, within scientific discourse, at least, there appear to be only two modes of justification for subjunctive assertions, namely: (i) those that can be sustained on *logical* grounds alone (as a function of meaning or of grammar), such as, "If John were a bachelor, then John would be unmarried", on the one hand, and (ii) those that can be sustained on *ontological* grounds (as a function of causal or of nomic necessity), such as, "If this were made of gold, then it would melt at 1063 °C", on the other. Frequentist analyses of statistical causality, however, are unable to secure support from either direction, for limiting frequencies, whether actual or hypothetical, are supposed to be logically contingent; and, as frequency distributions, with or without limits, they are supposed to forego commitments to the existence of non-logical necessities.

From an epistemological point of view, the actual limiting frequency construction seems to fare far better, since actual limiting frequencies, as features of the world's actual history, either occur as elements of that history or they do not, while hypothetical frequency hypotheses, as designating extensions of the world's actual history, are supposed to be true or false whether or not these occur as elements of that history. From an ontological point of view, moreover, actual limiting frequencies are purely descriptive and non-explanatory because they merely describe statistical correlations that happen to occur during the history of the world, which means that statements describing them are *extensional generalizations* as truth-functions of the truth-values of enormously numerous, purely descriptive singular sentences, in turn. Insofar as hypothetical limiting frequencies do not merely describe statistical correlations that happen to occur during the history of the world, however, statements describing them cannot be truth-functions of the truth-values of (even) enormously numerous, purely descriptive singular sentences, which means that, under the hypothetical frequency construction, these probabilistic hypotheses have to be *non-extensional generalizations* that are no longer truth-functional.

Although the hypothetical frequency approach provides *prima facie* support for subjunctive conditionals (concerning what limits would obtain if sequences were infinite, whether or not they actually are) and for counterfactual conditionals (concerning what limits would have obtained if sequences had been infinite, when they actually are not), therefore, the problem still remains of explaining why these statements — characterizing what these limiting frequencies would be or would have been if these sequences were or had been infinite — are true (cf. Fetzer 1974). In contrast with the actual frequency approach, which does not require them, in other words, the hypothetical frequency construction provides no theoretical principles or structural properties which might be invoked to explain the attribution of these hypothetical limiting frequencies to the physical world. This approach thus appears to be incapable, in principle, of affording a theoretical justification for its own probabilistic hypotheses — unless these limits are viewed as aspects of the world's "manifest destiny".

Indeed, there are several reasons to believe that any analysis of statistical causality that is based upon the hypothetical frequency interpretation will be essentially — and not merely incidentally — committed to distant teleological$_1$ causation. One such reason arises from the theoretical necessity to establish ordering relations within these classes of hypothetical events, since limiting frequencies are subject to variation with variations in ordering relations. An analysis based upon the actual frequency interpretation, by contrast, does not encounter this problem, because temporal relations between actual events serve to fix their order, which is a feature of the actual history **H** of the world **W**:

(J) *statistical causality as a function of actual frequencies*:

```
              X&-Y           X & Y            X& Y
H(W): ----+----+----+----+----+--+--+--+ ——→+t  lim = m/n
        . . .   t1    t2   t3   t4   t5   t6  . . .
              "PAST"        "NOW"        "FUTURE"
```

where m/n represents the conditional probability for Y given X as the limiting frequency with which Y-events occur relative to X-events during the world's history. Although a single infinite sequence is sufficient to exhibit the actual history of the world, an infinite sequence of infinite sequences is required to exhibit the enormous range of hypothetical extensions of the world's history up to the present, signified by "NOW", where, to guarantee these limits are unique, the same limits must obtain no matter how the world's history might "work out", i.e., the same attributes must occur with the same limits in each such history:

(K) *statistical causality as a function of hypothetical frequencies*:

```
               X &-Y          X & Y           X & Y
H1(W): ----+----+----+----+----+--+--+--+ ——→+t  lim = m/n;
         . . .   t1    t2   t3   t4   t5   t6  . . .
               X &-Y          X & Y    X &-Y
H2(W): ----+----+----+----+----+--+--+--+ ——→+t  lim = m/n;
  .      . . .   t1    t2   t3   t4   t5   t6  . . .
  .
Hi(W): . . . . . . . . . . . . . . . . . ——→+t  lim = m/n;
  .
  .
               "PAST"        "NOW"        "FUTURE"
```

thus, no matter what the specific features of those singular histories may be, the same attributes must occur with the same limiting frequency m/n in each of those hypothetical extensions of the world's actual history, H_i(**W**). Moreover, it is not difficult to establish that there must be an endless number of such hypothetical extensions by a diagonal argument, since, for any specific class thereof, simply construct a new sequence that differs from the first sequence in its first hypothetical "segment", from the second in its second, and so on.

If this conception guarantees that these limits are unique, it has to pay the price; for, since the hypothetical frequency approach provides no mechanistic properties, teleological$_2$ or not, that might be invoked to explain the ascription of these hypothetical extensions of the world's actual history, there appears to be no alternative explanation than to conclude that they occur as a reflection of the world's "manifest destiny", i.e., as manifestations of distant teleological$_1$ causation. Indeed, if these limiting frequencies themselves seem to reflect distant teleological$_1$ causation, consider how much worse things are for "the short run" and for "the single case", since finite frequencies and singular events are only "explainable", within the framework of this conception, through their assimilation as individually insignificant, incidental features of a single "grand design" whereby the future brings about the present and the past. Even without the added weight of the symmetrical properties that accompany any attempt to envision indeterministic causation as a species of conditional probability, these are grounds enough to embrace the conclusion that hypothetical frequency based theories of statistical causality are very unlikely to succeed.

4. PROBABILISTIC CAUSATION: LONG-RUN PROPENSITIES

The failure of frequency-based accounts of indeterministic causation as theories of statistical causality, no doubt, strongly suggests that alternative accounts must be considered, including, in particular, propensity-based analyses of "probabilistic causation". Propensity conceptions of objective probabilities as dispositional properties are built upon (a) the interpretation of dispositions as causal tendencies (to bring about specific outcomes under relevant test conditions), (b) where these causal tendencies are properties of (instantiations of) "probabilistic" experimental arrangements (or "chance set-ups"), (c) which are subject to formalization by "probabilistic" causal conditionals incorporating a primitive *brings about* relation (Fetzer 1981, Ch. 3). The most important difference between theories of statistical causality and theories of probabilistic causation is that, on propensity-based theories, probabilities *cause* frequencies, while on frequency-based theories, probabilities *are* frequencies. The long-run propensity construction thus identifies probabilities with "long-run" causal tendencies, while the single-case propensity construction identifies them with "single-case" causal tendencies, for relevant "trials" to bring about appropriate "outcomes".

The objective probability of an event of a particular type, *Y* (such as beta-particle emission, male-sex inheritance, . . .), therefore, is either identified with a dispositional tendency (of universal strength) for events of kind Y to be brought about with an invariable limiting frequency or with a dispositional tendency (of probabilistic strength) for events of kind Y to be brought about with a constant single-case propensity. The most important difference between the hypothetical frequency conception and these propensity conceptions, furthermore, is that long-run and single-case propensities, as long-run and single-case causal tendencies, qualify as mechanistic properties that might be invoked to explain the ascription of hypothetical extensions of the world's actual history, since frequencies are brought about by propensities. Because these properties are formalized as asymmetrical causal tendencies for events of kind Y to be brought about by events of kind X with propensity N,

(L) $Xat \ni\!\!- _N Yat^*$;

rather than as symmetrical conditional probabilities, it is not the case that whenever Xat "causes" Yat^*, Yat^* must also "cause" Xat, thereby conforming to our expectations about causal relations. In spite of these similarities, however, these propensity conceptions are dissimilar in several crucial respects.

The long-run propensity conception identifies probabilities with what the limiting frequencies for Y are or would be or would have been, had an appropriate chance set-up been subject to a suitable (infinite) sequence of trials, X, which may be characterized in relation to possible histories H_i of the world **W**:

(M) *probabilistic causation as a function of long-run propensities*:

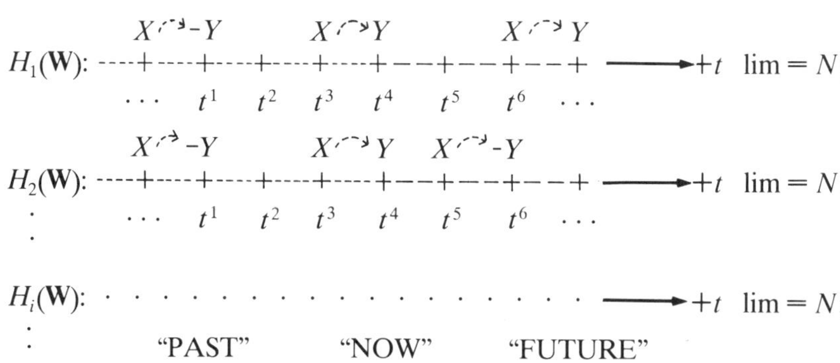

where, no matter what specific history the world's actual history **H** might have displayed or might yet display, the same specific "effects" *Y*, $-Y$, . . . , are invariably brought about with limiting frequencies *N* equal to their generating "long-run" propensities, whenever those histories reflect an infinite sequence of trials of the relevant *X*-kind. As a consequence, the "long-run" propensity conception ascribes a causal tendency of universal strength to bring about outcomes of kind *Y* with (invariable) limiting frequency *N* to every "chance set-up" possessing that dispositional property if subject to a trial of kind *X*, which, in the case at hand, itself consists of an infinite class of singular *X*-trials.

Since the "causes" *X* of these limiting frequencies for their "effects" *Y* themselves turn out to be infinite classes of spatially distributed and temporally extended singular trials, where "long run" dispositional tendencies serve as mechanistic properties, it should come as no surprise that this "long run" account exemplifies distant mechanistic causation (in relation to these infinite classes of trials). It is even more intriguing, therefore, to contemplate the contribution that each singular trial, *Xat*, must make to the attainment of that ultimate "effect"; for, unless each of these singular trials, so to speak, is able to "keep track" of the relative frequencies for outcomes of kind *Y*, $-Y$, . . . , it would be impossible to guarantee the same outcomes invariably occur with the same limiting frequencies during every possible history of the world. And if that, indeed, is the case, then, in effect, each singular trial member of these infinite trial sequences must not only "keep track" of how things are going but also "act" with the intention of insuring that things work out right, which means that this account also exemplifies proximate teleological$_2$ causation (in relation to each such singular trial). But surely no analysis of indeterministic causation incorporating both distant mechanistic and proximate teleological$_2$ species of causation could possibly satisfy the appropriate desiderata.

5. PROBABILISTIC CAUSATION: SINGLE-CASE PROPENSITIES

If an analysis of probabilistic causation can attain the reconciliation of mechanistic explanations with indeterministic theories, therefore, it must be based upon the single-case propensity conception. The most important difference between "long run" and "single case" propensities is that "single case" propensities are single-case as opposed to long-run causal tendencies, where their "effects" are the (variable) results (such as coming up heads or coming up tails, . . .) of singular trials (such as a single toss of a coin, . . .) rather than any (invariable) outcomes (in the form of limiting frequencies) of infinite sequences of trials (Fetzer 1971). Single-case propensities, as properties of chance set-ups, also qualify as mechanistic properties that might be invoked to explain ascriptions of hypothetical extensions of the world's actual history — not collectively (for infinite classes of trials), but distributively (for each specific individual trial), where collective consequences for finite and for infinite sets of trials follow by probabilistic calculations. Single case propensities, moreover, bring about specific outcomes Y, $-Y$, . . . , with constant (probabilistic) strength from trial to trial, but they generate only variable relative and limiting frequencies over finite and infinite sequences of trials, unlike any of the other interpretations we have considered before.

As we have already discovered, frequency-based constructions of statistical causality can be employed to generate frequency-based criteria of statistical relevance, such as (G), which can be utilized as the foundation for frequency-based accounts of statistical explanation (cf. especially Salmon 1971). Analogously, propensity-based conceptions of probabilistic causation can be employed to generate propensity-based criteria of causal relevance, such as:

(N) If $[(Xat \;\&\; Zat) \ni\!\!\rightarrow_N Yat^*]$ and $[(Xat \;\&\; -Zat) \ni\!\!\rightarrow_M Yat^*]$, where $N \neq M$, (the property) Z is causally relevant to the occurrence of (the property) Y, in relation to the occurrence of (the property) X;

which can be employed as the foundation for propensity-based analyses of probabilistic explanation (cf. Fetzer 1981, Part II). Thus, while frequency-based criteria entail the consequence that statistically-relevant properties are therefore explanatorily-relevant properties, propensity-based criteria entail the consequence that statistically-relevant properties may or may not be causally-relevant or explanatorily-relevant as well. Indeed, the most striking development in current work within this field has been the abandonment of a frequency-based model of statistical explanation in favor of a propensity-based model of causal explanation by Salmon (1984). [The extent to which Salmon has carried out this crucial conceptual exchange receives consideration in Fetzer (1987).]

The single-case propensity construction does not identify "probabilities" with what the limiting frequencies are or would be or would have been, therefore, because, in relation to the single-case conception, arbitrary deviations between the strength of a probabilistic "cause" and the limiting frequency of its "effects", even over the long run, are not merely logically possible, but nomologically expectable, where expectations of deviations are open to systematic computation. Unlike the hypothetical frequency and the "long run" propensity conceptions, in other words, the causal consequences of "single case" propensities now may be accurately represented by an infinite collection of infinite sequences, in which the same outcomes can occur with variable frequencies:

(O) *probabilistic causation as a function of single-case propensities*:

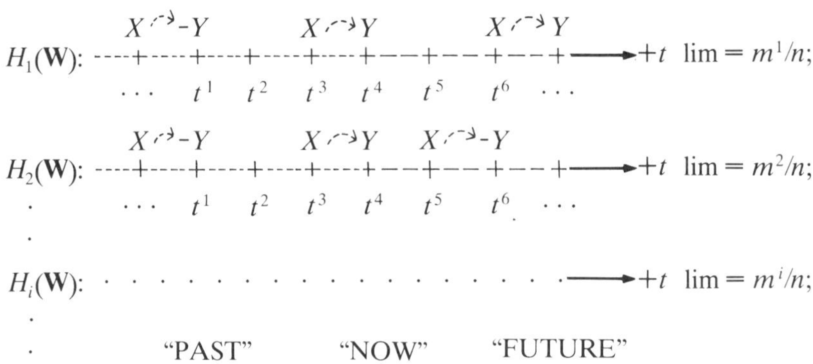

where, even though the same "effects", Y, $-Y$, . . . , are brought about by the same "causes", X, with constant propensity N from trial to trial, nevertheless, the limiting frequencies m^1/n, m^2/n, . . . , with which these outcomes in turn would occur or would have occurred, if sequences of X-trials were or had been infinite, are not at all invariable, where their expectations over finite and infinite sequences of trials can be calculated on the basis of classic limit theorems available for statistical inference, including the Borel theorem, the Bernoulli theorem and the central limit theorem (Fetzer 1981, Ch. 5 and Ch. 9).

Because these "causes" are single-case properties and single-case trials of individual chance set-ups, they exemplify proximate mechanistic causation, rather than any species of teleological causation (intentional or otherwise), thereby establishing an appropriate foundation for an acceptable analysis of probabilistic causation as the desired account of "indeterministic causation". Unlike frequency- or long-run propensity-based accounts, this interpretation can explain how it is possible for "different effects" to be brought about by "the same cause", i.e., under the very same conditions, within a purely mechanistic framework. Indeed, this approach also fulfills the additional probabilistic desideratum which Skyrms (1980) and Eells (1983) have endorsed that, insofar as there is "no end" to the variation in singular outcomes that might be displayed during any particular sequence of singular trials, there should also be "no end" to the variations in limiting frequencies displayed by those sequences themselves. That this result follows from the single-case propensity conception, moreover, should be viewed as very reassuring: if it were adopted as a "condition of adequacy" for an acceptable account of indeterministic causation, this requirement alone would enforce a sufficient condition for excluding both frequency-based and long-run propensity-based conceptions.

These reflections also bear upon the methodological desideratum that an adequate interpretation of single-case propensities ought to analyse physical probabilities by means of concepts that are understood independently of quantitative explications, which Suppes introduced and Eells (1983) has pursued. Although each of the four accounts we have considered would fulfill this condition in a different fashion, all of them seem to depend upon a prior grasp of the independent conception of relative frequencies in finite sequences (or short runs) of trials. The frequency theories add the notions of limiting frequencies and of hypothetical sequences, while the propensity theories add the notions of long-run and of single-case dispositions. The conception of a disposition, like that of a limit, however, was with us already; so, unless this requirement improperly implies that non-extensional notions must be given extensional definitions (a contradictory imposition), the concept of a single-case probabilistic disposition would appear to be (at least) as acceptable as that of a hypothetical long-run limiting frequency, relative to this condition.

In his recent discussion of Salmon's shift from frequency-based to propensity-based conceptions, Humphreys (1986) has objected to the idea of introducing the single-case notion of a probability distribution "which is impossible to construe in actualist terms". His position here appears to be related to the thesis that Armstrong (1983) refers to as "actualism", namely: the view that properties and relations exist only if they are instantiated. However, if this understanding of his position is correct, then it ought to be rejected: were a steel ball rolled across a smooth surface and allowed to come to rest only finite times during its existence, it would be silly to suppose it had no propensities (or causal tendencies) to come to rest upon any other of its nondenumerable surface points — and similarly for other sorts of phenomena. We typically assign truth-values to assertions concerning what would be or what might have been in probabilistic as well as non-probabilistic contexts with respect to actual and merely possible arrangements and contrivances. If all of these "effects" had to be displayed in order to "exist" (to be "real"), the world would be a vastly different — less threatening and promising — place.

Nonetheless, careful readers may discern a flaw in some of these examples, insofar as games of chance (such as tosses of coins) and other set-ups (such as rolls of balls) may be inadequate illustrations of indeterministic cases. While probabilistic causation appears to operate at the macro- as well as at the micro-level of phenomena, discriminating genuinely indeterministic cases from deterministic cases requires calibration for every property whose presence or absence makes a difference to the occurrence under consideration, a stringent epistemic condition that may remain unsatisfied for specific cases. Indeed, while our diagrams have been restricted to hypothetical extensions of the world's actual history in conveying the differences between these alternative conceptions, an adequate account of "indeterministic causation" carries with it the consequence that the history of an indeterministic world might be indistinguishable from the history of a deterministic world, insofar as they might both display exactly the same relative frequencies and constant conjunctions — where their differences were concealed "by chance"! (Cf. Fetzer 1983)

6. INDETERMINISTIC CAUSATION: CONCLUDING REFLECTIONS

Perhaps it should be clear by now why the comprehensive reconciliation of mechanistic explanations with indeterministic theories has not gone smoothly — perhaps even why the very idea of indeterministic causation has posed such an enormous conceptual problem. Indeed, the difference between the long-run and the single-case propensity conceptions is subtle enough to be invisible to the unaided eye, not only because neither diagram (M) nor diagram (K) incorporates its (implicit) distant "causal arrows", respectively, but also because diagram (M), like diagram (O), only exhibits (explicit) proximate mechanistic "causal arrows". But this is as it should be, for matters are more subtle than that: proximate teleological$_2$ causation, after all, is a species of proximate mechanistic causation! Still, this seems reason enough to sympathize with those who have experienced great difficulty in locating these differences (such as Levi 1977, 1979). Thus, by way of closing, it might be worthwhile to invite attention to three of the most important lessons that we have here to learn.

The first concerns the relationship between mathematics and the physical world; for, although frequency- and propensity-based conceptions of causation satisfy the "probabilistic conditions" of admissibility, of ascertainability, and of applicability, they do so in strikingly different ways. Frequency constructions formalize causal relations as a variety of conditional probability, while propensity conceptions formalize them as a species of causal conditional instead. The difficulties generated by symmetrical formulations reinforce the necessity to preserve the distinction between pure and applied mathematics; for if causation could be "probabilistic" only by satisfying inverse as well as direct probability relations, there could be no adequate conception of "probabilistic causation" (whether it were frequency-based or propensity-based in kind).

The second concerns the relationship between causation and the history of the physical world. What we have sought and what we have found is a conception of probabilistic causation that can be applied no matter whether the history of the physical world happens to be short or happens to be long. There appear to be inherent advantages in analysing "probabilistic causation" as a single-case conception that might be directly applied to finite and to infinite sequences, rather than as a long-run conception that might be applied indirectly to singular and to finite sequences instead. For it would be foolish to suppose that causal relations can obtain only if the world's history is infinitely long or that singular events can "cause" singular "effects" only if specific limiting frequencies happen to obtain during (actual or hypothetical) world histories.

The third and final (but by no means least important) concerns the relationship between indeterministic and deterministic theories and laws; for the considerations that have been brought to bear upon the analysis of indeterministic causation also apply to the analysis of deterministic causation. Deterministic causation, after all, is no more "symmetrical" than is indeterministic causation; and deterministic "causes" should no more be defined in terms of infinite classes of singular events than should indeterministic "causes": both should apply no matter whether the world's history is short or is long! The demise of statistical causality and the rise of probabilistic causation, therefore, should reap added dividends in analysing deterministic theories as well, further enhancing our understanding of the causal structure of the world.

University of Minnesota, Duluth

NOTE

I am grateful to Paul Humphreys and especially to Ellery Eells for criticism. I have also benefitted from the stimulating comments of an anonymous referee.

REFERENCES

Armstrong, D. M. (1983), *What is a Law of Nature*? Cambridge: Cambridge University Press, 1983.

Eells, E. (1983), "Objective Probability Theory Theory", *Synthese* **57** (1983), pp. 387—442.

Fetzer, J. H. (1971), "Dispositional Probabilities". In R. Buck and R. Cohen, eds., *PSA 1970*. Dordrecht: D. Reidel, 1971, pp. 473—482.

Fetzer, J. H. (1974), "Statistical Probabilities: Single Case Propensities vs. Long Run Frequencies". In W. Leinfellner and E. Kohler, eds., *Developments in the Methodology of Social Science*. Dordrecht: D. Reidel, 1974, pp. 387—397.

Fetzer, J. H. (1981), *Scientific Knowledge*. Dordrecht: D. Reidel, 1981.

Fetzer, J. H. (1983), "Transcendent Laws and Empirical Procedures". In N. Rescher, ed., *The Limits of Lawfulness*. Lanham: University Press of America, 1983, pp. 25—32.

Fetzer, J. H. (1987), "Critical Notice: Wesley Salmon's *Scientific Explanation and the Causal Structure of the World*", *Philosophy of Science* **54** (1987), forthcoming.

Good, I. J. (1961/62), "A Causal Calculus I—II", *British Journal for the Philosophy of Science* **11** (1961), pp. 305—318; and **12** (1962), pp. 43—51.

Hempel, C. G. (1962), "Deductive-Nomological vs. Statistical Explanation". In H. Feigl and G. Maxwell, eds., *Minnesota Studies in the Philosophy of Science*. Minneapolis: University of Minnesota Press, 1962, pp. 98—169.

Hempel, C. G. (1965), *Aspects of Scientific Explanation*. New York: The Free Press, 1965.

Hempel, C. G. (1968), "Maximal Specificity and Lawlikeness in Probabilistic Explanation", *Philosophy of Science* **35** (1968), pp. 116—133.

Humphreys, P. (1986), Review of Welsey Salmon's *Scientific Explanation and the Causal Structure of the World, Foundations of Physics* **16** (1986), pp. 1211—1216.

Levi, I. (1977), "Subjunctives, Dispositions, and Chances", *Synthese* **34** (1977), pp. 423—455.

Levi, I. (1979), "Inductive Appraisal". In P. Asquith and H. Kyburg, eds., *Current Research in Philosophy of Science*. East Lansing: Philosophy of Science Association, 1979, pp. 339—351.

Salmon, W. C. (1967), *The Foundations of Scientific Inference*. Pittsburgh: University of Pittsburgh Press, 1967.

Salmon, W. C. (1970), "Statistical Explanations". In R. Colodny, ed., *The Nature and Function of Scientific Theories*. Pittsburgh: University of Pittsburgh Press, 1970, pp. 173—231.

Salmon, W. C., ed. (1971), *Statistical Explanation and Statistical Relevance*. Pittsburgh: University of Pittsburgh Press, 1971.

Salmon, W. C. (1975), "Theoretical Explanations". In S. Korner, ed., *Explanation*. Oxford: Basil Blackwell, 1975, pp. 118—145.

Salmon, W. C. (1978), "Why Ask, 'Why?'?" *Proceedings and Addresses of the American Philosophical Association* **51** (1978), pp. 683—705.

Salmon, W. C. (1980), "Probabilistic Causality", *Pacific Philosophical Quarterly* **61** (1980), pp. 50—74.

Salmon, W. C. (1984), *Scientific Explanation and the Causal Structure of the World*. Princeton: Princeton University Press, 1984.

Skyrms, B. (1980), *Causal Necessity*. New Haven: Yale University Press, 1980.

Suppes, P. (1970), *A Probabilistic Theory of Causality*. Amsterdam: North-Holland, 1970.

van Fraassen, B. (1980), *The Scientific Image*. Oxford: Oxford University Press, 1980.

von Mises, R. (1964), *Mathematical Theory of Probability and Statistics*. Ed. by H. Geiringer. New York: Academic Press, 1964.

WAYNE A. DAVIS

PROBABILISTIC THEORIES OF CAUSATION

Probabilistic theories of causation have received relatively little attention. This is understandable, perhaps, since presentations are usually rather technical. But the neglect is unfortunate. The basic idea is quite simple, and very attractive. Moreover, competing theories all have serious problems that have been discussed *ad nauseam*. This paper is a critical exploration of the two main types of probabilistic theory, represented by Suppes (1970) and Cartwright (1979) on the one hand, and Giere (1980) on the other. I shall explain the theories' appeal, defend them against some outstanding objections, and improve them. Alas, the probabilistic approach is not without serious flaws of its own. These will be presented too.

I. SUPPES AND THE HUMEAN BACKGROUND

It is natural to think that causes in some way *necessitate* their effects. Suppose I strike the table and cause a noise; then the noise *had* to occur, it seems, given that I struck the table. Hume, of course, vigorously opposed this alleged necessary connection between cause and effect. For necessary connections can only be known *apriori*, he thought, whereas causal connections must be verified *aposteriori*. Hume saw no way for experience to tell us what must occur as opposed to what does occur, so he concluded that causation cannot entail necessitation. Instead, Hume emphasized *constant conjunction*. Striking the table caused the noise, on Hume's analysis, because striking it is always followed by the noise.

The problems for Hume's theory are legion. First, causes are not *always* followed by their effects. Flipping the switch this morning caused the light overhead to come on. But flipping the switch is not followed by the light coming on when the bulb is burned out, or when the power is off. Similarly, operating a slot machine sometimes produces a stream of money, but not always. I call this the *background conditions problem*: cause and effect are constantly conjoined only in certain circumstances. While the light does not come on whenever the

James H. Fetzer (ed.) Probability and Causality. 133–160.

switch is flipped, it does come on whenever the switch is flipped in conditions like those of this morning.

Mill responded to the background conditions problem by sticking to the Humean theory, and insisting that the cause "philosophically speaking" had to be the entire set of conditions sufficient to produce the effect in all cases. But Mill's approach amounts to legislating a new sense for the word "cause." Flipping the switch caused the light to come on, in the conventional sense of "caused," even though flipping the switch does not completely explain why the light came on. A more appropriate response is to say that one event caused another only if there are conditions which, together with the cause, always lead to the effect. This strategy, the Hempelian "covering law" approach, is hard to fault at the level of "ordinary" events. But many have thought that it comes to grief at the quantum level. Bombarding a uranium atom with a neutron typically splits it into barium and krypton atoms; but sometimes other atoms are produced, or no fission occurs. Some believe that nuclear fission is irreducibly probabilistic, that there simply are no conditions in which bombarding uranium with neutrons always results in barium and krypton. If so, then no form of constant conjunction is necessary for causation.

The constant conjunction condition implies that the probability of the effect given the cause is maximal: $P(E/C) = 1$. Suppes avoided the background conditions problem by requiring only that the cause must *raise the probability of the effect*. Only in special cases would the cause make the effect certain. More precisely, the requirement is that the conditional probability of the effect given the cause must be greater than the unconditional probability of the effect: $P(E/C) > P(E)$. This implies that the the effect must be more probable given the presence rather than the absence of the cause: $P(E/C) > P(E/-C)$. In other words, Suppes requires that the cause is *statistically relevant* to the effect, and *positively* so.[1] This positive relevance condition is faithful to the epistemological principles motivating the Humean analysis, since probabilities can be verified empirically, by observing relative frequencies. So replacing constant conjunction with positive relevance represents a natural weakening of the Humean theory.

This move does not eliminate all the difficulties of the Humean theory. Suppose that firing a certain cannon produces a characteristic flash and noise. Then the flash always precedes, but does not cause, the noise. Hume's theory, however, rules that the flash does cause the noise.

This is the *common cause problem*: the effects of a common cause will often be constantly conjoined without one causing the other. The same example shows that positive relevance alone is not sufficient for causation. For the probability of the noise given the flash is greater than the probability of the noise without the flash. Suppes observed that while the flash increases the probability of the noise, firing the gun increases it just as much. That is, the probability of the noise given the flash and the firing equals that of the noise given just the firing. Whenever $P(E/CF) = P(E/F)$, F is said to *screen off* C from E.[2] The flash does not cause the noise, Suppes suggests, because the firing is a screening factor.

Omitting refinements discussed later, Suppes' theory says that C causes E provided C is positively relevant to E, and provided there are no screening factors. In short, a cause is a positively relevant unscreened condition.

The Basic Conditional Probability Theory: *C causes E iff $P(E/C) > P(E)$ and there is no factor F such that $P(E/CF) = P(E/F)$.*

In some cases, "C causes E" is true at one time, false at another. Thus some diseases that used to cause death no longer do so. This presents no problem, since probabilities change over time too. When generalized beyond the present, the basic theory will read: C causes E *at t* iff $P(E/C) > P(E)$ *at t* and there is no factor F such that $P(E/CF) = P(E/F)$ *at t*. The basic theory can also be relativized to specific conditions or populations, by conditionalizing on them: C causes E *in B* iff $P(E/CB) > P(E/B)$ and there is no factor F such that $P(E/CFB) = P(E/FB)$. This allows a cause to have different effects in different circumstances or populations, which is another aspect of the background conditions problem. Suppes himself placed temporal restrictions on the events, requiring that F occur before C and C before E. We will examine these restrictions later.

The positive relevance and screening conditions together seem to eliminate the third and perhaps greatest problem plaguing the Humean theory, the *accidental generalization problem*. Hume's theory is too broad because constant conjunctions include accidental generalizations as well as lawlike regularities, whereas only the latter give rise to causal connections. Whenever a rain dance is performed, rain may always follow, just because it always rains (eventually); but the rain dance would not cause the rain. And whenever there is a tornado near

Topeka, there may by sheer coincidence be an earthquake in Southern California; but tornados near Topeka would not cause earthquakes in Southern California. Innumerable constant conjunctions like these just happen to hold, and so do not constitute lawlike connections between events. This presents a serious problem for the Humean theory, because no one has yet succeeded in distinguishing laws from accidental generalizations except in terms of natural necessity. Suppes' theory, however, rules out the problem cases without violating Humean epistemological scriptures. The rain dance example violates the positive relevance requirement: the probability of rain given the dance is no greater than the unconditional probability of rain. And the earthquake case violates the screening condition: tectonic motion along the San Andreas fault screens off the tornados near Topeka.

Despite its success with the Humean problem cases, Suppes' theory has aroused severe criticism. Some object that it contradicts determinism. This is a mistake. "Determinism" is ambiguous, denoting two logically independent principles. In one sense, determinism says that every event has a cause ("the principle of universal causation"). In another, it says that if an event has a cause, then it has a sufficient cause ("the principle of sufficient causation"). It is possible that for every event E another event C can be found satisfying the relevance and screening conditions (when the latter is more explicitly formulated). So the probabilistic theory does not deny universal causation. It is also possible that for some event E no other event C can be found satisfying those conditions; so Suppes' theory does not entail universal causation either. Indeed, this is the proper result: a *definition* of causality should entail nothing about the *existence* of causes. Suppes' theory similarly leaves it open whether or not all causes are sufficient causes. If C is a sufficient cause of E, then $P(E/C) = 1$. It is fully possible that the relevance and screening conditions are satisfied only when $P(E/C) = 1$; but it is also possible for those conditions to be satisfied when $P(E/C) < 1$ for every E and C.

It has been argued, however, that a definition of causation should *not* leave open the principle of sufficient causation. Unlike the Humean theory, a probabilistic theory allows that C may cause E on some occasions, and fail to cause E on others, *when there are no other relevant differences between the occasions*. But this seems absurd. For causation is connected with explanation: C causes E only if C explains E. But in the case imagined, we do not fully understand *why* E

occurred. How then could *C* have caused *E*?[3] I believe this argument can be countered by observing that we may have a *partial* explanation of something without having its *complete* explanation. Suppose a rock caused a window to break. Then we know that the window broke because the rock hit it. We do not, however, have a complete explanation of why the window broke. For that, we need the rock's momentum, the glass's strength, and so on. Similarly, the knowledge that a uranium atom split as a result of neutron bombardment does not enable us to fully understand why the fission occurred. In the window case, of course, we think a complete explanation is possible "in principle." In the fission case, though, it is argued that a complete explanation is not possible. The claim is that even if we know all the explanatory factors there are, we will not fully understand why the atom split.[4] The principle of sufficient causation is tenable, therefore, only if a complete explanation is possible whenever a partial explanation exists. But this has not been proven and does not seem self-evident. Probabilistic theories of causation cannot be faulted, therefore, for leaving open the principle of sufficient causation.

II. GIERE'S THEORY

Like Suppes, Giere starts with the idea that a cause must raise the probability of its effect. Unlike Suppes, Giere interprets this as requiring an inequality between *counterfactual* rather than *conditional* probabilities. Consider smoking and cancer. In fact, some but not all people smoke. But consider two counterfactual cases, one in which everybody smokes, and one in which nobody smokes, where in other respects the two hypothetical populations are as much like the actual population as possible. If smoking causes cancer, the frequency of cancer should be greater in the first group than in the second. Any failure of this inequality could be attributed to chance factors. That is, smoking causes cancer, on Giere's view, provided the probability of cancer would be greater if everyone smoked than if no one did. In general, Giere begins with the actual population of objects to which (or situations in which) events *C* and *E* may occur. He than considers what the probability of *E* would be if *C* occurred to every member, and what it would be if *C* occurred to no member. *C* causes *E*, according to Giere, provided the first probability is greater than the second — that is, provided $P(E)$ would be greater if *C* always occurred than if *C* never occurred.

Conspicuously absent from Giere's theory is any sort of screening condition. How then does it deal with the problems motivating that condition in Suppes' theory? Giere has no trouble with accidental generalizations, because they do not remain true under the indicated counterfactual assumptions. Consider the tornado-earthquake case. We imagined that, coincidentally, a tornado in Topeka is always followed by an earthquake in Southern California. Giere's theory denies that the tornadoes cause the earthquakes, because the probability of an earthquake in Southern California would be no greater if tornadoes always struck Topeka than if they never did.

The common cause problem remains, though. Once again imagine a type of cannon with a characteristic flash and sound. Does the flash cause the sound on Giere's theory? It would seem that if the flash occurred, the noise would occur. For if its characteristic flash occurred, a cannon of that type would have to have fired. And if the cannon had fired, it would have produced its characteristic noise. If this reasoning is correct, then the probability of the noise occurring would be greater if the flash were produced by every cannon of that type than if it were produced by none. Hence Giere's theory rules, incorrectly, that the flash causes the noise.

A similar problem confronts Lewis's counterfactual analysis of causation. His solution is "flatly to deny the counterfactuals that cause the trouble" (1973a, p. 566). I do not think they can be denied, at least not in all cases of the common cause problem. Admittedly, if that particular flash were to occur, it is conceivable that some other type of cannon (with a different noise) would have produced it, in which case that flash would no longer be characteristic of the original cannon. But it is also possible that no other type of cannon could have produced it, which is the case I am imagining. We frequently move counterfactually from one effect of a common cause to another. Consider: "If one pellet from the shotgun had hit the bull's-eye, others would have at least hit the target (because at that range there is little scatter)"; "If the cheese had spoiled, the milk would have spoiled (because milk spoils before cheese)." The fact that such counterfactuals are readily asserted and accepted as true would seem to constitute strong linguistic evidence against Lewis's position. One drawback of Giere's theory is clear, however: counterfactual probabilities are more difficult to evaluate than conditional probabilities.

Giere can avoid the common cause problem by imposing a screening condition of his own, requiring that no factor F exist such that $P(E)$ would be the same if C and F always occurred as if F always occurred. Then the cannon's flash would not cause the noise because firing the cannon would screen off the flash.

Giere's theory has other problems. Consider whether firing a *particular* cannon causes a certain noise. The population should be the class of all times, since the firing and the noise could occur at any time. But while a cannon can be fired at any time, it cannot be fired at all times (continuously). Consequently the probability of E if C always occurred will be undefined, and Giere's theory will assign no truth value to "Firing the cannon causes that noise." Consider next the hypothesis that having below average income leads to crime. To evaluate this hypothesis, on Giere's theory, we must determine the probability of crime if the suspected cause were universal. But it is logically impossible for everyone to have below average income. The probability of E if C always occurred is again undefined, and the hypothesis is left without a truth value.

These difficulties are related to the problem of "frequency dependent causation" discussed by Sober (1982). In most situations, nudity attracts attention.[5] This depends, however, on the rarity of nudity. If everyone practiced nudity, it would attract no special attention. So attracting attention would not necessarily be more likely if everyone went nude than if no one did. Giere's theory, then, incorrectly rules that nudity does not attract attention. It also erroneously rules that trying to withdraw money from a bank causes a run. For the probability of a run would be much greater if everyone attempted a withdrawal than if no one did.

The basic problem seems to be this: the hypothesis that the cause *always* or *never* occurs represents too great a departure from the actual state of affairs. Fortunately, we are not limited to such all-or-nothing hypotheses. It may be better to consider a *marginal* increase or decrease in the frequency of the cause, and determine what differential that would produce in the frequency of the effect. Giere's basic insight is that we should expect the effect to occur more often if the cause occurred more often. Let $P_{+C}(E)$ designate what the probability of E would be if C occurred a little more often, and let $P_{-C}(E)$ designate what it would be if C occurred a little less often. If we wanted our modification of Giere's theory to be as close as possible to the original while avoiding the above problems, we would formulate the positive relevance requirement as $P_{+C}(E) > P_{-C}(E)$. But why consider the case in which C occurs less often as well as that in which C occurs more often? It would suffice, as far as I can see, to require only that $P_{+C}(E) > P(E)$, which says that the probability of E would be greater than it actually is if C occurred a little more often than it actually does. This gives us:

The Basic Counterfactual Probability Theory: *C causes E iff $P_{+C}(E) > P(E)$ and there is no factor F such that $P_{+CF}(E) = P_{+F}(E)$.*

The fact that everyone cannot have below average income is no longer a problem. For more people could have low income. The fact that nudity attracts attention is secured because at current levels of nudity, more people would attract attention if more people went nude. Finally, the result that withdrawing money from a bank causes a run is avoided because a few extra withdrawals would not not make a run more likely.

I will leave the counterfactual theory vague in certain respects. For example, the theory does not specify exactly how the members of the population are to be selected for the hypothetical addition of the cause. Presumably, the selection should be representative, to keep "other things equal." Nor does the theory specify exactly how many more causes are to be added. Presumably, the number should be large enough to minimize "chance differences," while small enough to avoid frequency dependence effects. This vagueness may not be a defect. For causal statements are themselves vague, as we shall see below.

The basic conditional and counterfactual theories are very similar, and seem to give the same results in all cases. To avoid duplication, I shall henceforth concentrate on the conditional probability theory. What I say will apply, *mutatis mutandis*, to the counterfactual theory.

III. CARTWRIGHT'S THEORY

We have seen some advantages of the probabilistic theory over Hume's. The probabilistic theory has its own difficulties, however. Indeed, both the positive relevance and screening clauses are too strong. The positive relevance condition is too strong because a cause may *lower* the probability of its effect. Hesslow (1976) observed that this would happen whenever one cause of an event lowers the probability of other more effective causes. Thus contraceptive pills and pregnancy both cause thrombosis. The probability of thrombosis given pregancy may be greater than its probability given the pills. But since contraceptive pills lower the probability of pregnancy, the probability of thrombosis given the pills may be less than its probability without them. Giere stipulated that his counterfactual populations are to be as much like the actual population as possible given the different distributions of C. But any differences attributable to that difference have to be allowed for Giere's theory to work at all. It might be replied that in certain populations, such as women who never get pregnant, contraceptive pills do increase the chances of thrombosis. Thus the probabilistic theory correctly rules that contraceptive pills cause thrombosis in women who don't get pregnant. But contraceptive pills also cause thrombosis in the population of women at large, a result not forthcoming from the theory[6]. Hesslow's objection seems especially serious, since it questions the key idea of the probabilistic theory.

The screening clause as stated is also too strong, as Suppes himself observed. Flipping the switch (C) causes the light to come on (E). One screening factor is C itself: $P(E/CC) = P(E/C)$; another is E itself. Clearly, it must be understood that *F is neither C nor E*. But screening factors still abound. F might be: some event in the causal chain between C and E, such as electrons flowing through the circuitry; or: certain effects of E, such as the room being illuminated, or someone observing that the light came on. For these choices of F, it is quite possible that $P(E/CF) = P(E/F)$. To avoid all this, Suppes required that *F precede C*. But such a temporal restriction may leave many spurious causes unscreened. Suppose that purely by coincidence, a pitcher only tips his cap before throwing his fearsome fastball. The probability of his striking out the batter may thus be greater if he tips his hat than if he does not. With Suppes' temporal requirement, though, tipping his hat would not be screened off by his fastball.

Cartwright solved the problem with both conditions by focusing on Hesslow's example.

> All the counter examples I know to the claim that causes increase the probability of their effects work in this same way. In all cases, . . . the cause is correlated with some other causal factor which dominates in its effects. This suggests that the condition as stated is too simple. A cause must increase the probability of its effects — but only in situations where such correlations are absent. The most general situations in which a particular factor is not correlated with any other causal factors are situations in which all other causal factors are held fixed . . . (Cartwright, 1979, p. 423).

Thus contraceptive pills increase the probability of thrombosis in women who do not get pregnant, and also in those who do.

Cartwright's Theory: *C causes E iff* $P(E/CF) > P(E/F)$, *where F is any alternative causally relevant factor.*

"Alternative causally relevant factor" means any event that causes either E or $-E$ other than C or its effects, or any combination of such events.[7] A restricted definition of screening is implicit: An alternative causally relevant factor F is screening when $P(E/CF) \leqslant P(E/F)$. Cartwright's theory entails that C causes E only if there is no screening factor. A similar modification could, of course, be made in the counterfactual theory: C causes E iff $P_{+CF}(E) > P_{+F}(E)$ for any alternative causally relevant F.

Cartwright avoids many of the problems facing Suppes' theory. But she pays a price. Hume and Suppes were trying to *define* causal statements. Their goal was to explain what causation is, and to reduce the concept to other clearer concepts. Since Cartwright's theory is circular, it cannot serve these purposes, as she recognized. Her necessary and sufficient conditions are far from tautological, however, so they still constitute a significant theory. It will not create an understanding of causation in someone with no previous understanding. But the theory can increase our understanding of causation by relating it to probability.

Cartwright's theory fails to overcome all the problems confronting Suppes', however, and has some of its own. Cyanide gas is one thing that causes death. Another is the absence of any gases. But the presence of cyanide gas is incompatible with the absence of any gases. Since the conditional probability of death given cyanide gas plus no gases is *undefined*, Cartwright's theory as stated does not assign a truth value to "Cyanide gas causes death." We shall accordingly require that alternative causally relevant factors be *compatible* with C.

More seriously, the probability of an *inevitable* event is unaffected even by its causes. Death is one such event: all human beings die eventually of one thing or another. The probability of death is 1 given either the presence or the absence of cyanide gas, for example, even when alternative causes are held fixed. Thus our probabilistic theories rule that death has no causes. Note that a conditional probability $P(E/C)$ does not express just the likelihood that E will result from C, nor does it impose any constraints on when E occurs in relation to C. The same goes for the counterfactual probability $P_{+C}(E)$. For similar reasons, an inevitable event never raises the probability of another event. $P(E/C) = P(E)$ when C is inevitable, and $P_{+C}(E)$ is undefined. Hence our probabilistic theories also rule that death has no effects.

It might be denied that a problem exists here, on the grounds that cyanide clearly does *not* cause people to be *mortal*.[8] While this is true, cyanide *does* cause people to *die*. And while dying is different from being mortal, the theory as stated cannot explain why cyanide causes one but not the other. It might also be observed that the theory correctly rules that cyanide causes people to *die within a few minutes*. But even though this entails that cyanide causes people to die, the theory denies that cyanide causes people to die.

The latter observation suggests a remedy: relativize all probabilities to "test situations" of short duration. But how can we restrict the length of the test situation without ruling out delayed effects? Some consequences of radiation exposure, for example, take decades to show up. The proper solution, I believe, is to *existentially* quantify over the time interval. Cartwright's theory would then become: C causes E iff $P(E$ in i/CF in $i) > P(E$ in i/F in $i)$ *for some interval* i *in the immediate future*,[9] and for all alternative causally relevant factors F. Thus let i be the next 200 years. The probability that one will die in the next two hundred years given that one receives cyanide is no greater than the probability that one will die in the next two hundred years. Now let i be the next 2 minutes. The probability that one will die in the next two minutes given that one receives cyanide is considerably greater than the unconditional probability that one will die in the next two minutes. Since there is *some* immediately future interval in which cyanide raises the probability of death, our modification of Cartwright's theory rules that cyanide causes death.

IV. THE DISTINCTNESS CONDITION

One of Hume's most widely accepted principles is that cause and effect must be *distinct* events. Our probabilistic theories do not entail this, however. Indeed, they rule that any event causes itself, as long as it has no other sufficient cause. So we must add that *C and E are not the same event.*

But numerical distinctness is not enough. Let A be drawing a spade, and let B be drawing the ace of spades. Then $P(A/B) > P(A)$, and $P(B/A) > P(B)$. There are no screening factors, if the draw is random.

So the conditional theory rules that drawing a spade causes drawing the ace of spades, and that drawing the ace of spades causes drawing a spade. Both rulings are incorrect since drawing an ace of spades logically implies drawing a spade. Our probabilistic theories therefore need an additional clause requiring that *C and E are logically independent.* It is not enough to require simply that C does not logically imply E. That would prevent the result that drawing the ace of spades causes drawing a spade, but not the result that drawing a spade causes drawing the ace of spades.

Logical connections are not the only noncausal connections, though, as Kim (1973a) observed in his critique of Lewis's counterfactual analysis of causation. Imagine a light that is turned on by flipping a switch. Let S be flipping the switch, and let T be turning on the light. Then $P(T/S) > P(T)$, and $P(S/T) > P(S)$. But turning on the light does not cause the act of flipping the switch. And while flipping the switch causes the light to come on, it does not cause the act of turning on the light. We might be inclined to say that flipping the switch and turning on the light are not distinct acts. But they are numerically[10] and logically distinct. One may flip the switch without turning on the light, for example. Following Goldman (1970, p. 21), let us say that one action *generates* another when the latter is done by, or in, doing the former. Then an additional requirement is that *C and E are generationally independent.* Turning on the light is commonly said to *supervene* on flipping the switch. In these terms, our new requirement is that causes and effects must not supervene on each other.

The definition of screening must be similarly restricted. Let C be administering cyanide and E death. Any screening factor F must be logically independent of C, otherwise C will be screened off by all of the following: the conjunction of C with the act of bringing about the effect, or with some irrelevant action like wearing a hat; the disjunction of C with some act whose probability is zero, such as getting a man pregnant, or with some equally effective cause, such as administering cyanate; and acts such as correctly predicting C. Screening factors must furthermore be generationally independent of C, otherwise turning on the light will screen off flipping the switch as a cause of the light going on. When Cartwright defined "alternative causally relevant factor," she required only that F be *different* from C or its effects. We should add that *F must be logically and generationally independent of C or its effects.*

V. SINGULAR VS. GENERAL CAUSAL STATEMENTS

The variables in our analysandum "*C* causes *E*" represent events. But there is a familiar distinction between *general events* ("event-types," "event-universals") and *singular events* ("event-tokens," "event-particulars"). Death is a general event, occurring to different people in different places at different times. Socrates' death was a singular event, happening at a unique spatio-temporal location. A related distinction exists among statements. A *general causal statement*, like "Drinking hemlock causes death," asserts that one general event causes another. A *singular causal statement*, like "Socrates' drinking hemlock caused his death," asserts that one singular event caused another. The relationship between singular and general causation is not simple. From the fact that being poisoned causes death, we cannot infer that Alan's being poisoned caused his death (he might have died of a bullet wound first). And even though Jim Fixx's last run caused his death, it is too strong to say that going for a run causes death.[11]

So the question naturally arises: Do the variables *C* and *E* represent general events or singular events in our probabilistic theories? Are the theories stating truth conditions for general or singular causal statements? Suppes said "Both." "A deliberate equivocation in reference between events and kinds of events runs through the earlier systematic sections of this monograph. It is intended that the formalism can be used under either interpretation" (1970, p. 79ff). Giere and Cartwright, however, restrict their variables to general events. This means that Giere and Cartwright fail to provide a complete theory of causation. But there are good reasons for their restriction.

One concerns the appropriate interpretation of the probability function. Three interpretations are commonly distinguished. *Empirical* probability is a measure of the relative frequency with which events tend to occur in the long run. *Logical* probability measures the evidential support for a statement, and so represents degree of rational belief. *Subjective* probability is perceived probability or degree of actual belief. Both logical and subjective probability apply to statements describing singular events. But since causation is objective, it cannot be defined in terms of subjective probability. The subjective probability of an event depends on the individual whose beliefs are being measured; but whether or not one event causes another is generally independent of the individual. The fact that pilot error caused a plane to crash depends in no way on how certain I am (or you are) that the plane crashed or that it crashed if the pilot made an error. Similarly, logical probability is relative to the available evidence; but the existence of causal connections, unlike our knowledge of them, does not depend on our evidence. The plane may have crashed because of pilot error even though the evidence indicates instrument malfunction.

Since causation is neither subjective nor relative to evidence, any plausible probabilistic theory must define it in terms of empirical probability. The empirical interpretation is eminently suitable, since causation and empirical probability both concern the occurrence and nonoccurence of events. The interpretation is natural, moreover, since the probabilistic theory is viewed as a weakening of the Humean constant conjunction theory. However, it is notoriously difficult to assign empirical probabilities to singular events. The notion of "long run relative frequency" does not even apply to singular events, since they occur only once.[12]

There are independent reasons for restricting probabilistic theories to general causation. It seems self-evident, for example, that causation among singular events is transitive, asymmetric, and irreflexive. Among general events, however, causation is not asymmetric. Suppose Jack and Jill regularly give each other colds. Then Jack's getting a cold causes Jill to get one, and Jill's getting a cold causes Jack to get one. This is so even though in no particular case could Jack and Jill's colds cause each other.[13] General causation is also not transitive: an inability to run or walk causes one's leg muscles to become flaccid; losing one's legs causes an inability to run or walk; but losing one's legs does not cause them to become flaccid. Obviously, such a case could not arise with singular events. Irreflexivity may seem even easier to refute. Thus world history shows that aggression causes aggression. But this statement is ambiguous. As intended, it means that aggression causes *further* aggression, and is not an instance of irreflexivity. "Aggression causes aggression" can mean that aggression causes *itself*, though; but in this sense it is false.

Our probabilistic theories fit general causation better because none entails that causation is asymmetric or transitive. None entails that causation is irreflexive either without the distinctness condition. Henceforth, upper-case variables will range over general events, lower-case variables over singular events.

While probabilistic theories are better interpreted as defining general causation, problems remain. General causal statements do appear *cumulatively transitive*, like indicative and subjunctive conditionals.[14] That is, while "A causes C" does not follow from "A causes B" and "B causes C," it does seem to follow from "A causes B" and "AB causes C." The case used to refute simple transitivity does not disprove cumulative transitivity. For while losing one's legs causes an inability to walk or run, and an inability to walk or run causes one's legs to become flaccid, an inability to walk or run together with the loss of one's legs does not cause one's legs to become flaccid. Unfortunately, our probabilistic theories do not guarantee even cumulative transitivity. Let A be drawing a spade, let B be drawing the ace of spades, and let C be drawing an ace. Then $P(B/A) > P(B)$, and $P(C/AB) > P(C)$. There are no screening factors. Nevertheless, $P(C/A) = P(C)$. For an example that does not violate the distinctness condition, consider: (1) Having only a small hole in a tire (A) causes the driver to hear a characteristic hissing noise immediately (B); (2) Having only a small hole and hearing a characteristic hissing noise immediately (AB) causes the driver to have the tire repaired immediately (C); and (3) Having only a small puncture (A) causes the driver to have the tire repaired immediately (C). While seldom heard, a small leak does raise the probability of hearing the noise characteristic of a small hole; so $P(B/A) > P(B)$. And obviously, $P(C/AB) > P(C)$. So assuming no screening factors, our theories rule that (1) and (2) are true. Nevertheless, since other types of damage are more noticeable, having only a small puncture actually reduces slightly the likelihood that the tire will be repaired immediately; hence $P(C/A) < P(C)$. So (3) comes out false.

It might be argued that the probabilistic theory rules correctly in this case, and that general causation is not cumulatively transitive.[15] But (3) will be false only if the hole is so small that a hissing noise will *seldom* be heard immediately. And in that event, (1) will be false as well. Conversely, if the hole is large enough so that a hiss is normally heard immediately, then (3) will be true as well as (1).

My analysis of this alleged counterexample relies on the fact that while singular and general causation are different, they are not unrelated. Indeed, general causal statements are just vague quantifications of singular causal statements. That is, "C causes E" means something like "C *always* or *sometimes* causes E." The proper quantifier, however, is hard to specify. "Cancer causes death" would be counted as true even though cancer does not always cause death; so "C always causes E" is too strong. That transitivity fails for general causation while holding for singular implies the same thing. On the other hand, running sometimes causes death, but "Running causes death" would not be considered true; so "*C* sometimes causes *E*" is too weak.[16] "Smoking causes cancer" thus resembles "Americans like football." The "intermediate" quantifiers *frequently* and *commonly* are also too weak. Going for a drive frequently and even commonly causes death; but "Going for a drive causes death" would not be counted true. It seems, therefore, that "*C* causes *E*" means "*C normally* or *typically* causes *E*"; that is, *when C occurs, it normally or typically causes an occurrence of E.* As the frequency with which *C* causes *E* changes, the truth value of "*C* causes *E*" will change. Thus surgery *used to* cause peritonitis; but due to modern antiseptic methods it no longer does so.

Note that "normally causes" does not mean the same as "is a normal cause of." Legionnaires' disease normally causes death, since death is the usual outcome when that disease strikes. But being rare, Legionnaires' disease is not a normal cause of death. A *common location* is understood in general causal statements when *C* and *E* do not specify one. Thus "Cuts cause pain" means "When cuts occur *to a person*, they normally cause pain *in that person.*" And "Lightning causes thunder" means "When lightning occurs *in a place*, it normally causes thunder *there*". When *C* and *E* do specify a location, however, it need not be common. Thus, "Sunspots cause electrical disturbances on earth" is true even though sunspots occur on the sun and the electrical disturbances on earth.

If I am right, general causal statements are inherently vague. For it is impossible to specify how often something must occur for it to be normal or typical. Consequently, it is hard to see how precise probabilistic truth conditions of the sort Suppes, Giere, and Cartwright proposed could possibly be provided. Moreover, any attempt to analyze general causation independently of singular would seem unlikely to succeed. At least three problems confront the probabilistic approach.

First, since "*C* causes *E*" means something like "*C* normally causes *E*," it implies that *C* and *E* at least occasionally occur.[17] But none of our probabilistic analyses guarantees that *C* or *E* ever occur. *A fortiori*, none guarantees that occurrences of *C ever* cause occurrences of *E*. "*C* causes *E*, but *C* never occurs and never causes occurrences of *E*" sounds quite self-contradictory, though. Rosen has argued against requiring the occurrence of *C* or *E* becauses of the "many theoretical contexts where predictions are for fanciful situations, future contexts, or, in general, scenarios one hopes to avoid realizing" (1980, p. 82). However, without a requirement of occurrence, one may have defined "*C would* cause *E*," or something else perhaps, but not "*C causes E*." Thus we might say that "All-out nuclear war would cause unimaginable destruction" but not "All-out nuclear war causes unimaginable destruction." Some may be content to provide truth conditions for "*C* would cause *E*," or to stipulate some new and more precise sense of "*C* causes *E*." But my goal is to determine whether probabilistic theories correctly answer the age-old question "What is it for one event to cause another?" as it has usually been intended. A successful analysis of "*C* would cause *E*," or a redefinition of "*C* causes *E*," however laudable, would not be a solution to that problem.

Second, even if we add a requirement that *C* and *E* occur, our theories will not guarantee that occurrences of *C normally* cause occurrences of *E*. Since the probability of drowning is greater if you go for a swim than if you do not, the theories rule that "Going for a swim causes you to drown" is true. Yet this statement is too strong. While swimming occasionally causes drowning, it does not normally do so.

Third, cuts have caused pain in millions of particular cases. This observation should at least *support* the generalization that cuts cause pain no matter what its precise interpretation. (The fact should even support "Cuts would cause pain.") But the observation provides *no* support on any of our theories. For all require a *comparison* of frequencies.

VI. THE TEMPORAL PRECEDENCE CONDITION

Hume thought that causes must precede their effects. The absence of a temporal precedence condition in our probabilistic descendents of Hume's theory generates some absurdities. The theories correctly rule that getting pregnant results in having a baby. For the probability of having a baby is much greater if you get pregnant than if you do not. But the theories also allow the backward result that *giving birth causes pregnancy*. For the probability of your getting pregnant is also greater if you will have a baby than if you won't. Of course, if pregnancy has a sufficient cause, it would screen off birth. But whether birth causes pregnancy should not depend on whether pregnancy has a sufficient cause.

Suppes himself insisted on temporal precedence in addition to positive relevance and screening. Despite his claim of formal neutrality, though, any temporal requirement must be very different for singular and general causation. Singular events occur at particular times. So it is relatively clear what temporal precedence is for them: *c* precedes *e* iff the time at which *c* occurs is earlier than the time at which *e* occurs.[18] General events like lightning, however, occur at many different times. We cannot say, therefore, that one general event precedes another iff the time at which the one occurs is earlier than that at which the other occurs.

We do have a concept of temporal precedence among general events, though.[19] We think getting pregnant precedes having a baby, lightning precedes thunder, and day precedes night. These are generalizations about the normal temporal relations among occurrences of the general events mentioned. With singular events, *c precedes e* is equivalent to *c is followed by e.* With general events, surprisingly, this equivalence fails. Of course, "Getting pregnant precedes having a baby" and "Getting pregnant is followed by having a baby" are both true. But while "Getting pregnant precedes *losing* a baby" is also true, "Getting pregnant is followed by losing a baby" is false. There are, therefore, two kinds of general temporal precedence. *C is followed by E* means something like *When C occurs, E normally occurs later.* We might call this *forward looking* temporal precedence, since we focus on the earlier event and see what normally occurs later. In contrast, *C precedes E* means *When E occurs, C normally occurs earlier.* This is *backward looking* temporal precedence, since we focus on the later event and see what typically happens earlier. General causation requires forward rather than backward looking temporal precedence. That is, "*C causes E*" *entails* "*C is followed by E*" *rather than* "*C precedes E.*" Legionnaires' disease causes death. But while the disease is normally followed by death, it does not normally precede death.

Normally is required in the analysis of general temporal precedence rather than *frequently.* "Getting pregnant is followed by losing a baby" is false even though miscarriages are not uncommon. My initial examples suggest that general temporal precedence requires some *regular interval* between the occurrences of *C* and *E*. Thus nine months normally elapse between conception and birth. But pregnancy precedes losing a baby even though there is no due date for a miscarriage, and birth is followed by death even though there is no normal age for death.

Note finally that a *common location* is also understood in statements of general temporal precedence when C and E do not specify one. Thus "Getting pregnant is followed by having a baby" is true because when getting pregnant occurs *to a woman*, having a baby occurs *to her* later.

Given that temporal precedence must be defined so differently for singular and general events, it should be no surprise that the two relations have different logical properties. Thus like causation, temporal precedence is asymmetric among singular events, but not among general events. In any particular case, Mary's getting a cold cannot both precede and follow Bill's getting one. But in general, Mary's getting a cold may be followed by Bill's getting one, and Bill's getting a cold may be followed by Mary's getting one. It should not be inferred, of course, that general temporal precedence is symmetric. Being born is followed by dying, but dying is not followed by being born. Indeed, the temporal precedence condition helps with the pregnancy example only because having a baby is not followed by getting pregnant. Of course, having a baby is often followed by getting pregnant *again*, but not by getting pregnant with the very same baby.

Suppes' formulation of the temporal precedence condition presupposes that events have a unique time associated with them. While this is appropriate for singular events, it often fails for general events. *Dying*, for example, is *temporally unindexed*: it can occur at any time. *Dying at 10:00 AM today*, in contrast, is *temporally indexed*: it can only occur at 10:00 AM today. Dying at 10:00 is still a general event, since it can occur to different individuals. Suppes' theory can only be applied to indexed general events. Let $F^{t''}$, $C^{t'}$, and E^{t} be temporally indexed general events, and let t'', t', and t be variables ranging over times.

Suppes' Theory: *$C^{t'}$ causes E^{t} iff t' precedes t, $P(E^{t}/C^{t'}) > P(E^{t})$, and there is no $F^{t''}$ such that t'' precedes t' and $P(E^{t}/C^{t'}F^{t''}) = P(E^{t}/F^{t''})$.*

The temporal restriction in the third clause restricts screening factors to events occurring before the suspected cause. This restriction was criticized earlier.

Suppes' temporal precedence requirement is weaker than mine: t' may precede t even though $C^{t'}$ is not followed by (and does not precede) E^{t}. Although 10:00 precedes 11:00, "Dying at 10:00 is followed by swimming at 11:00" is false (as conventionally interpreted)[20]. Suppes' requirement is strong enough, though, to avoid ruling that birth causes pregnancy. But this represents little improvement over

Cartwright. For Suppes' theory does not rule that "Giving birth causes pregnancy" is *false* either: it assigns *no* truth value to statements about temporally unindexed general events. The theory is also inapplicable when the temporal indexes are sufficiently broad. Thus "Giving birth in the 20th century causes getting pregnant in the 20th century" is not ruled false, nor is its converse ruled true.

We might extend Suppes' theory by defining causation for unindexed general events in terms of causation for indexed events. Thus pregnancy results in having a baby because getting pregnant *at one time* produces a baby *about nine months later.* And cyanide causes death because ingesting cyanide *at one time* causes death *a minute or so later.* A similar relationship holds when the general events are broadly indexed. Thus pregnancy in the 20th century causes birth in the 20th century because getting pregnant *at one time* in the 20th century results in giving birth *about nine months later* in the 20th century. Let C^0 and E^0 range over unindexed or broadly indexed general events, and let C^t and E^t be the result of indexing them with "at t." Then we might supplement Suppes' theory as follows: C^0 causes E^0 iff for some i, C^t causes E^{t+i}. The increment i may well be an interval like "9 months ± 2 weeks." The increment must be existentially quantified, since the time between cause and effect varies, as the pregnancy-birth and cyanide-death cases show.

This extension of Suppes' theory avoids the problem of inevitable events discussed earlier. Even though the probability of death is 1 because death will inevitably occur *at some time*, the probability of death at $t + 1$ minute is less than 1 for any t. So the probability of death at $t + 1$ minute given cyanide at t may well be greater. However, the fact that i is existentially quantified allows C^0 to cause both E^0 and $-E^0$ in some cases, which seems impossible.[21] Whereas an i of 9 months leads to the ruling that getting pregnant results in having a baby, an i of 1 month leads to the ruling that getting pregnant results in *not* having a baby. The probability of not having a baby at $t + 1$ month given conception at t is higher than the unconditional probability of not having a baby at $t + 1$ month, for any t. "Getting pregnant results in not having a baby" seems patently false, however, and seems incompatible with "Getting pregnant results in having a baby." I see no way to extend Suppes' theory without generating such absurdities.

Unlike Suppes, Giere and Cartwright omitted the temporal precedence clause. While Giere simply ignored temporal considerations,

Cartwright wanted to leave open the question of temporally backward causation. But Cartwright's theory does more than leave the question *open*: it wrongly implies that temporally backward causation is common and widespread, as the pregnancy-birth case illustrates.

The possibility of temporally backward causation is usually defended by an appeal to modern physics. Thus Sayre cites the annihilation of an electron and a positron in the production of an X-ray: "It remains an acceptable interpretation of the Feynman equations that the positron is actually an electron moving in the reverse temporal direction, in which case it cannot be ruled out *a priori* that the effect precedes at least part of its cause in time" (1977, p. 209ff). The idea of motion in "reverse temporal direction" is curious. Suppose a particle is at position p at t and at p' at t', where p and p' are different, and where t is earlier than t'. Is it an open question, to be settled empirically, whether we describe the particle as moving from p to p' (regular temporal direction) or as moving from p' to p (reverse direction)? Hardly. Given what the words *from* and *to* mean, we have to describe the particle as moving from its earlier position to its later position. We do not need further evidence. Describing the particle as moving from p' to p would be like describing a growing child as shrinking. My position here might be criticized on the grounds that it is illegitimate "*apriori* physics" to argue for or against a physical claim on the basis of what certain English words mean. But note that the particle's motion — its change in position over time — was *given*. Obviously, such a change can only be known *aposteriori*. But the question at issue was how to describe the given motion using the English words "from" and "to." Since that is the issue, the words' meaning will obviously be critical.

In general, causing something to occur that has already occurred is impossible, in my opinion, for the same sort of unprofound reason that it is impossible to move to a place one is already at. A more serious problem, I think, is that the temporal precedence condition rules out *simultaneous causation*. Consider an airplane climbing. The motion of air over and under the wings causes the wings (and the rest of the plane) to rise. But the motion of the air does not precede the upward motion of the wings.[22] The temporal condition could be weakened, by requiring only that C is followed *or accompanied* by E. But since simultaneity is symmetric, our probabilistic theories will have problems with the direction of causation among simultaneous events. If the theories correctly rule that the air moving around it causes the wing to rise, they will also incorrectly rule that the wing rising causes the air to move around it. So if a theory requires strong temporal precedence, it rules out simultaneous causation; but if a theory requires only weak precedence, it allows backward simultaneous causation. The probabilistic theorist can take some comfort, however, in the fact that all other approaches face the same dilemma.

VII. FORKS, CHAINS, AND WEDGES

Three events may be causally related in several ways. Some common configurations are *forks, chains*, and *wedges*.

Forks *Chains* *Wedges*

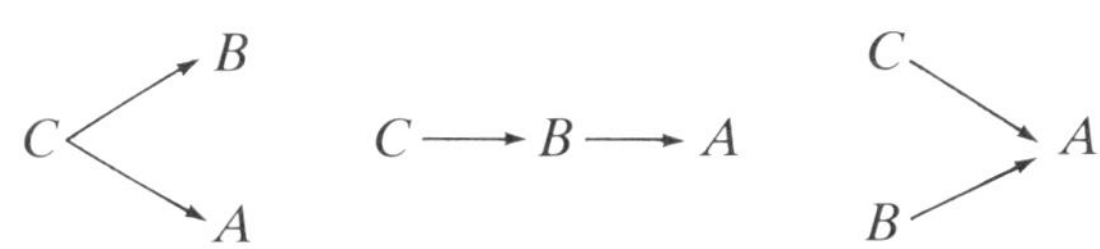

In a fork, *A* and *B* have *C* as a *common cause*. Equivalently, *C* has *A* and *B* as *multiple effects*. But there is no causation between *B* and *A*. In a wedge, *C* and *B* have *A* as a *common effect*, and *A* has *B* and *C* as *multiple causes*. But there is no causation between *C* and *B*. In a chain, *C* causes *B*, which causes *A*.

Forks, chains, and wedges are possible among both general and singular events. General causal chains may be either transitive or nontransitive, depending on whether *C* causes *A*. All singular causal chains are transitive. As for wedges, it is important to distinguish between singular causes that act *independently* (as when a building burns because of a short and coincidentally because of an arsonist's torch) and those that do not (as when a building burns because of a short and because flammable materials were nearby). Independent multiple causation of a singular event is called *overdetermination*; its existence is controversial.[23] Many would argue (unsuccessfully in my opinion) that when singular events are described finely enough, or understood properly, the appearance of overdetermination disappears. This controversy does not carry over to general events. It cannot be denied, for example, that both decapitation and electrocution cause death.

Let us focus on general causation. We saw that one advantage of probabilistic theories over Hume's is their success with the common cause problem, which involves forks. The firing of a cannon may produce a characteristic flash and noise, but the flash does not cause the noise even though it is always followed by the noise. Suppes and Cartwright accounted for this by observing that the firing screens off the flash: the noise's probability given the firing and the flash does not exceed its probability given just the firing. Unfortunately, the Suppes-Cartwright treatment for the common cause problem has side-effects: the screening condition often rules that *B* is a spurious cause of *A* in chains and wedges as well.[24] Consider transitive chains. Whenever $P(A/BC) = P(A/C)$, *C* screens off *B*. This happens when (but not only when) *C* is sufficient for *B*, or when *B* is necessary for *A*. Thus in ample quantities, cyanide always causes asphyxiation, which causes death. The probability of death given asphyxiation and enough cyanide equals the probability of death given enough cyanide. So the conditional probability theory denies that asphyxiation causes death. Rosen somehow finds this "not an unattractive result." "If a prior event has already completely determined a later event," she says, "then all intervening events are mere lackeys and deserve the spurious epithet" (1980, p. 78). But asphyxiation is no spurious cause of death. Nor is it a mere "lackey": cyanide causes death only because it causes asphyxiation.

Remaining with chains, *B* screens off *C* on the basic theory whenever $P(A/CB) = P(A/B)$. This obtains when *B* is sufficient for *A*, when *C* is necessary for *B*, or when the events form part of a Markov chain. Suppes' restriction of screening factors to earlier events prevents *B* from screening off *C* in these situations, as does Cartwright's exclusion of effects from the set of alternative causally relevant factors.

Our probabilistic theories also rule out many wedges. Whenever the probability of *A* given both *C* and *B* does not exceed the probability of *A* given just *C*, *C* screens off *B*. This happens when (but not only when) *C* is a sufficient or maximal cause of *A*. Since decapitation is a sufficient cause of death, the probability of death given decapitation is 1. Since this probability cannot be exceeded, decapitation screens off all other causes of death. The probability of death given cyanide gas and decapitation will be no greater than the probability of death given just decapitation, for example. As long as $P(A/C) = 1$, *C* screens off *B* even when *C* is only a potential cause of *A*. This occurs in wedge-like cases where *C* *would* cause *A* if *B* should not cause *A*. *C* might be a *back-up* for *B*, or be *preempted* by it.[25]

Cartwright recognized the difficulty for her theory when the probability of effect E given some alternative cause F is 1 (1979, p. 428). She suggested that C causes E in that case as long as $P(E/CF) = 1$. But this makes almost anything a cause of death, even a kiss. For the likelihood of death given a kiss plus any sufficient cause of death is 1 too. Moreover, the same problem arises when E has a *maximal* cause, even if that cause is not sufficient. Let 10% of all humans be invulnerable. Then the probability of death given decapitation would be just 90%. Adding cyanide would not make death any more likely.

We might try weakening the positive relevance condition, requiring only that a cause *sometimes raises* and *never lowers* the probability of its effect.[26] Cartwright's theory would then say that C causes E only if $P(E/CF) > P(C/F)$ for *some* F and $P(E/CF) < P(C/F)$ for *no* F, where F ranges over alternative causally relevant factors. But such a weakened theory would have as much trouble with the common cause problem as Hume's, ruling that the flash of the cannon causes the noise.

We have seen that the Suppes-Cartwright treatment of the common cause problem has serious side-effects. It is also not completely effective. Their theories correctly rule that B does not cause A in a fork as long as $P(A/BC) \leqslant P(A/C)$. But this condition need not obtain. When $P(A/BC) > P(A/C)$, in what Salmon has called an *interactive fork*, C does not screen off B.[27] Imagine a television set with a balky switch: it usually turns the set on, but not always. When the set is on, it produces both sound and picture. Then the probability of a picture given that the switch is on and given sound is greater than the probability of a picture given just that the switch is on. But the sound does not cause the picture. They are causally independent effects of a common cause even though they are not statistically independent. As Shrader (1977, p. 141) observed, probabilistic theories share one of the most serious defects of Hume's theory: the failure to distinguish completely between *causes* of an event and mere *indications* that it occurred.

VIII. CONCLUSIONS

Probabilistic theories of causation are extremely attractive given that, *pace* Hume, causes do not *always* produce their effects. Causes more commonly just raise their effects' probability, whether "raising" involves differences in conditional probabilities (Suppes and Cartwright) or counterfactual probabilities (Giere). The notion of screening solves the standard common cause and accidental generalization problems of the Humean theory. The complaint that probabilistic theories violate determinism is groundless.

Probabilistic theories do face serious problems. First, a distinctness condition is needed to guarantee that causes and effects, as well as screening factors and spurious causes, are numerically, logically, and generationally independent. Second, probabilistic theories are only plausible for general causal statements, since causation among singular events is transitive and asymmetric, and since empirical probability is difficult to assign to singular events. But the cumulative transitivity of general causal statements remains unaccounted for. And it seems impossible to analyze general causation independently of singular. Third, a temporal precedence condition is needed to rule out widespread backward causation. But it would either eliminate all simultaneous causation or fail to eliminate backward simultaneous causation. Fourth, while the screening condition solves some common cause problems (involving forks), it creates severe problems with chains and wedges. Earlier links of a chain screen off later links, and a maximal cause screens off all other causes. Furthermore, interactive forks remain problematic, which means that probabilistic theories do not distinguish completely between causes and mere indicators.

In short, probabilistic theories of causation are attractive, but have major drawbacks. I would sound a death-knell for the approach except for the fact that the alternatives' vital signs are equally weak.[28]

Georgetown University

NOTES

[1] Probabilistic theorists commonly distinguish between *positive* and *negative* causation, defining the former in terms of positive relevance, the latter in terms of negative relevance. We will focus exclusively on positive causation, since that is what is expressed by English sentences of the form "*C* causes *E*."

[2] As Suppes observes, "screening factor" should be defined more strictly. *F* should satisfy $P(E/CF) \geqslant P(E/C)$ as well as $P(E/CF) = P(E/F)$. Nothing will hang on this, so I am omitting it for simplicity.

[3] This argument has been advanced by Stegmüller (1973), Hesslow (1976, p. 292), von Fraassen (1978), and Brand (1979, p. 273). It has been rejected by Krüger (1976),

Rosen (1978, p. 613), Fetzer (1981, p. 134ff), Humphreys (1980, p. 310), and Railton (1981, pp. 238, 248).

[4] Thus it is no help to define a complete explanation as one citing all the explanatory factors there are, as some do.

[5] My example is a modification of Eells' (1986).

[6] Cf. Rosen (1978, pp. 606—607), Salmon (1980, pp. 62—64), and Otte (1981, pp. 182—84).

[7] This formulation is based on Cartwright's applications of her theory (e.g., see 1979, p. 427). Her "official" formulation is somewhat different (see 1979, p. 423). There she requires that F be a *state description* over all events that cause either E or $-E$ other than C or its effects. Nothing I say below hinges on the precise formulation of Cartwright's theory.

[8] This was urged by Ellery Eells (in a discussion of his paper at the 1986 Pacific APA meeting) and by an anonymous reader.

[9] I picked the *immediate* future because "C causes E" is in the present tense, and because all probabilities are assumed to be current. In general, we can say that C causes E at t iff C raises the probability of E in some interval beginning with t.

[10] The Davidsonian theory of event identity does not apply here since we are dealing with act-*types*.

[11] Cf. the "pairing problem" discussed by Kim (1973b) in connection with the Humean theory. See also Hesslow (1976, p. 291).

[12] See Fetzer (1981) for a spirited defense of the "single case propensity theory of probability."

[13] Anglin (1981) argues that circular causation is possible even among singular events. As an example, he claims that "William went to the party because Marilyn did" and "Marilyn went because William did" may both be true. I believe Anglin misdescribes the causal facts. It is not one person's *actual* going that causes the other to go: that would be hard to imagine. Rather, William goes because *he believes Marilyn is going*. Similarly, Marilyn goes because *she believes William is going*.

[14] See Stalnaker (1968, p. 106) and Lewis (1973b, p. 32).

[15] This was argued by an anonymous reader.

[16] Cf. the familiar "birdie" example: Jones hit a wild shot, which struck a tree, and miraculously deflected into the hole for a birdie. Then hitting a wild shot resulted in a birdie in this case; but in general, a wild shot does not result in a birdie. See Suppes (1970, p. 41), Rosen (1978, pp. 607—608), Humphreys (1980, pp. 307—309, 311), Salmon (1980, pp. 63—64), and Otte (1981, pp. 183—84).

[17] Once is not enough. Suppose I have only taken LSD and hallucinated once. Then "My taking LSD causes me to hallucinate" and "My taking LSD used to cause me to hallucinate" are incorrect general causal statements even if the singular statement "My taking LSD caused me to hallucinate" is true.

[18] Actually, things are more complex, since some singular events occur over *intervals* rather than at *points* of time. Cf. Pollock (1982, p. 287).

[19] Contrast Kim (1973b, p. 217).

[20] It would be quite natural to define temporal precedence for indexed general events as follows: $C^{t'}$ precedes E^{t} iff t' precedes t. On this definition, general temporal precedence is just like singular. What I am pointing out in the text is that this sense of

general temporal precedence is not what is expressed by "precedes" in conventional English.

[21] Again, there is a parallel for conditionals. The law of conditional noncontradiction states that "If P then Q" and "If P then $-Q$" are incompatible except when P is contradictory. There is no similar exception to the incompatibility of "C causes E" and "C causes $-E$," because both statements entail that C has occurred, which guarantees that it is a possible event.

[22] Rosen (1980, p. 76ff) argues against simultaneous causation on the grounds that "the speed of light is a constant which places an upper limit on the speed of propagation of energy." I see no justification for Rosen's assumption that causation necessarily involves the propagation of energy.

[23] Skyrms (1980, p. 109) and Rosen (1980, p. 79) both deny the possibility of overdetermination, for example.

[24] Otte (1981, pp. 173—75, 180) observes that when all the causes in question are necessary and sufficient for their effects, Suppes' theory fails to distinguish among forks, wedges, and chains. For all the probability relations are the same.

[25] Otte (1981, p. 174) and Ehring (1984) describe excellent cases of preemption, and show that they are not allowed on Suppes' or Cartwright's theories.

[26] This "pareto-dominance condition" was suggested by Skyrms (1980, p. 108) as a plausible weakening of the Suppes-Cartwright theory.

[27] See Salmon (1980, p. 66ff), Otte (1981, pp. 180—81), and Shrader (1977, pp. 143—44).

[28] I would like to thank Ellery Eells, two anonymous readers, and the philosophy department of the University of Cincinnati for helpful comments on an earlier version of this paper. One of those readers deserves special mention for an incredibly conscientious and useful review.

REFERENCES

Anglin, W. S. (1981) "Backwards Causation," *Analysis* **41** (1981), pp. 86—91.

Brand, M. (1979) "Causality," in *Current Research in Philosophy of Science*, P. D. Asquith, ed. (Ann Arbor, MI: Edwards Brothers, 1979), pp. 252—281.

Cartwright, N. (1979) 'Causal Laws and Effective Strategies,' *Nous* **13** (1979), pp. 419—37.

Eells, E. (1985) 'Probabilistic Causal Interaction,' (1985), *Philosophy of Science* **53** (1986), pp. 52—64.

Ehring, D. (1984) 'Probabilistic Causality and Preemption,' *British Journal of Philosophy of Science* **35** (1984), pp. 55—57.

Fetzer, J. (1981) *Scientific Knowledge* (Dordrecht: Reidel, 1981).

Giere, R. (1980) 'Causal Systems and Statistical Hypotheses,' in *Applications of Inductive Logic*, L. J. Cohen and M. Hesse, eds. (Oxford: Clarendon, 1980) pp. 251—70.

Goldman, A. I. (1970) *A Theory of Human Action* (Englewood Cliffs, NJ: Prentice-Hall, 1970).

Hesslow, G. (1976) 'Two Notes on the Probabilistic Approach to Causality,' *Philosophy of Science* **43** (1976), pp. 290—92.

Humphreys, P. (1980) 'Cutting the Causal Chain,' *Pacific Philosophical Quarterly* **61** (1980), pp. 305—314.

Kim, J. (1973*a*) 'Causes and Counterfactuals,' *Journal of Philosophy* **70** (1973), pp. 570—72.

Kim, J. (1973*b*) 'Causation, Nomic Subsumption, and the Concept of Event,' *Journal of Philosophy* **70**, (1973), pp. 217—36.

Krüger, L. (1976) 'Are Statistical Explanations Possible?' *Philosophy of Science* **43** (1976), pp. 129—146.

Lewis, D. (1973*a*) 'Causation,' *Journal of Philosophy* **70** (1973), pp. 556—67.

Lewis, D. (1973*b*) *Counterfactuals* (Cambridge MA: Harvard, 1973).

Otte, R. (1981) 'A Critique of Suppes' Theory of Probabilistic Causality,' *Synthese* **48** (1981), pp. 167—90.

Pollock, J. L. (1982) 'Causes, Conditionals, and Times,' *Pacific Philosophical Quarterly* **63** (1982), pp. 275—88.

Railton, P. (1981) 'Probability, Explanation, and Information,' *Synthese* **48** (1981), pp. 233—56.

Rosen, D. (1978) 'In Defense of a Probabilistic Theory of Causality,' *Philosophy of Science* **45** (1978), pp. 604—13.

Rosen, D. (1980) 'A Probabilistic Theory of Causal Necessity,' *Southern Journal of Philosophy* **18** (1980), pp. 71—86.

Salmon, W. C. (1980) 'Probabilistic Causality,' *Pacific Philosophical Quarterly* **61** (1980), pp. 50—74.

Sayre, K. M. (1977) 'Statistical Models of Causal Relations,' *Philosophy of Science* **44** (1977), pp. 203—14.

Shrader, D. W. Jr. (1977) 'Causation, Explanation, and Statistical Relevance,' *Philosophy of Science* **44** (1977), pp. 136—45.

Skyrms, B. (1980) *Causal Necessity* (New Haven: Yale, 1980).

Sober, E. (1982) 'Frequency-Dependent Causation,' *Journal of Philosophy* **79** (1982), pp. 247—53.

Stalnaker, R. (1968) 'A Theory of Conditionals,' *American Philosophical Quarterly Monographs*, No. 2 (1968), pp. 98—112.

Stegmüller, W. (1973) *Personelle und Statistische Wahrscheinlichkeit* (Springer-Verlag, 1973).

Suppes, P. (1970) *A Probabilistic Theory of Causality* (North Holland Publishing Co., 1970).

van Fraassen, B. (1978) 'Review of Stegmüller's *Personelle und Statistische Wahrscheinlichkeit*,' *Philosophy of Science* **45** (1978), pp. 158—63.

BRIAN SKYRMS

CONDITIONAL CHANCE[1]

By *Chance* I mean physical probability as opposed to degree of belief. Good subjectivists do not believe in *Chance*, but this does not free them from the obligation to give a theory of chance. Ordinary and scientific discourse is full of chance talk, and if its function is not to be given straightforward naive analysis in terms of reference to real chances, then a more sophisticated analysis is in order. Subjective Bayesians have a pragmatic story to tell which is, by now, well known. Putative beliefs about chance are given a reading in terms of ordinary beliefs about the world. The sort of pragmatic analysis of meaning offered is of general philosophical interest, and invites close study by the philosophical community.

Chance enters into a number of philosophical analyses not as an unconditional probability, but rather as a conditional probability: *Conditional chance*. In this paper, I focus on two related examples: the theory of subjunctive conditionals and the theory of probabilistic causation. In certain approaches to each of these areas the availability of conditional chance is assumed and it is made to play a fundamental role. Does the subjective Bayesian story about chance together with the standard definition of conditional probability generate an adequate Bayesian story about the functioning of conditional chance in these contexts or do these uses of conditional chance generate additional problems which demand the attention of honest Bayesians?

I believe the latter to be the case, and this paper is intended to be a contribution towards the development of such a Bayesian theory. The theory will throw some light on the relation between theories which use conditional chance, and those which don't: in particular, between Adams and Stalnaker-Lewis types of analysis of conditionals and between Good and Suppes-Salmon-Granger analyses of probabilistic causation. And difficulties in the theory of conditional chance will predict corresponding difficulties in the theories of conditionals and probabilistic causation.

In Section 1 I will introduce the leading ideas of the subjective Bayesian treatment of chance in a mathematically simple setting that

James H. Fetzer (ed.) Probability and Causality. 161–178.

should be accessible to most philosophers. Readers familiar with these ideas can skip this section. Section 2 will indicate how the ideas of Section 1 are generalized to a mathematically and physically more interesting theory. Readers who already know, and readers who do not want to know about these refinements can skip this section. Section 3 will sketch a theory of conditional chance which is a natural generalization of the foregoing theory of chance. Section 4 will discuss the import of this account of conditional chance for theories of subjunctive conditionals. Section 5 will discuss its relevance to theories of probabilistic causation.

1. CHANCE

Typically, when one is uncertain about the chance of an outcome, one assigns it a degree of belief that is a weighted outcome of the possible chances. Suppose that you are uncertain about the bias of a coin, and suppose for simplicity you believe that there are only two possibilities: 2 to 1 in favor of heads or 2 to 1 in favor of tails. Perhaps you know that the coins came from a joke shop that only stocks these two kinds. Suppose your degree of belief in bias towards heads is somewhat greater than your degree of belief in bias towards tails, say 0.6. The coin is about to be flipped. You assign degree of belief to it coming up heads of $0.6(2/3) + 0.4(1/3)$: an average of the possible chances weighted by your degrees of belief that they are the chances. We say that *degree of belief* is the expectation of chance. It is essential to a Bayesian story to the effect that ordinary degrees of belief behave *as if* we had degrees of belief about chance that this principle be preserved.

In the sort of simple situations we have in mind here, the appropriate mathematical trick is to define *Chance* as degree of belief conditional on a partition. Suppose that we have a space of possible situations (or worlds) with degrees of belief defined over appropriate subsets of them (or Carnap propositions). Suppose you are to learn whether a certain proposition, S, which determines the chance of another proposition, T, is true or false and nothing more. Then your new degree of belief in T, after learning the truth about S will be either your old degree of belief conditional on S, $Pr(T \mid S)$, or your old degree of belief conditional on not-S, $Pr(T \mid \text{not-}S)$. If you are uncertain about S, you can still say what the "chance" of T is in any possible situation, w. In any w in the set S, $\text{chance}(T) = Pr(T \mid S) = Pr(T \,\&\, S)/Pr(S)$ and in any w in the set (not S), chance(T) = $Pr(T|\text{not-}S)$. In so doing you have defined the value of chance(T) for every situation, w, and have therefore given a kind of *truth definition* for statements of the form chance(T) = x. Notice that the definition has the consequence that chance(T) is the same for every situation in S and is the same for every situation in not-S, as it should be if S is to be interpreted as the proposition whose truth value determines the value of chance(T), and that it is a consequence of the definition that the probability of T is the expectation of the chance of T.

The propositions [S, not-S] in the preceding example are said to form a *partition* of the space of possible situations; i.e. they are mutually exclusive and their union is the whole space. The technique of the example can be applied quite generally to finite partitions, $S = [S_1, \ldots, S_n]$. We construe chance (T) as a function $Pr[T||S]$ which takes at situation w the value $Pr(T|S_i)$ where S_i is the member of the partition containing w. I say *a* function rather than *the* function because some members of the partition may have zero probability in which case $Pr(T|S_i)$ is undefined. In this case any value constant over situations in S_i will do, since the degree-of-belief expectations will not be affected. Different choices of ways to fill in the undefined cases lead to different functions, all of which are called *versions* of the probability of T conditional on the partition, S. Any version of $Pr[T||S]$ gives a Kripke style semantics for statements of the form "chance(T) = x" with the desired formal properties: chance(T) is constant for situations in the same element of the partition and $Pr(T)$ is the expectation of chance(T).

The subjective Bayesian story now goes something like this: "All this talk about so-called chances is really just the talk about probability conditional on a partition for some partitions that we have found to be of continuing usefulness. Usefulness of a partition may have something to do with physical science and may have something to do with the mathematics of convergence, but a reification of "Chance" contributes nothing to the analysis of these questions."

2. BELIEF CONDITIONAL ON A σ-FIELD

In order to tell the Bayesian story in any generality, we need to generalize the notion of belief conditional on a partition to belief conditional on a σ-field. A σ-field of propositions is a set of propositions closed under not only finite truth functions, but also countable conjunction and disjunction. We assume that the set of propositions that our degree of belief measure, *Pr*, is defined over is a σ-field. We will also assume without discussion here that *Pr* is countably additive, i.e. that for a countable disjunction of pairwise incompatible propositions, its probability is equal to the limiting sum of the probabilities of the disjuncts.

Now if the determination of chance involves a limiting process, one may need something fancier than a finite partition to specify the factors which determine chance. A subset of the propositions which is itself a σ-algebra is the appropriate vehicle. The determination of chance in a world depends on the truth values of all the members of the σ-algebra in this world. We want to define chance of a given proposition, *T*, as a random variable taking a definite value on [0, 1] in every situation, in such a way that degree of belief is the expectation of chance and such that this principle is preserved under conditioning on members of the σ-algebra. For every proposition, *S*, in the σ-algebra *S*, we want:

$$\int_S \text{Chance}(T)\, \delta Pr = Pr(A \,\&\, S)$$

Such a random variable always exists, and it is called the probability of *T* conditional on the σ-field *S*: $Pr(T \mid S)$. It is again determined only up to a set of measure zero, the various functions in the family again being called *versions* of the probability of *T* conditional on *S*.

For example, suppose that our probability space is the unit square, with our "propositions" being the Borel sets and our initial probability measure being the uniform Lebesgue measure. Suppose we want to consider probability of a proposition, *T*, conditional on the σ-subfield generated by the Borel sets on the *x* axis: i.e., the sets $b\ X\ [0, 1]$ where *b* is a Borel set of the *x* axis. To specify for a point, *w*, whether it is in or out for each member of this subfield is to specify its *x* coordinate. At almost every point, the probability of *T* conditional on this subfield is just the length of the portion of the vertical line with this coordinate which lies in *T*. To make the mathematics fit the motivation, think of a chance process on vertical location with unknown horizontal location.

If we can specify an appropriate σ-field for chance (*T*) and an appropriate σ-field for chance (*S*), can we specify an appropriately fine σ-field which gives them all at once? In abstract measure theory technical difficulties may arise which preclude an affirmative answer. I don't want to enter into details here. It should suffice to say that these difficulties do not arise if our underlying space of states has a reasonable topological structure (as the space in our example does).

Probability conditional on a σ-field makes contact with the great convergence theorems of probability, and thus explains the intimate connection of the notions of chance and limiting relative frequency while avoiding the absurdities of the crude philosophical theory which identifies them. The problem of identifying an appropriate σ-field can be attacked from two directions. One way, using sufficient statistics, is to try directly to isolate what is relevant to the chances. The complimentary approach, involving groups of measure preserving transformations, seeks the maximal symmetry which leaves the structure of the chances invariant. (There is a sketch of these approaches for philosophers in chapter 3 of my *Pragmatics and Empiricism*.) Whether approached *via* partial exchangeability or ergodic theory, this Bayesian conception of chance explains why degree of belief is the expectation of chance; why and how degrees of belief based on knowledge of chances are resilient; why in typical settings consistency forces us to belief that ideal evidence would almost surely lead our opinions to converge to the sure chances; and why in many cases chances are almost surely equal to the appropriate limiting relative frequencies. It also explains how a subjective theory of chance can make fruitful contact with objective physics.

This is the real theory, standing in the background, when we discuss our toy examples with small finite partitions. In the balance of the paper I will return to toy examples in order to avoid obscuring the philosophical problems that I want to raise with mathematical complexity. But the reader should remember that the issues involved apply to the general theory as well.

3. CONDITIONAL CHANCE

When we have given a theory of chance, have we not also given a theory of conditional chance, at least for conditions of positive probability, by virtue of the standard definition of conditional probability in terms of unconditional probabilities? The answer depends on the kind of condition we have in mind.

Let us consider an example that I have discussed elsewhere[2] in this connection. A coin is loaded one way or another with iron and a magnet is on or off so that the chances of heads are:

Chance(*H*)	*BH*	*BT*
OFF:	2/3	1/3
ON:	5/6	1/6

Suppose that tosses are independent in the chance distributions and that the coin is to be flipped 10 times under constant conditions. Then some conditional chances, e.g. chance (at least 5 heads given at least 4 heads), have indeed been adequately dealt with. In any situation we simply consider whether the bias is *BH* or *BT* and whether the electromagnet is on or off to fix the unconditional chances of "at least 5 heads" and "at least 4 heads" and then apply the definition of conditional probability.

But what if the condition includes a partial specification of the determinants of chance? What is the chance of heads on toss 1 conditional on *BH* in a situation where we have *BT* & *OFF*? Here a mechanical application of chance as probability conditional on the four-celled partition and the ratio definition of conditional probability does not work, because the stated condition is incompatible with some cells of the partition. It is intuitively clear, however, how this example should be treated. If we are interested in the value of $Ch(H \mid BH)$ in a situation in which we have *BT* & *OFF*, we should take the partial determinant of chance supplied by the stated condition, *BH*, and let the situation supply the rest, i.e. *OFF*. The appropriate value for conditional chance here is then 2/3.

Then $Ch(H \mid BH)$ has the same value in situations with *BT* & *OFF* as in those with *BH* & *OFF*; and similarly it has the same value in those with *BT* & *ON* as in those with *BH* & *ON*. Our intuitive procedure is equivalent to evaluating conditional chance in the straightforward way as conditional "objective" probability where the "objective" probabilities are probabilities conditional on the coarser partition: $\{OFF, ON\}$.

This suggests that for a general theory of conditional chance we need not one, but a family of partitions. That is, we need a function which maps each consistent condition onto a partition each of whose members is consistent with that condition. If we assume a background in which each consistent proposition has positive probability, we can then simply define the value of the conditional chance, $Ch(q \mid p)$ in a situation as $pCh(p \ \& \ q)/pCh(p)$ in that situation where pCh is probability conditional on the partition, πp, which the family assigns to p. Such a theory is sketched in Chapter 5 of my *Pragmatics and Empiricism*.

Such a family of partitions is *deterministic* iff for every consistent proposition, p, each member, b, of πp is such that b & p contains only one point. Deterministic families of partitions are extreme cases which give conditional chances always equal to zero or one. At the other end of the spectrum, is the *trivial* family of partitions which has only one member, i.e. the trivial partition which itself has only one member, the whole space. One can investigate natural properties of families. For instance, suppose that p entails q and that the partition appropriate to q, $\{b^1, b^2, b^3\}$, is too fine for p because one of its members, b^3, is incompatible with p. Then a natural way to move to a coarser partition appropriate to p is to weaken the background conditions corresponding to b^1 and/or b^2 to $b^{1'}$ and $b^{2'}$ so that these cover all the cases originally in b^3. A family of partitions where coarsenings are always done in this way is called *omonotonic*. And we want our theory of conditional chance to mesh correctly with our theory of unconditional chance. The stated condition together with the background factors specified by the cells of the partition assigned to that condition should jointly be sufficient to specify the unconditional chances. Let πc be the partition for unconditional chance and πp be the partition assigned by the family for conditional chance to p. The treatments of conditional and unconditional chance are said to *mesh* if for every consistent p, and every cell bpi in πp, bpi & p is a subset of c for some c in πc.

It would be of some interest to develop the technical properties of various kinds of families of partitions, and to generalize the account to families of σ-algebras. Here, however, I want to turn to philosophical considerations raised by Robert Stalnaker[3] which suggest the need for an even more general account.

The theory works naturally for chance conditional on some conditions. Questions of chance conditional on *BT* call up the partition $\{$*ON, OFF*$\}$; questions of chance conditional on *OFF* call up the partition $\{$*BH, BT*$\}$; questions of chance conditional on a tautology call up the finer partition {*BH* & *ON*, *BH* & *OFF*, *BT* & *ON*, *BT* & *OFF*}; questions of chance conditional on *BH* & *ON* call up the trivial partition. But what, asks Stalnaker, is the chance of heads on toss 1 conditional on *BH* or *ON* in a situation where we in fact have *BT* and *OFF*?

If we make up the rest of the story in the most favorable way, things are not that bad. There is another factor, *S*, taking on values S_1, S_2, S_3. All values are compatible with *BT* and *OFF*; but only S_1 is compatible with *BH* and *ON*, only S_2 with *BH* and *OFF*, only S_3 with *BT* & *ON*. Specification of bias of the coin, state of the magnet and compatible *S*-factor value gives us a partition with six cells, with *BT* & *ON* being split into three parts according to the value of the *S* factor. This partition is finer than need be to determine chance, but that does no harm. Now, to answer Stalnaker's question we look to the 3 celled partition generated by the *S*-factor. Then the chance of heads on toss one conditional on *BH* or *ON* is 5/6, 2/3, 1/6 in a situation according to whether we have S_1, S_2, or S_3 holding.

Must there always be such an *S*-factor available to save the day? There must be on Stalnaker's theory of subjunctive conditionals: "If *BH* or *ON* were the case then *BH* and *ON* would be"; "If *BH* or *ON* were the case then *BH* and *OFF* would be"; "If *BH* or *ON* were the case then *BT* and ON would be"; are S_1; S_2; S_3 and on Stalnaker's theory exactly one of them must be true at each point.

However, I think that I agree with Stalnaker's intuition that there need not be such an *S*-factor available. Suppose that we limit ourselves to the factors of bias of the coin and state of the magnet. Then what can we say about the problem case. The formal machinery allows us to take the trivial partition whose only member is the whole space as the relevant partition making the conditional chance equal to the subjective conditional probability, but this is an unintuitive answer. In fact *any* answer here is an unintuitive answer. The disjunctive condition is inhomogeneous in such a way that there is no natural way to make it cooperate with background conditions to give an unambiguous specification of chance. (If we changed the example so that the probability of heads on toss one had the same value conditional on *BH* & *ON*, *BH* & *OFF*, *BT* & *ON* then we could take this as the natural value of chance conditional on the disjunction, *BH* or *ON*. In this case we would say that the disjunctive condition is homogeneous with respect the outcome, heads on toss one.) The only natural thing to say here is that conditional chance is indeterminate. This means that the theory should be generalized to allow partial families of partitions. In the case of inhomogeneous conditions, an appropriate correlative partition may fail to exist. From this perspective, the case of a family which always yields an appropriate partition is rather special and deserves the special name of a *total* family.

A case much less problematic that of a total family is that of *the natural partial family generated by a list of factors*. Let a *factor* be a variable taking a finite number of mutually exclusive and jointly exhaustive values[4], and consider a finite list of factors. A natural condition will be the specification of a value for each of some subset of the variables. The natural partial family generated by a list of factors assigns to each natural condition the partition generated by joint assignments of values to the remaining variables. The natural partial family generated by a list of factors is *monotonic*: If q entails q and πp and πq both exist, then πp is a coarsening of πq.

This discussion of Stalnaker's problem leads to a prediction for theories which use conditional chance, which we will test shortly. Natural partial families should be unproblematic, but difficulties should arise where the conditional chance called for by the theory depends on an inhomogeneous disjunctive condition.

4. CONDITIONALS

On one theory of subjunctive conditionals, that of Robert Stalnaker, a subjunctive conditional, $p \Rightarrow q$, always has an objective truth value in a possible world, w: i.e. the truth value of q in the possible world, w', selected by the operative Stalnaker selection function when w and p are given as arguments. According to a rival theory, due mainly to Adams[5], a subjunctive conditional has an objective value in every possible world, but that value is not a truth value. Rather it is the conditional chance of the consequent of the conditional on its antecedent. These ideas appear to be quite different from one another, but given the foregoing account of conditional chance we can see that they are intimately related.

Consider a selection function which maps pairs consisting of a possible world, w, and a consistent proposition, p, to possible worlds. (We neglect conditionals with inconsistent antecedents here, since the treatment of them is separable from the issues to be discussed.) We say that such a function is van Fraassen iff it satisfies: (i) $f(w, p) \in p$ and

(ii) If $w \in p$ then $f(w, p) \in p$ and that it is Stalnaker iff it is van Fraassen and, in addition, it satisfies (iii) If $f(w, p) \in q$ and $f(w, q) \in p$ then $f(w, p) = f(w, q)$.

Consider now the conditional chance account, when the family of partitions is total and deterministic. In this case we can think of the family of partitions, p, as inducing a selection function: i.e. that function which maps (w, p) onto that point, w', which is the sole member of the intersection of p with the member of πp which contains w. This selection function is van Fraassen, and the truth value that it yields for a conditional, $p \Rightarrow q$, is equal to the conditional chance of q on p. The van Fraassen selection function account is equivalent to the Adams account in the special case where conditional chance is given by a total deterministic family of partitions.

If, in addition, a total deterministic family of partitions is Omonotonic then the selection function it induces is Stalnaker. Conversely, we can think of a selection function, f, as inducing a family, π, of partitions where πp has as elements the set of inverse images under $f(\cdot, p)$ of points in p. If a selection function is Stalnaker, then the family of partitions it induces in this way is a total deterministic Omonotonic family. Since the requirement of Omonotonicity does not seem to be transparently compelling, the conditional chance account may appear to support van Fraassen's intuition that Stalnaker's condition (iii) is not well-motivated.

However, the problem that Stalnaker raised for the conditional chance account suggested that we should consider natural partial families of conditional chances. This suggests that Stalnaker's own account should be generalized to cover partial selection functions. A partial selection function will be said to be quasiStalnaker iff it meets Stalnaker's three conditions where defined. A deterministic partial family of partitions generates a partial selection function in the obvious way. Remember that the natural partial family of partitions generated by a list of factors is Monotonic (hence, *a fortiori*, Omonotonic). Here Stalnaker's third condition finds a natural justification from the perspective of the conditional chance theory:

The partial selection function induced by a deterministic natural partial family of partitions generated by a list of factors is quasiStalnaker.

The move to partial families of partitions when coupled with the theory that the objective value of a conditional is its conditional chance has an additional insight to offer. It predicts trouble for subjunctive conditionals whose antecedents are disjunctions, and everyone who has worked on conditionals knows that this is an accurate prediction! More precisely, it predicts that the objective value of a subjunctive conditional whose antecedent is an inhomogeneous disjunction is undefined. It is plausibly a presupposition or implicature of the use of a such a conditional on this theory that its objective value *is* well-defined. Thus, the assertion of a subjunctive conditional with disjunctive antecedent should presuppose or imply that the disjunction is approximately homogeneous with respect to the chance of the consequent, and that the epistemic expectation of the conditional chance is high. Then, in the ordinary vague style of everyday speech, the speaker who asserts a subjunctive conditional with a disjunctive antecedent, $(p \text{ or } q) \Rightarrow r$, presupposes or implies the warranted assertability of $p \Rightarrow r$ and $q \Rightarrow r$.

Everyone has noticed that this is so in natural language, and that it is so is what is known as the problem of disjunctive antecedents. The perceived problem is that none of the contending accounts of subjunctive conditionals make $p \Rightarrow r$ and $q \Rightarrow r$ logical consequences of $(p \text{ or } q) \Rightarrow r$. The account offered here does not make them *logical consequences* either, but it provides a reasonable explanation for the observed phenomenon. That explanation is a direct result of the interaction of the theories of conditional chance and subjunctive conditionals.

Much of that interaction remains to be explored. Obviously, the case of indeterministic families of partitions invites interaction with Lewis style semantics where there can be ties for most similar possible world in which p is true; and with the closely related set selection function account.

There is another consequence of the conditional chance account that I think has not been generally noticed. That is the asymmetry between possibilities of iteration of conditionals in the antecedent and consequent. The conditional chance account lends itself naturally to iteration in the consequent. It says that the objective value of $p \Rightarrow q$ in situation w, where p and q are propositions, is the conditional chance of q on p in w. It would be equivalent to say that the objective value of $p \Rightarrow q$ in w is the conditional chance expectation of the truth value of q conditional on p in w, where truth value of q is a random variable taking on values 1 in worlds where p is true and 0 in worlds where p is false. But now objective value of $p \Rightarrow q$ has been defined as a random variable, taking a definite numerical value in the closed interval [0, 1], and there is no reason why we should not take its conditional expectation. So we can generalize the theory as follows:

E: *The objective value of* $p \Rightarrow q$ *in* w *is the chance expectation of the objective value of* q *conditional on* p.

This gives a general Adams type treatment for conditionals such as $p \Rightarrow (q \Rightarrow r)$, $p \Rightarrow (q \Rightarrow (r \Rightarrow s))$, etc. [with p, q, r, s all propositions.] (I leave it as an exercise to verify that in this theory $(p \,\&\, q) \Rightarrow r$ is not equivalent to $p \Rightarrow (q \Rightarrow r)$. In this connection see Adams (1975, p. 33.) However, in general a conditional will not do as the antecedent of a conditional because the theory requires that the antecedent be a proposition; that it determine a measurable set of situations. [As I pointed out elsewhere[6] we can *impose* a reading for $(p \Rightarrow q) \Rightarrow r$ by reinterpreting it as something like [Objective value of $(p \Rightarrow q) = 1] \Rightarrow r$, but that *is* a reinterpretation.] Of course, in the case of determinism the conditional chances can also be interpreted as truth values; iterations as well as truth functional composition are unproblematic, and the extension of the conditional chance approach embodied in definition E is entirely consistent with the selection function semantics.

5. PROBABILISTIC CAUSATION

In a deterministic setting, two main families of accounts of causation make the cause (i) a sufficient condition or (ii) a necessary condition for the effect. A cause may or may not be required to satisfy other physical principles, such as temporal precedence to the effect, or locality of action. If we wish to modify such theories to operate under conditions of indeterminism, it is natural to weaken causation to causal tendency. Thus (i) C has a tendency towards sufficiency for E iff the conditional chance of E on C is greater than the conditional chance of E on not-C and (ii) C has a tendency towards necessity for E iff the conditional chance of not-C on not-E is greater than that of not-C on E. If one takes the appropriate conditional chances as primitive, one can simply choose (i) or (ii) as the appropriate qualitative notion and add whatever additional physical principles of action that seem appropriate. The theory of I. J. Good (1961—62) can be interpreted as a development of option (ii) wherein *the weight of evidence against C if E does not happen*:

$$\log[PR(\text{not-}E \mid \text{not-}C)/PR(\text{not-}E \mid C)]$$

is offered as a measure of the strength of the causal tendency.

On the other hand, one might start with degrees of belief (which in the case of large surveys may approximate sample relative frequencies) and attempt to tease out of them degrees of belief about probabilistic causation. Positive probabilistic relevance of C to E (or equivalently of not-E to not-C) in one's degrees of belief need not indicate that one believes that C has a causal tendency to produce E. It merely indicates that C is evidence for E. This is consistent with a number of alternative causal pictures, the simplest of which has C and E being effects of a common cause, CC. To see whether this is the case, we control for the potential common cause; i.e. we see if the positive relevance remains when we condition on elements of the partition $\{CC, \text{not-}CC\}$. The idea can be pushed further by seeking a finer partition which controls for the elements in more complicated possible causal networks.

Thus Reichenbach (1956) says that C is causally relevant to E iff (i) $Pr(E \mid C) > Pr(E)$ and (ii) there is no set of events earlier than or simultaneous with C such that conditional on these events E and C are probabilistically independent. Suppes (1970) replaces Reichenbach's second clause with the requirement that there is no partition of events earlier than C such that conditionally on each element of the partition, C and E are independent. I am taking some liberties in transplanting Reichenbach's frequentist remarks to subjective ground and it should be noted that Suppes intends his account to be neutral among interpretations of probability, but I believe that subjective probability assignments are the natural home for these theories.

It should be fairly obvious that the subjectivist theory of conditional chance is the bridge between theories of the kind proposed by Good, and those more in the spirit of Reichenbach and Suppes. Controlling for the other relevant causal factors in order to eliminate spurious causal relevance is just the process of identifying a sufficient partition for the conditional chances.

This point of view suggests that the spirit of clause (ii) in Reichenbach and in Suppes might be better captured in another way. Let πc be the partition appropriate to chance conditional on c. Then we say that E *has a tendency of cause C in a situation* iff in that situation the chance of E conditional on C is greater than the chance of E conditional on not-C. If we assume here that $\pi c = \pi$ not-c, that means that conditional on the member of πc with contains the situation in question, E has positive probabilistic relevance to C. If this is so conditional on *every* member of the partition we say we have *unanimity* with respect to the partition. Then E has a tendency to cause C in any situation, and reference to the situation may be dropped. [Of course other physical requirements for proximate causation can be added to such a theory, and proximate causes can be strung together in causal chains.] The accounts offeréd by Cartwright (1979), Eells and Sober (1983), Granger (1980) and myself (1980) (1984) are more or less of this kind.

While the basic spirit of the Reichenbach and Suppes approaches is maintained, the revision motivated by conditional chance has some important technical advantages. In the first place clause (i) in the Reichenbach-Suppes accounts can be dropped; a real cause need not be, in Suppes terminology, a *prima facie* cause. This is all to the good, because it is possible to give examples where a genuine statistical cause is not a *prima facie* cause because mixing has created *spurious independence* in the degrees of belief.[7] In the second place, the existential quantifier in clause (ii) of these accounts makes clause (ii) much too strong. One indication of this is that in a deterministic universe *nothing* gets past clause (ii). Clearly these accounts need be some restriction on the kinds of events or partitions considered in (ii).[8] The theory of conditional chance helps put this requirement into focus.

So far so good. But the foregoing assumes that conditional chance is well-defined. We have seen that may not be a natural assumption in the case of conditions which are inhomogeneous disjunctions. So, taking the "tendency towards sufficient condition" version of the theory, we need the conditional chance of E on C and of E on not-C to be well defined. But if C is chosen so as to be a natural factor, its negation is liable to be an inhomogeneous disjunction. Similar problems arise for the "tendency towards necessary condition" variant of the theory. Negation introduces inhomogeneous disjunctive causal factors. The upshot is that it appears that probabilistic causation as treated above may be very rarely well-defined.

Of course, appearance may be misleading and conditional chance may be well-defined after all. There may be underlying factors which overdetermine chance in such a way that they allow us to say to any situation how it would be located *vis a vis* salient factors which determine chance if it were in the disjunction. We saw earlier that if the appropriate Stalnaker subjunctive conditionals are always well-defined, they supply such factors. Eells explores a solution to the problem of disjunctive causal factors along these lines in his contribution to this volume.

For certain cases, the "hidden factors" approach is plausible but not, I think, in all cases. In many causes the reasons which lead us to believe that there is no natural conditional chance lead us to believe that there is no natural truth value for the subjunctive conditionals which would save the day. What then, do we say for these cases? Well, if we cannot compare the chance of E conditional on C with the chance of E conditional on not-C; we can compare it with chance of E conditional on C' and conditional on C'', where C' and C'' are incompatible with C and specific enough so that the requisite conditional chances *are* well-defined. Then statements of causal tendency (of the sufficient condition variety) should really include another index. They should be of the form: *C as an alternative to C′* has a tendency to produce E in situation *w*. (In cases where the negation of C is intended as C', the explicit reference can be suppressed.) The tendency towards being a necessary condition version needs to specify alternatives on the effect end. "Had E not happened" gives way to "If E' rather than E had happened". The explicit statement of alternatives in causal claims is advocated by Good (1961—62), by Holland (1986) and by Glymour (1986).

In cases in which the chances are determined by variables whose values have a natural order, there may be a natural choice of alternatives. Suppose that chance(E) is determined by two variables, V_1; V_2 which take on the values Low, Medium, High. Then to see whether V_1 = high has a tendency to produce E, it is natural to take the partition $[V_2 = L, V_2 = M, V_2 = H]$ as the background partition, and to compare the conditional chance of E on $V_1 = H$ with the conditional chance of $V_2 = M$ rather than with the conditional chance of $V_2 = L$.[9]

The natural order of values of the variables helps even more when the chance of E is a differentiable function of real variables. For simplicity, assume that there are only two relevant variables so that Chance(E) can be pictured as a surface above the $V_1 - V_2$ plane. Then we can say that variable V_1 is a *positive factor for E* at a point in the plane iff the partial derivative of the chance of E at that point is positive. By this we mean that keeping V_2 constant, a little less V_1 gives us a little less Chance(E) and a little more V_1 gives us a little more Chance(E). This can be seen as consistent with the foregoing theory (plus a little terminological shorthand) with the background σ-algebra being that generated by the statements of the form $V_2 \in I$, where I is a real interval.

CONCLUSION

A generalization of the Bayesian theory of chance to conditional chance is not quite so trivial as one might expect. It requires not one, but a family of partitions (or σ-algebras), and the natural notion is that of a partial rather than a total family. This theory, including the aspect of partiality, has something to offer both the study of subjunctive conditionals and the analysis of probabilistic causation. In each of these areas it is a bridge between theories with different points of view and in each it suggests generalizations and modifications of these theories that are of interest in their own right.

University of California at Irvine

NOTES

[1] This paper further develops ideas presented in Chapter 5 of my *Pragmatics and Empiricism*. I am indebted to Robert Stalnaker for forcing me to face the problem of partiality. This research was partially supported by NSF grant SES-8605122.

[2] *Pragmatics and Empiricism* Ch. 5.

[3] In comments on my paper "A Bayesian Theory of Subjunctive Conditionals" at the Pacific Division Meetings of the American Philosophical Association.

[4] Generalization to factors which are continuous variables would use the sort of machinery referred to in section 2.

[5] But developed here in a way to which differs in some respects with Adams' interpretation of his theory. See Adams (1965), (1966), (1970), (1975), (1976).

[6] *Pragmatics and Empiricism* p. 105.

[7] See Skyrms (1987) for examples.

[8] As noted in Suppes (1984).

[9] This may be thought of as a version of the subjunctive strategy "If it hadn't been high, it would have been medium", but I am not sure that is the best way to think of it.

REFERENCES

Adams, E. (1965), "On the Logic of Conditionals", *Inquiry* **8**, 166—197.

Adams, E. (1966) "Probability and the Logic of Conditionals", in *Aspects of Inductive Logic*, eds. J. Hintikka and P. Suppes (North-Holland: Amsterdam), pp. 265—316.

Adams, E. (1970), "Subjunctive and Indicative Conditionals", *Foundations of Language* **6**, 89—94.

Adams, E. (1975), *The Logic of Conditionals* (D. Reidel: Dordrecht).

Adams, E. (1976), "Prior Probabilities and Counterfactual Conditionals", in *Foundations of Probability Theory, Statistical Inference, and Statistical Theories of Science* eds. W. Harper and C. Hooker (D. Reidel: Dordrecht), pp. 1—21.

Cartwright, N. (1979), "Causal Laws and Effective Strategies", *Nous* **13**, 419—437.

Eells, E. (this volume), "Probabilistic Causal Interaction and Disjunctive Causal Factors", pp. 189—209.

Eells, E. and Sober, E. (1983), "Probabilistic Causality and the Question of Transitivity", *Philosophy of Science* **50**, 35—57.

Ellis, B. (1968), "Probability Logic", manuscript.

Ellis, B. (1976), "Epistemic Foundations of Logic", *Journal of Philosophical Logic* **5**, 187—204.

Gibbard, A. (1981), "Two Recent Theories of Conditionals", in *Ifs* eds. W. Harper, R. Stalnaker and G. Pearce (D. Reidel: Dordrecht), pp. 211—247.

Glymour, C. (1986), "Comment: Statistics and Metaphysics", *Journal of the American Statistical Association* **81**, 964—986.

Good, I. J. (1961—62), "A Causal Calculus", *British Journal for Philosophy of Science* **11**, 305—328; **12**, 43—51; **13**, 88.

Granger, C. (1980), "Testing for Causality: A Personal Viewpoint", *Journal of Economic Dynamics and Control* **2**, 329—352.

Granger, C. (1988), "Causality Testing in a Decision Science", in *Causation, Change and Credence: Proceedings of the Irvine Conference Vol. 1*, ed. B. Skyrms and W. Harper (D. Reidel: Dordrecht), pp. 3—21.

Holland, P. (1986), "Statistics and Causal Inference" (with comments by Rubin, Cox, Glymour, and Granger), *Journal of the American Statistical Association* **81**, 945—970.

Jeffrey, R. (1964), "If" (abstract), *The Journal of Philosophy* **61**, 702—703.

Lewis, D. (1973), "Causation" *Journal of Philosophy* **70**, 556—567.

Lewis, D. (1974), *Counterfactuals* (Harvard University Press: Cambridge, Mass.).

Lewis, D. (1976), "Probabilities of Conditionals and Conditional Probabilities", *Philosophical Review* **85**, 297—315.

Reichenbach, H. (1956), *The Direction of Time* (University of California Press: Berkeley and Los Angeles).

Salmon, W. (1971), *Statistical Explanation and Statistical Relevance* (University of Pittsburgh Press: Pittsburgh).

Salmon, W. (1984), *Scientific Explanation and the Causal Structure of the World* (Princeton University Press: Princeton, N.J.).

Skyrms, B. (1980), *Causal Necessity* (Yale University Press: New Haven).

Skyrms, B. (1984) *Pragmatics and Empiricism* (Yale University Press: New Haven).

Skyrms, B. (1987), "Review of Suppes' *Probabilistic Metaphysics*" *Philosophical Review.*

Skyrms, B. (forthcoming), "Probability and Causation", *Journal of Econometrics.*

Stalnaker, R. (1968), "A Theory of Conditionals", in *Studies in Logical Theory*, ed. N. Rescher (Blackwell: Oxford).

Stalnaker, R. (1970), "Probability and Conditionals", *Philosophy of Science* **37**, 68—80.

Stalnaker, R. (1984), *Inquiry* (Bradford Books: Boston).
Suppes, P. (1970), *A Probabilistic Theory of Causality* (North-Holland: Amsterdam).
Suppes, P. (1984), *Probabilistic Metaphysics* (Blackwell: Oxford).
van Fraassen, B. C. (1976), "Probabilities of Conditionals", in *Foundations of Probability Theory, Statistical Inference, and Statistical Theories of Science* W. Harper and C. Hooker (D. Reidel: Dordrecht), pp. 261—308.

PART II

PROBABILITY, CAUSALITY, AND DECISION

NANCY CARTWRIGHT

HOW TO TELL A COMMON CAUSE: GENERALIZATIONS OF THE CONJUNCTIVE FORK CRITERION

1. INTRODUCTION

For much of his career Wesley Salmon has defended Reichenbach's principle of the common cause, and in particular he has endorsed and developed Reichenbach's statistical characterization of common causes in terms of conjunctive forks: common causes screen off joint effects from each other — that is, given the common cause, the correlation between joint effects that have no direct causal influence on each other will disappear. But in his recent work on causal processes, Salmon has given up this view.[1] He now maintains that where there is a common cause there will be either a conjunctive fork or an interactive fork. This is a considerable weakening of his position, for the implication does not go the other way around. So long as interactive forks are characterized purely statistically, it is not the case that any factor which produces either a conjunctive or an interactive fork will be a common cause. Interaction needs some more robust, non-statistical characterization, and this is just what Salmon tries to provide in his work on causal processes.

There are a variety of reasons for thinking that purely statistical characterizations of causality won't do, and that the notion of an individual causal process is more fundamental than that of a causal regularity, no matter whether the regularity is deterministic or purely probabilistic. But one motivation for weakening his position is a mistake: that is an argument by Bas van Fraassen using a quantum mechanical example of correlated spin systems to show that the principle of the common cause is not always satisfied.[2] This paper will defend Salmon against van Fraassen. Quantum mechanical results about correlated spins — and, more interestingly, the purely classical macroscopic analogues of these which van Fraassen describes — do not violate the principle of the common cause. I say "more interestingly" because the macroscopic cases are indeterministic, but they lack the peculiar quantum characteristics, due to the principle of superposition,

James H. Fetzer (ed.) Probability and Causality. 181–188.

that lead one to say that quantum systems have no properties until we look at them. Should it turn out that these peculiar quantum systems fail to satisfy the common cause principle, that would not be very disturbing for Salmon's general project. But it is disturbing if the principle fails for straightforward cases of macroscopic indeterminism. I will argue that it does not fail for these.

Although I am going to defend the principle of the common cause, I am not going to defend conjunctive forks. For the conjunctive fork is too narrow. The statistical conditions that characterize the conjunctive fork are just a special case: they are the statistical conditions that are appropriate for marking out a common cause under special assumptions — in particular under the assumption that the common cause produces each of its effects independently of each of the others. This particular assumption is violated in the van Fraassen examples. Given other assumptions about how causes operate, other statistical characterizations for a common cause will be appropriate. Even though the conjunctive fork fails, common causes can indeed be postulated in van Fraassen's examples, common causes that satisfy the statistical conditions that are appropriate for the special conditions he formulates.

My discussion is an exercise in linear causal modelling theory, for causal modelling theory provides the one principled way I know to derive the conjunctive fork as the correct statistical criterion for the operation of a common cause. My argument is thus subject to all the limitations that beset causal modelling theory. On the other hand, I know no other way to derive the conjunctive fork condition, and hence think this is the correct context in which to set the discussion.

2. HOW TO DERIVE THE CONJUNCTIVE FORK

Causal modelling theory assumes that where there are causal relations among variables, the variables will be functionally related as well, and that the functional relationship will be linear. Variables can sometimes be redefined to make this true. For example, if x and y cause z, where $z = xy$, this can be rewritten as $z' = x' + y'$, where $z' = \log z$, $x' = \log x$, and $y' = \log y$. Nevertheless, this is still a very restrictive condition. I will also assume that all variables contain a time index, and that causes precede their effects. It will then be possible to arrange a linear array of equations with effect variables on the left and causes on the right.

Considering just three variables, the array looks like this:

(CM) a. $x_1 = u_1$
b. $x_2 = a_{21}x_1 + u_2$
c. $x_3 = a_{31}x_1 + a_{32}x_2 + u_3$

There may be some special conditions under which an array of equations like this force a causal interpretation. It is less controversial to assume that the causal information is added on, that is, that a causal model consists not just of a set of linear equations, but of a set of linear equations written in a certain way — with causes as independent variables and effects as dependent. That is what I will assume here. Besides the x's, the equations above contain u's as well. These are supposed to represent variables which play a definite causal role, but which have not been identified by us. (They are unknown.). Standard treatments assume that they have some nice properties. In particular, they are statistically independent of everything that occurs simultaneous to or earlier than themselves. It is convenient if they are scaled to have mean 0.

Imagine that x_2 and x_3 are statistically dependent on each other. Under what conditions will the dependence be entirely due to the operation of the shared cause, x_1, with x_2 having no direct influence on x_3? That situation is represented by setting $a_{32} = 0$ in the equations of the model, with $a_{31} \neq 0$ and $a_{21} \neq 0$. So the question reduces to that of finding statistical conditions under which $a_{32} = 0$. Standard treatments[4] proceed thus. Multiply (CM)c by x_2 and take the conditional expectation with respect to x^1; also multiply by x_3, and do the same. Using the fact that $\text{Exp}(u_3/x_1) = \text{Exp}(u_3) = 0 = \text{Exp}(u_3/x_1)\text{Exp}(x_2/x_1) = \text{Exp}(u_3x_2/x_1)$, this yields

1. $\text{Exp}(x_2x_3/x_1) = a_{31}x_1\text{Exp}(x_2/x_1) + a_{32}\text{Exp}(x_2^2/x_1)$

2. $x_1\text{Exp}(x_3/x_1) = a_{31}x_1^2 + a_{32}x_1\text{Exp}(x_2/x_1)$.

Solving for a_{32} gives

$$\text{(CF)} \quad a_{32} = \frac{\text{Exp}(x_2x_3/x_1) - \text{Exp}(x_2/x_1)\text{Exp}(x_3/x_1)}{\text{Exp}(x^2/x_1) - (\text{Exp}(x_2/x_1))^2}$$

So $a_{32} = 0$ if and only if $\text{Exp}(x_2x_3/x_1) = \text{Exp}(x_2/x_1)\text{Exp}(x_3/x_1)$. This is just the characterization of a conjunctive fork.

Consider now a van Fraassen kind of example, where there is a joint cause, but it produces its effects in tandem. A particle collides with an atom and the atom emits two new particles as a consequence. Particle 1 may be emitted either at angle θ or at angle θ'; and the probabilities are 50—50. Particle 2 may be emitted at angle $-\theta$ or at $-\theta'$; and again the probabilities are 50—50. But momentum must be conserved. So particle 1 is emitted at θ if and only if particle 2 follows the path defined by $-\theta'$.

It is obvious in this situation that a cause may be at work: λ, when it is present in the atom produces motions along θ, $-\theta$; otherwise the atom emits the particles at angles θ' and $-\theta'$. λ may be a deterministic cause: Prob (the angle of emission for particle 1 = θ/λ) = 1 = Prob (the angle of emission for particle 2 = $-\theta/\lambda$),d with Prob (λ) = 0.5. Or, it may be a purely probabilistic cause: Prob (the angle of emission for particle 1 = θ/λ) = r = Prob (the angle of emission for particle 2 = $-\theta/\lambda$); in which case Prob (λ) = 1/2r. If λ is totally deterministic in producing θ, and $-\theta$, it will form a conjunctive fork: Prob (the angle of emission for particle 1 = θ and the angle of emission for particle 2 = $-\theta/\lambda$) = 1 = Prob (the angle of emission for particle 1 = θ/λ) × Prob (the angle of emission for particle 2 = $-\theta/\lambda$) = 1 × 1. But if the probabilities depart from 1 at all, the conjunctive fork will no longer obtain: Prob (the angle of emission for particle 1 = θ and the angle of emission for particle 2 = $-\theta/\lambda$) = 1/2r ≠ Prob (the angle of emission for particle 1 = θ/λ) × Prob (the angle of emission for particle 2 = $-\theta/\lambda$) 1/2r × 1/2r.

But in this case it is not reasonable to expect the probabilities to factor, conditional on the common cause. Momentum is to be conserved, so the cause produces its effects in pairs. Particle 1 follows θ just in case particle 2 follows $-\theta$, so the probability for the conjunction of the two events is bound to be the same as that for either on its own. Clearly the conjunctive fork criterion is not appropriate here. That is because it is a criterion tailored to cases where the cause operates independently in producing each of its effects: whether one of the effects is produced or not has no bearing on whether the cause will produce the other. In effect, van Fraassen and Salmon both agree to this diagnosis. The question I want to raise is this: Can the equations of (CM) be modified to produce a criterion more appropriate for cases in which one action of the cause constrains its others?

3. MORE GENERAL CRITERIA FOR COMMON CAUSES AND FOR CAUSAL INDEPENDENCE

Yes, and at the same time the equations can be improved in another way as well. As they stand, the equations of (CM) are entirely deterministic in the relation between cause and effect — x_i contributes the value $a_{ji}x_i$ to x_j with probability 1. There is no possibility for the cause to take on the value x_i, but then contribute to x_j only in some fixed percentage of cases. This is particularly striking where the causal variables are 2-valued, representing the presence or absence of the cause. In that case the effect is bound to occur if a cause does, and there is no space for a purely probabilistic cause. But it is easy to model probabilistic causes, by the simple device of including information about whether the cause operates or not. For each x_i on the right hand side of the equation for x_j, introduce the random variable $\dot{a}_{ji}$. $\dot{a}_{ji}$ takes on two values: it has value 1 if x_i operates to produce an effect on x_j, and 0 if it does not. $\dot{a}_{ij}$ is meant to represent a genuine physical occurrence, though in cases where the causes are supposed to be purely probabilistic, it will coincide with no further physical state or property of the system that determines whether the cause will operate or not. But in general one can expect that there will be further empirical signs, beyond the mere occurrence of the effect itself, by which one can detect whether the cause has operated or not.

Before considering what kinds of statistical relations the $\dot{a}_{ij}$'s have to other factors, look first at the u's. The u's are commonly called "error terms," and it is often assumed that their interpretation can be left deliberately ambiguous. On the one hand, they may represent genuine quantities or states, about which we are ignorant. On the other hand, they may serve as a way to represent a genuine element of indeterminism in a deterministic format. In this case the expectation of u_j gives the background rate with which x_j occurs on its own for no cause whatsoever. In this case, the u_j's — like the $\dot{a}_{ij}$'s — represent genuine physical happenings — the spontaneous occurrence of x_j — but again there need be no physical state or property which determines whether a u-type event happens or not, although, just as with the $\dot{a}$'s, in many cases there may be independent ways to confirm that it has done so.

This interpretation has considerable advantage over the 'ignorance' interpretation of the u's, for now the independence assumptions about them are perfectly natural. In general it is a very fortunate accident — an accident which one has little reason to hope for — should the causes unknown to us distribute themselves in just the right way to guarantee conditional independence of their expectations. But if u stands for an event of spontaneous production, it is reasonable to suppose that this event occurs "totally randomly." The independence assumptions are one way to formulate this supposition.

The same reasoning leads to analogous independence assumptions for the $\dot{a}_{ij}$. If the hypothesized causes $x_1, \ldots, x_{j-1}$ are pure stochastic causes, acting entirely 'randomly' and subject to no constraints, for each i, $\dot{a}_{ij}$ should be independent of all earlier or simultaneous factors, in all combinations. In this case, the conjunctive fork condition can be derived just as before.

The proposal, then, is to begin with equations which are one step back from (CM).

(CM)* a. $x_1 = u_1$
b. $x_2 = \dot{a}_{21}\alpha x_1 + u_2$
c. $x_3 = \dot{a}_{31}x_1 + \dot{a}_{32}x_2 + u_3$

Since x_3 is the effect of concern, for simplicity I am assuming a rescaling of the variables so that each unit of x_1 and x_2 contributes one unit to x_3. In this case the parameters a_{31} and a_{32} from equations (CM) will turn out to be the expectations of $\dot{a}_{31}$ and $\dot{a}_{32}$ from (CM)*. The condition that $a_{32} = 0$ will mean that Exp $\dot{a}_{32} = 0$, which, since $\dot{a}_{32}$ is always either 0 or 1, means that $\dot{a}_{32}$ "never" occurs. Hence x_2 never operates to produce x_3, and the condition that $a_{32} = 0$ remains as the appropriate way to express x_2's lack of effect on x_3.

The normal assumption which gives rise to the conjunctive fork is that each $\dot{a}_{ij}$ is independent of all earlier factors, in all combinations. In this case, multiplying (CM)*c as before, first by x_1, then by x_2, and taking expectations conditional on x_1, gives, using the independence assumptions on the u's:

1′. $\text{Exp}(x_2x_3/x_1) = x_1\text{Exp}(\dot{a}_{31}x_2/x_1) + \text{Exp}(\dot{a}_{32}x_2^2/x_1)$
2′. $x_1\text{Exp}(x_3/x_1) = x_1^2\text{Exp}(\dot{a}_{31}/x_1) + x_1\text{Exp}(\dot{a}_{32}x_2/x_1)$.

But given the independence assumptions on the $\dot{a}$'s the expectations of the $\dot{a}_{ij}$ will factor out, and (CM) will follow exactly as before, setting $a_{32} = \text{Exp}\ \dot{a}_{32}$.

Where the operation of the joint cause is constrained, as by the need for conservation of momentum, the assumption of total independence among the $\dot{a}$'s will the mistaken. In the simple example above, the joint cause, λ, operates to produce an effect in particle 1 if and only if it operates to produce an effect in particle 2. Assume an analogous constraint in (CM)*: $\dot{a}_{21} \equiv \dot{a}_{31}$. In this case on the right hand side of equation 1′, $\dot{a}_{31}$ will no longer be independent of x_2, conditional on x_1. Rather,

$$x_1\text{Exp}(\dot{a}_{31}x_2/x_1) = x_1\text{Exp}(\dot{a}_{31}\dot{a}_{21}\alpha x_1/x_1) = \alpha x_1^2 a_{31}$$

where $a_{31} = a_{21} = \text{Exp}\ \dot{a}_{31} = \text{Exp}\ \dot{a}_{21}$. This follows because $\dot{a}_{31}$ and $\dot{a}_{21}$ are either 1 or 0 together, and are independent of x_1. So equation 1′ yields

$$\text{Exp}(x_2x_3/x_1) = \alpha x_1^2 a_{31} + a_{32}\text{Exp}(x_2^2/x_1).$$

(CF) no longer follows. Instead,

$$a_{32} = 0 \text{ iff } \text{Exp}(x_2x_3/x_1) = x_1^2 a_{31}$$

or, using equation 2′ as well,

$$\text{(CI)} \qquad a_{32} = 0 \text{ iff } \text{Exp}(x_2x_3/x_1 = \text{Exp}(\alpha x_1x_3/x_1).$$

I have labelled this condition (CI) for it is the appropriate condition to express that x_3 is causally independent of x_2 — that x_2 does not cause x_3. It is equally, in a three variable model, a *common cause* condition, since any statistical dependencies between x_2 and x_3 must be entirely due to the common cause x_1 if x_2 itself has no effect on x_3; and this is exactly what (CI) says: for any x_1, the joint expectation of x_2 and x_3 is exactly the same as the joint expectation between x_1's contribution to x_2 (αx_1) and its contribution to x_3. There is nothing left over that could be contributed by a separate direct connection between x_2 and x_3.

In the case just considered, x_1's operation to produce x_2 is correlated 100% with its operation to produce x_3. Obviously the two operations may be correlated without being coextensive. More generally then, let $\text{Exp}(\dot{a}_{21}\dot{a}_{31}) = \gamma$. It follows that, in the right hand side of equation 1′,

$$x_1\text{Exp}(\dot{a}_{31}x_2/x_1) = \alpha x_1^2\,\gamma$$

and (CI) becomes

$$\text{(CI)}' \qquad a_{32} = 0 \text{ iff } \text{Exp}(x_2x_3/x_1) = \alpha x_1^2\,\gamma.$$

Like (CI), this condition asserts that, for a given x_1 the correlation between x_2 and x_3 is exactly the same as that between x_1's contribution to x_2 and its contribution to x_3, only this time the correlation is reduced by the factor γ, which measures how often x_1 acts jointly to produce its two contributions.

Now that the idea is clear, it is apparent that there is nothing necessary about the conditions that give rise to (CF), or to (CI), or to (CI)′. A variety of different causal structures can be imagined, for which very different kinds of conditions on the a's and u's will be appropriate, and these in turn will imply very different criteria for the causal independence of x_2 and x_3. On reflection, this is exactly what one should expect: what statistical conditions mark specific causal features of a structure will depend intimately on how the causes in that structure are supposed to work. Van Fraassen style counterexamples do not show that there is anything fundamentally mistaken about the common cause idea, but rather that Salmon, following Reichenbach, has not been employing sufficiently general characterizations of it. Reichenbach hypothesized that correlated events must share a common cause if neither is a direct cause of the other. Effects produced in tandem do not show that this principle is false, but rather that its statistical formulation requires some care. I do not think that quantum mechanical correlations are an exception either, but that is a longer story.[4]

Stanford University

NOTES

[1] Salmon, Wesley, 1984, *Scientific Explanation and the Causal Structure of the World*. Princeton, NJ: Princeton University Press.
[2] Cf., van Fraassen, Bas, 1982, "Rational Belief and the Common Cause Principle" in McLaughlin, Robert, *What? Where? When? Why?* Dordrecht: D. Reidel.
[3] Cf., Simon, Herbert, 1954, "Spurious Correlation: A Causal Interpretation," *JASA* **49**, 467–479. Reprinted in Simon, H., 1977, *Models of Discovery*. Dordrecht: D. Reidel.
[4] For the longer story, see Cartwright, N. 1988, *Causes and Capacities: Empiricism Reclaimed*. Oxford: Oxford University Press.

ELLERY EELLS

PROBABILISTIC CAUSAL INTERACTION AND DISJUNCTIVE CAUSAL FACTORS*

The basic idea in probabilistic theories of causality is that causes raise the probability of their effects. Of course, it is necessary to control for other causes of the effect in question in order to avoid being misled by "spurious correlations", which arise in cases of "Simpson's paradox". But, even when we control for other causes, it is possible for a causal factor to raise the probability of a second factor in some situations and lower that probability in other situations, where these situations do *not* correspond to the presence or absence of other *causes* of the second factor. When this happens, the reason may be that the causal factor "*interacts*" with other factors. In this paper, I will explore the phenomenon of probabilistic causal interaction in detail and in general, and I will suggest a general way of accommodating the possibility of causal interaction in one common current understanding of probabilistic causation.

I will be especially concerned with causal factors that are themselves or whose negations are disjunctive in nature, so that the presence or absence of the causal factor need not confer a unique causally significant probability on the effect factor in question, and the *average* probabilistic impact of the presence or absence of such a factor on the effect factor in question will depend on the base probabilities of the disjuncts. I will argue that these are cases of causal interaction and can be accommodated in the probabilistic theory of causality on that basis.

In Section 1, I briefly present the common current understanding of probabilistic causation alluded to above. In Section 2, I present some simple examples of interaction and briefly discuss some ways in which the phenomenon has previously been dealt with in the theory of probabilistic causality. Here I point out some limitations of these ways, which involve disjunctive causal factors (in this section, I draw from my (1986)). Section 3 shows how the problem described in Section 2 can be generalized. And in Sections 4—6, I offer solutions.

1. PROBABILISTIC CAUSALITY

As mentioned above, the basic idea is that causes X raise the proba-

James H. Fetzer (ed.) Probability and Causality. 189—209.

bility of their effects Y. But the probability increase idea is, strictly speaking, neither necessary nor sufficient for causation. To see that it is not necessary, we need only imagine a situation in which X causes Y but in which X is strongly correlated with a negative cause, Z, of Y. If, for some reason, proper exercise and diet (X) were strongly correlated with smoking (Z), then proper exercise and diet may be a cause of coronary health (Y) even though, on average, it lowers the probability of coronary health — because of the correlation between the factor of exercise and diet and the factor of smoking. This could happen, for example, if there were a genetic common cause of smoking and proper exercise and diet. To see that the probability increase idea is not sufficient for causation, we need only imagine a situation in which X is *not* a cause of Y but in which X is strongly correlated with a strong positive cause, Z, of Y. For example, falling barometers (X) do not cause rain (Y), but they are strongly correlated with, because caused by, the approach of cold fronts (Z). In this case, Z is a common cause of X and Y, which can give rise to a positive correlation between X and Y.

In order to control for other causes Z of the effect factor Y in question in assessing the causal role of X for Y, Nancy Cartwright (1979) says that such Z's should be "held fixed", positively and negatively, and that separate probability comparisons should be made for each way of holding them all fixed. Let $Z_1, \ldots, Z_n$ be all factors, other than X and effects of X, that are positive or negative causes of Y; these are all the factors that are causally relevant to Y "independently of X". Then there are 2^n conjunctions in which each of these factors appears either unnegated or negated. Of these 2^n conjunctions, let K_i's be those that, both in conjunction with X and in conjunction with $\sim X$, have nonzero probability. The K_i's are called "causal background contexts". Then, according to Cartwright's (1979) formulation of the theory, X is a positive causal factor for Y if and only if, for each i, $Pr(Y/K_i \,\&\, X) > Pr(Y/K_i \,\&\, \sim X)$. Negative causal factorhood is defined by changing the "$>$" to "$<$", and causal neutrality is defined by changing the "$>$" to "$=$".[1]

It is easy to see how this theory should deliver the right answers in the two examples given above. In the first example, we must hold fixed the factor of smoking, since it is independently (of exercise and diet) causally relevant to coronary health. And for both smokers and nonsmokers, proper exercise and diet should increase the probability of a healthy heart. In the second example, we have to hold fixed the factor of an approaching cold front. And both in cases in which a cold front is approaching, and when no cold front is approaching, a falling barometer will not affect the probability of rain.

Note that positive, negative, and neutral causal factorhood are not exhaustive of the possible causal significances a factor X can have for a factor Y. These three significances each require what is called "context unanimity", i.e., that the relevant inequality, or the equality, hold for *all* causal background contexts. There is also the possibility of various kinds of "mixed" causal significance, in which inequalities get reversed or changed to equalities across contexts.

Another thing to note is that it is crucial that judgments of probabilistic causation be understood as *relative to a given population*. For example, the Surgeon General says that smoking is a positive causal factor for heart attacks. Clearly, what is meant is that smoking causes heart attacks *in the human population*, and the intent of the claim is perfectly consistent with the possibility that there are creatures somewhere in the universe for whom smoking is a crucial part of daily nutrition, important specifically for the maintenance of coronary health. Also, of course, probability assignments are always relative to populations, on many interpretations of probability, so that what increases the probability of what will differ from population to population.

I think that Cartwright's formulation of the theory is on the right track, but that an important further refinement of the theory must be made. On Cartwright's formulation, we must "hold fixed" in the background factors all and only independent *causes* of the effect factor in question. We shall see, however, that there is another important kind of factor that must be held fixed if the theory is to deliver the right answers in cases of interaction.

2. PROBABILISTIC CAUSAL INTERACTION

As mentioned above, it is possible for a causal factor to raise the probability of a second factor in some situations while lowering it in others, even when these situations do not correspond to the presence or absence of independent *causes* of the second factor. For example, John Dupré (1984) discusses the following possibility (for purposes somewhat different from those of this paper). Suppose that for most of the human population smoking increases the probability of lung cancer but that there is a rare physiological condition under which smoking actually lowers the probability of lung cancer. That physiological condition need not be a negative or positive *cause* of lung cancer. Consistent with the description of the example, that condition could be causally positive, negative, neutral, or mixed for lung cancer.[2]

Cartwright suggests that when a causal factor raises the probability of a second factor in some situations and lowers it in others, an "interaction" may be the reason. She illustrates this idea with the following example. Ingesting an acid poison increases the probability of death, unless you have also ingested an alkali poison; and ingesting an alkali poison increases the probability of death, unless you have also ingested an acid poison. If you ingest one without the other, you increase the probability of death; and if you ingest none or both, you do not.

The question now arises of what kind of causal significance a proper theory of probabilistic causality should assign to smoking for lung cancer in Dupré's example, and to the acid and alkali poisons for death in Cartwright's example. Should we deny that smoking is a positive causal factor for lung cancer in Dupré's example just because there is a rare kind of individual for whom smoking lowers the probability of lung cancer? And should we deny that ingesting acid (alkali) poison is a positive causal factor for death just because there is a kind of individual — namely those who have just ingested an alkali (acid) poison — for whom ingesting acid (alkali) poison lowers the probability of death? Affirmative answers to these questions may seem to be implied by the "unanimity" feature of the formulation of the theory of probabilistic causality given above, but more on this below.

My answer to each of the questions above is, indeed, in the affirmative: smoking is *not*, in the general population, a positive causal factor for lung cancer in Dupré's example; and acid and alkali poisons are *not*, in the general population, positive causal factors for death in Cartwright's example. In the remainder of this section, I will (i) explain my answers, (ii) explain why the Cartwright formulation of probabilistic causality *does not* deliver these answers, (iii) show how to refine that formulation so as to make it give these answers, and (iv) consider Cartwright's way of dealing with interaction and point out two difficulties with it, *one* of which seems also to confront the refinement of the theory I suggest, but will be dealt with in later sections.

It may seem that by giving affirmative answers to the questions above, we would be ignoring important parts of the causal truth, and, indeed, be positively misleading. In fact, however, just the opposite is true, and it is a negative answer that would ignore important causal truth and be misleading. The fact that probabilistic causation is a relation between three things — a causal factor, an effect factor, and *a population* within which the former is a cause of the latter — is crucial here. To express the whole causal truth in Dupré's example, I think we should say that smoking is causally positive for lung cancer *in the subpopulation of the human population in which that physiological condition is absent*, that it is negative for lung cancer *in the subpopulation in which that condition is present*, and that it is causally mixed for lung cancer *in the whole, combined population.* And in Cartwright's example, acid poison is causally negative for death *among those who have just ingested an alkali poison*, positive for death *among those who have not*, and, *in the combined population*, ingesting acid poison is causally mixed for death (and similarly for the causal role of alkali poison for death). Thus, if properly formulated so as to give all these answers, the probabilistic theory of causality will neither miss important parts of the causal truth nor be misleading.

It is if we give *negative* answers to the questions above (and affirm that smoking is simply causally positive for lung cancer in the general population in Dupré's example and that acid and alkali poisons are simply causally positive for death in Cartwright's example) that judgments of probabilistic causal connection can mislead and ignore important parts of the whole causal truth. This tack would ignore reversals of causal roles across subpopulations; and it would mislead some decision makers, for example, if the difference of causal roles across subpopulations is ignored. Furthermore, there are serious, I believe insurmountable, difficulties with formulating, in a principled way, an acceptable theory of probabilistic causation that would yield these answers. For example, just *how* rare does that physiological condition of Dupré's example — or the ingestion of the interacting poison of Cartwright's example — have to be for it to be correct to ignore reversals across the relevant subpopulations?[3]

Unfortunately, the theory of probabilistic causality as formulated above does not deliver the correct answers, given above, for the Dupré and Cartwright examples. The correct answers relative to the relevant combined populations are that smoking is causally *mixed* for lung cancer in Dupré's example, and that ingesting acid (alkali) poison is causally *mixed* for death. But in order for the theory to deliver these answers, we have to hold fixed the factor of that rare physiological condition in Dupré's example and the factor of having already ingested alkali (acid) poison in Cartwright's example. But on the formulation of probabilistic causation given above, we are only allowed (and required) to hold fixed independent positive and negative *causes* of the effect factor in question. I have already pointed out that in Dupré's example the rare physiological condition need not count as a positive or negative cause of lung cancer: it may be mixed or neutral. The same goes for the factor of having ingested alkali (acid) poison in the assessment of the causal role of acid (alkali) poison for death in Cartwright's example.

Here is an argument to the effect that, *on the formulation of the theory given above*, the factor of having just ingested an alkali poison should *not* be held fixed when assessing the causal role of acid poison on death. (Similar considerations apply to Dupré's example; see also Note 2 above.) Suppose that, on the formulation of the theory given above, we *were* supposed to hold fixed this factor of whether or not one has just ingested an alkali poison. Then on the formulation of the theory given above, this means that ingesting *alkali* poison must be a *positive or negative cause* of death. Also, if we hold it fixed, we get the (right) answer that ingesting *acid* poison is *causally mixed* for death. But the example is completely symmetrical in acid and alkali, so it cannot be that ingesting the alkali poison is causally positive, and ingesting acid poison causally mixed, for death. So the supposition that we should hold fixed the factor of having just ingested an alkali poison must be false (or else the theory is *trivially* false, for the very simple reason of implying an asymmetry in this example where there is none).

The solution to the problem is, of course, that we must include in the theory the requirement that another kind of factor be held fixed. Let us say that causal factor X *interacts with* a factor F, *with respect to* Y as the effect factor, if in the presence of F, X raises the probability of Y, and in the absence of F, X lowers the probability of Y — or vice versa, exchanging "raises" and "lowers".[4] Then in assessing X's causal role for Y, we should hold fixed in causal background contexts not only all independent positive and negative *causes* of Y (as in the formulation of the theory given above), but also all *factors with which X interacts, with respect to Y*. On this revised theory, the factor of that rare physiological condition of Dupré's example, and the factor of having just ingested the interacting poison in Cartwright's example, will have to be held fixed, so that the theory will deliver the right answer of mixed causal significance in the whole, combined populations.

Here is a full statement of the revised theory. Let $F_1, \ldots, F_n$ be all and only those factors that are, in population P, either (i) a positive or negative cause of Y independently of X or (ii) a factor with which X interacts with respect to Y. Let K_i's be those maximal conjunctions of the F_i's that, relative to population P, have nonzero probability both in conjunction with X and in conjunction with $\sim X$. Then X is a positive causal factor for Y in population P if and only if, for each i, $Pr(Y/K_i \;\&\; X) > Pr(Y/K_i \;\&\; \sim X)$. Negative causal factorhood and causal neutrality are given by changing the "$>$" to "$<$" and "$=$", respectively.[13]

Cartwright (1979) deals with interaction in a different way. First, she characterizes interaction differently: "two causal factors are interactive if in combination they act like a single causal factor whose effects are different from at least one of the two acting separately" (pp. 427—428). Second, for the acid/alkali example, her solution is to focus on the *combined* factors determined by whether or not one has ingested acid poison and whether or not one has ingested alkali poison:

> In this case, it seems, there are three causal truths: (1) ingesting acid without ingesting alkali causes death; (2) ingesting alkali without ingesting acid causes death; and (3) ingesting both alkali and acid does not cause death. All three of these general truths should accord with [the theory]. (p. 428)

Of course there is also the fourth causal truth that ingesting neither acid nor alkali does not cause death.

I find two difficulties with this approach to interaction — difficulties that are independent of the difference between Cartwright's conception of interaction and the one given above.[5] First, it fails to deal with the question of the causal roles of the *uncombined* factors — for example, the ingestion of acid poison (without saying whether or not alkali poison has been ingested). It fails to deliver the right answer that the factor of ingesting acid poison is causally mixed for death in the whole population. Indeed, it fails to deliver *any* answer in connection with the causal roles of the "uncombined" factors. And second, the approach is not very general, as I shall now explain.

The success of the combined factors approach for the acid/alkali example turns, I think, on a kind of symmetry in the example. Figure 1 depicts how I think we picture the probabilistic relations among ingesting acid poison, ingesting alkali poison, and death, in the example (X can be ingesting acid poison, F ingesting alkali poison, and Y death). But there are other, less symmetric, ways in which two factors can be interactive. Figure 2 and 3 depict other possibilities. (In these three

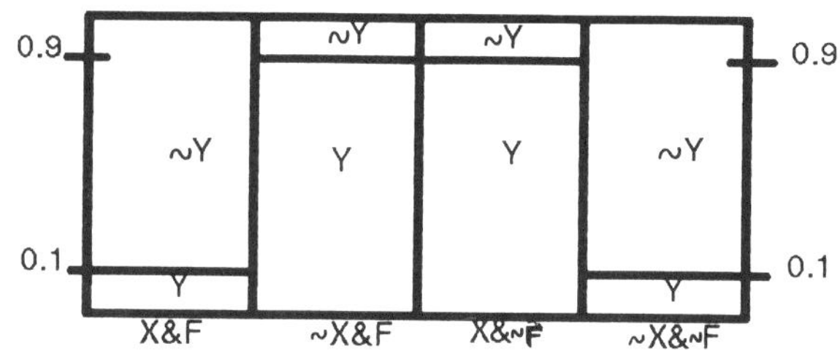

Figure 1.

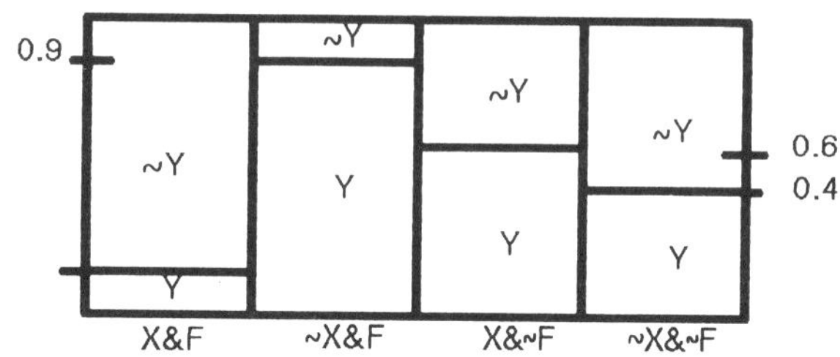

Figure 2.

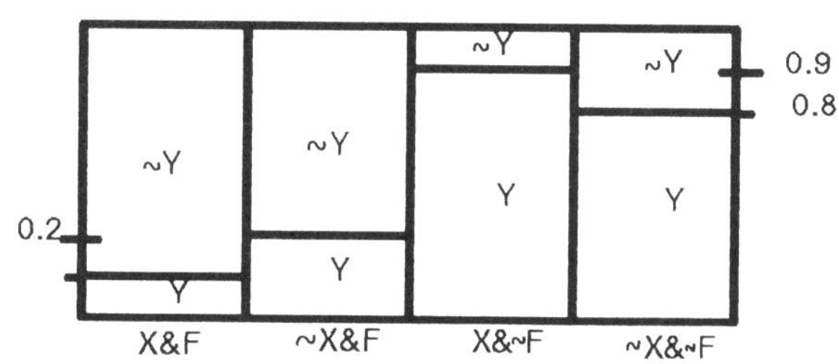

Figure 3.

figures, the spacing of the vertical lines is not meant to indicate the frequencies, or probabilities, of the combined factors in the relevant population.) In each of these three kinds of cases, *X* and *F* are interactive for *Y* on Cartwright's characterization, and *X* interacts with *F* with respect to *Y* on the characterization of interaction given earlier.

Consider the Figure 2 possibility. Clearly *X* & *F* is negative for *Y* and ~ *X* & *F* is positive — so in combination *X* and *F* have a different effect on *Y* than at least one of them (*F*) acting alone does, and *X* and *F* are interactive in Cartwright's sense. But we do not have four definite causal truths as in the acid/alkali example. For example, whether X & ~ F is *probabilistically* positive, negative, or neutral for Y will depend on the probabilities of the other combined factors: if ~ X & F is probable enough, then X & ~ F will be negative for Y; and if X & F, or ~ X & ~ F, is probable enough, then X & ~ F will be positive for Y. But it seems that the question of the *casual* significance of X & ~ F for Y should not turn on the overall probabilities or frequencies of the other combined factors. This is because different individuals who are X & ~ F may have different propensities to distribute themselves among the other combined factors *were they not X* & ~ *F*'s.[6]

The same problem arises, of course, for the assessment of the causal significance of ~ F & ~ X for Y in Figure 2 type cases. And for Figure 3 cases (another kind of case of interaction of X with F in both senses of interaction discussed), the same problem arises in assessing the causal roles of ~ X & F and ~ X & ~ F for Y.

Thus, the first problem with the "combined factors" approach is that it does not provide an answer to the question of the causal role of the *uncombined* factors, which, of course, is a question we should be allowed to ask. And the second problem is that it does not provide a principled answer even to the question of the causal role of all the *combined* factors. I conclude that we should reject the combined factors approach as a general solution to the problem of causal interaction.

However, the second problem seems to be shared with the approach to interaction advocated above, involving holding fixed factors with which the causal factor in question interacts. For while this approach handles the uncombined causal factors just fine, it has not been explained how to evaluate the causal significance of the combined factors. In addition to the uncombined factors, it should *also* be allowable to ask what the causal roles of the combined factors are. But it seems that all the considerations advanced above against Cartwright's approach on this count apply equally well to the revised theory. In the next section, I will formulate this problem in a very general way. And in the sections that follow, I will offer solutions, exploiting a more general understanding of interaction.

3. THE PROBLEM OF DISJUNCTIVE CAUSAL FACTORS

According to the theory of probabilistic causation, to judge the causal role of a factor X for a factor Y, we must have at hand the probabilities of Y both in the presence and in the absence of X (in each background context). We must have at hand these two *single* probabilities. What was problematic in the previous section about combined causal factors was the probability of the effect factor in the *absence* of a combined causal factor: in the examples, there were more than one causally significant probability of the effect factor given the absence of the combined factor (and no principled way of averaging them). This is because the absence of a combined, i.e., *conjunctive*, causal factor is a *disjunctive* factor. A natural generalization of these problem cases would be cases in which there are multiple causally significant probabilities of the effect factor both in the presence and in the absence of the causal factor, i.e., cases in which the causal factor and its negation are both disjunctive.

Of course the negation of a disjunction is a conjunction of the negations of the disjuncts; but it will nevertheless be a disjunction of other factors. Suppose, for example, that factors A, B, and C are all the factors that may be truth-functional ingredients of any factor causally relevant to the effect factor Y in question, and consider the disjunctive factor $X = A \vee (B \,\&\, C)$. Expressed as a disjunction of maximally specific causally relevant factors (i.e., in disjunctive normal form in terms of A, B, and C), X is equivalent to $(A \,\&\, B \,\&\, C) \vee (A \,\&\, B \,\&\, {\sim}C) \vee (A \,\&\, {\sim}B \,\&\, C) \vee (A \,\&\, {\sim}B \,\&\, {\sim}C) \vee ({\sim}A \,\&\, B \,\&\, C)$. And its negation, ${\sim}(A \vee (B \,\&\, C))$, is equivalent to ${\sim}A \,\&\, {\sim}(B \,\&\, C)$, and to $({\sim}A \,\&\, B \,\&\, {\sim}C) \vee ({\sim}A \,\&\, {\sim}B \,\&\, C) \vee ({\sim}A \,\&\, {\sim}B \,\&\, {\sim}C)$.

If the probabilities of Y conditional on the various disjuncts of X are not *all* greater than (or *all* less than or *all* equal to) *each* of the probabilities of Y conditional on the various disjuncts of ${\sim}X$, then whether or not the probability of Y given X is greater than (and whether or not it is less than, and whether or not it is equal to) the probability of Y given ${\sim}X$ will turn on the base probabilities of the disjuncts of X and of ${\sim}X$. However, the question of the *causal* significance of X for Y should not turn on this, it seems. Suppose, for example, it were true of all individuals who have X, that were they not X, they would all have the factor ${\sim}A \,\&\, {\sim}B \,\&\, {\sim}C$. Then for *these* individuals, the relevant probability of Y in the absence of X would be $Pr(Y/{\sim}A \,\&\, {\sim}B \,\&\, {\sim}C)$, and not $Pr(Y/{\sim}X)$, the latter being an *average* whose value depends on the base probabilities of *all* the disjuncts of ${\sim}X$.[7]

Here is the problem of disjunctive causal factors at the level of generality at which I will discuss it in the following sections. Let $Z_1, \ldots, Z_M$ be the consistent, exhaustive, maximal (hence mutually exclusive) conjunctions of all factors under discussion in a given problem that may be truth-functional ingredients of any factor causally relevant to a factor Y. Then when X is any disjunction, $Z_{i1} \vee \ldots \vee Z_{im}$, of a subset of $\{Z_1, \ldots, Z_M\}$, the question may arise of the causal significance of X for Y. Let the negation of X be $Z_{j1} \vee \ldots \vee Z_{j(M-m)}$ (so that $\{Z_{i1}, \ldots, Z_{im}, Z_{j1}, \ldots, Z_{j(M-m)}\} = \{Z_1, \ldots, Z_M\}$, an M-membered set).

In order to state the solution to the problem succinctly in terms of interaction in Section 5, it will be convenient notationally to reformulate the problem as follows. Let the set $\{X_{i1}, \ldots, X_{in}\}$ be the result of disjoining any factors Z_{ik} that confer any given equal probability on Y into a single factor X_{ik}, so that $n \leqslant m$, $X_{i1} \vee \ldots \vee X_{in}$ is equivalent to $Z_{i1} \vee \ldots \vee Z_{im}$, *no two of the* X_{ik}'s *confer the same probability on* Y, and no X_{ik} is a disjunction of Z_{ik}'s that confer different probabilities on Y. Similarly, let $X_{j1} \vee \ldots \vee X_{j(N-n)}$ be equivalent to $Z_{j1} \vee \ldots \vee Z_{j(M-m)}$, *such that* $N - n \leqslant M - m$, *no two of the* X_{jl}'s *confer equal probability on* Y, and no X_{jl} is a disjunction of Z_{jl}'s that confer different probabilities on Y.

Thus, we have X equivalent to the disjunction $X_{i1} \vee \ldots \vee X_{in}$, each of whose disjuncts confers a different probability on Y (and each of which cannot, using factors under discussion in the given context, be decomposed into a disjunction whose disjuncts themselves confer different probabilities on Y). And $\sim X$ is equivalent to the disjunction $X_{j1} \vee \ldots \vee X_{j(N-n)}$, each of whose disjuncts confers a different probability on Y (and each of which cannot, using factors under discussion in the given context, be decomposed into a disjunction whose disjuncts themselves confer different probabilities on Y). So there are multiple causally significant probabilities of Y both in the presence of X (n such probabilities) and in the absence of X ($N - n$ such probabilities).

The problem of disjunctive causal factors, then, is how to identify, within background contexts, a *single* causally significant probability for Y in the presence of X and a *single* causally significant probability for Y in the absence of X. Another way of putting the problem is this. How can we appropriately identify causal background contexts within which the presence of X and the absence of X each yield single causally significant probabilities for Y? Note, incidentally, that this problem can arise *within* situations that have been identified, using rules and techniques developed henceforth, as causal background contexts (so that $Z_1, \ldots, Z_M$ would be specifications of all the factors under discussion that have not already been specified in such a context). But for simplicity in what follows, let us suppose that no contexts have been identified as yet; the application of what follows to the other case is straightforward.

4. PART OF A SOLUTION

In this section I offer *part* of a solution to a *special case* of the problem. The result in this section will be: *if*, in the special case in question, we hold fixed in background contexts factors of a certain kind, *then* we can identify the two single probabilities we need within each background context. In the next section, I will explain why we *should* hold those factors fixed; that will involve a more general conception of interaction than the one explained above.

The special case handled in this (and the next) section is the one in which for each individual who has X, there is a determinate factor X_{jl} (a disjunct of $\sim X$) that that individual *would* have if it were not X, and for each individual who has $\sim X$, there is a determinate factor X_{ik} (a disjunct of X) that that individual would have if it were not $\sim X$. (Of course, which X_{jl} an X would be if not X can depend on more than just which of the X_{ik}'s the individual actually has; and which X_{ik} a non-X would have if it were an X can depend on more than just which of the X_{il}'s the individual actually has.) In the more general case, handled in Section 6 below, there may not be a *determinate* X_{jl} that a given X would have if not X, but rather there may be *various* X_{jl}'s that it might have if not X, with different *probabilities*; and similarly for actual non-X's, were they not non-X.

Let us begin with the simple kind of case diagrammed in Figure 2. (Here the notation will diverge from that used to describe the general case.) Suppose we are interested in the causal significance of X & $\sim F$ for Y. Given the assumption of the special case just described, there are the following three kinds of individuals in the example: (i) those who would be X & F if not X & $\sim F$ (these include all the actual X & F's and possibly some of the X & $\sim F$'s), (ii) those who would be $\sim X$ & F if not X & $\sim F$ (these include all actual $\sim X$ & F's and possibly some of the X & $\sim F$'s), and (iii) those who would be $\sim X$ & $\sim F$'s if not X

& $\sim F$ (these include all actual $\sim X$ & $\sim F$'s as well as possibly some of the X & $\sim F$'s).

Call these three kinds of individuals K_1, K_2, and K_3. These three kinds of individuals are mutually exclusive and (in our special case) collectively exhaustive: holding each of them fixed positively or negatively gives us back just the three factors K_1, K_2, and K_3 as consistent maximal conjunctions of them and their negations. Holding each of them fixed, then, gives us three background contexts in the example. And we have the following three two-way probability comparisons:

$$Pr(Y/K_1 \,\&\, (X \,\&\, \sim F)) = 0.6 > 0.1 = Pr(Y/K_1 \,\&\, \sim(X \,\&\, \sim F));$$
$$Pr(Y/K_2 \,\&\, (X \,\&\, \sim F)) = 0.6 < 0.9 = Pr(Y/K_2 \,\&\, \sim(X \,\&\, \sim F));$$
$$Pr(Y/K_3 \,\&\, (X \,\&\, \sim F)) = 0.6 > 0.4 = Pr(Y/K_3 \,\&\, \sim(X \,\&\, \sim F)).$$

The assignment of probabilities on the left hand sides above relies on the plausible assumption that the probability of Y depends just on the way an individual *is* with respect to all factors in any way causally relevant to Y, and not, beyond that, on the way it *would be* if it were different. We have already identified all the causally relevant factors for Y in X and F, so that the K_i's are not causally relevant to Y once X and F have each been specified positively or negatively; hence, the K_i's should not be *probabilistically* relevant to Y once we have specified X and F in one way, as we have in the left hand sides in X & $\sim F$.[8] On the right hand sides, however, if we ignore the K_i's, then neither X nor F is specified positively or negatively: $\sim(X$ & $\sim F)$ is consistent with each of X, $\sim X$, F, and $\sim F$. But $\sim(X$ & $\sim F)$ together with a K_i determines a specification of each of X and F, which gives us a unique causally significant probability for Y.

Note also that if the underlying population is diverse enough, there is no need to postulate nonactual individuals: each of the six conditioning factors in the six probabilities above determines a set of actual members of any actual population in question. For example, K_1 & $(X$ & $\sim F)$ is a subset of the actual X & $\sim F$'s and K_1 & $\sim(X$ & $\sim F)$ is the set of the actual X & F's.

So, referring to the probability comparisons above, we see that if we *should* hold fixed the K_i's, and if there are individuals of all six kinds determined by specifying a K_i and whether or not X & $\sim F$ (so that all six probabilities are defined), then in this example, X & $\sim F$ is not unanimous for Y: X & $\sim F$ has *mixed* causal relevance for Y.

Let us now consider a somewhat more complicated example. Turning back to the notation used to describe the general case, and using an example from the previous section, let

$$
\begin{array}{ll}
X_1 = A \,\&\, B \,\&\, C & X_5 = {\sim}A \,\&\, B \,\&\, C \\
X_2 = A \,\&\, B \,\&\, {\sim}C & X_6 = {\sim}A \,\&\, B \,\&\, {\sim}C \\
X_3 = A \,\&\, {\sim}B \,\&\, C & X_7 = {\sim}A \,\&\, {\sim}B \,\&\, C \\
X_4 = A \,\&\, {\sim}B \,\&\, {\sim}C & X_8 = {\sim}A \,\&\, {\sim}B \,\&\, {\sim}C
\end{array}
$$

$$X = A \vee (B \,\&\, C) = X_1 \vee X_2 \vee X_3 \vee X_4 \vee X_5$$

$${\sim}X = X_6 \vee X_7 \vee X_8$$

Now there are eight counterfactual factors that have to be held fixed in background contexts. For $i = 1, \ldots, 5$, and $j = 6, 7, 8$, let $K_{i,j}$ be the conjunctive counterfactual factor: If X then X_i, and if ${\sim}X$ then X_j. The $K_{i,j}$'s are 15 mutually exclusive and collectively exhaustive kinds of individuals; and assuming nothing else needs to be held fixed, these are our 15 causal background contexts for this example. So to determine the causal significance of X for Y, we must compare $Pr(Y/K_{i,j} \,\&\, X)$ with $Pr(Y/K_{i,j} \,\&\, {\sim}X)$, for each of the 15 $K_{i,j}$'s. These will all be comparisons involving just two causally significant probabilities, since, for each i and j, $K_{i,j} \,\&\, X$ is a subset of X_i and $K_{i,j} \,\&\, {\sim}X$ is a subset of X_j: just note that $K_{i,j}$ is a subset of $X_i \vee X_j$, for each i and j. Also, $Pr(Y/K_{i,j} \,\&\, X) = Pr(Y/X_i)$ and $Pr(Y/K_{i,j} \,\&\, {\sim}X) = Pr(Y/X_j)$, since, again, the X_i's and X_j's are assumed to specify all factors causally relevant to Y in this example.

In the general case described at the end of the last section, there will be $n + (N - n)$ counterfactual factors to be held fixed in background contexts. For $k = 1, \ldots, n$ and $l = 1, \ldots, N - n$, let $K_{k,l}$ be the conjunctive counterfactual factor: If X then X_{ik}, and if ${\sim}X$ then X_{jl}. Again, the $K_{k,l}$'s are $n(N - n)$ mutually exclusive and collectively exhaustive kinds of individuals; and assuming that nothing else needs to be held fixed, they are the causal background contexts. The probabilities $Pr(Y/K_{k,l} \,\&\, X)$ and $Pr(Y/K_{k,l} \,\&\, {\sim}X)$ must be compared in each of these $n(N - n)$ background contexts. And these will all be comparisons of two causally significant probabilities, since $Pr(Y/K_{k,l} \,\&\, X) = Pr(Y/X_{ik})$ and $Pr(Y/K_{k,l} \,\&\, {\sim}X) = Pr(Y/X_{jl})$, for each k and l. Recall that the X_{ik}'s and X_{jl}'s are assumed to specify all factors causally relevant to Y in the general case (except that some of them may be disjunctions of such specifications Z_{ik} and Z_{jl} when more than one confers the same probability on Y).

This procedure gives us the intuitively correct answer in cases of the kind depicted in Figure 2. And the way causal factors of the form $A \vee (B \& C)$ were handled, and the general case just described, are just straightforward generalizations of the Figure 2 kind of case. But what else can be said in the way of justifying or motivating this way of dealing with the problem of disjunctive causal factors. As far as justification goes, Elliott Sober and I (1983) have already argued that on any interpretation of probability suitable for developing probabilistic theories of causality, it cannot hurt to hold fixed factors that may not absolutely need to be held fixed: once enough has been held fixed to get the right answer, the theory should give the same right answer if more and more factors are held fixed in the background contexts. If this is right, then we cannot be doing anything wrong in holding fixed the counterfactual factors described here. In the next section, further positive motivation will be provided for holding these factors fixed.

5. PROBABILISTIC CAUSAL INTERACTION GENERALIZED

In Section 2, a causal factor *X*'s interacting with a factor *F* with respect to an effect factor *Y* was characterized as there being a *reversal*, across *F* and $\sim F$, of an *inequality* between the probability of *Y* given *X* and the probability of *Y* given $\sim X$. Intuitively, this roughly corresponds to *X*'s being a positive cause of *Y* in either the presence or absence of *F* and a negative cause in the other case. If this is the rough intuitive basis for our understanding of interaction, then clearly the characterization given earlier needs to be broadened. For one thing, there are, besides mixed causal factorhood, *three* kinds of causal significance that a factor *X* can have for a factor *Y*: positive, negative, and neutral. So the first natural generalization of probabilistic causal interaction would be to say that *X* interacts with *F* with respect to *Y* if the *probabilistic significance — positive, negative, neutral —* of *X* for *Y* is just *different* in the presence of *F* from what it is in the absence of *F*, where *X* has positive, negative, or neutral probabilistic significance for *Y* according to whether the probability of *Y* given *X* is greater than, less than, or equal to, the probability of *Y* given $\sim X$.

Another natural generalization would allow a causal factor to interact not just with an "on/off" factor *F* (contrasted simply with $\sim F$), but with a three-way partition $\{F_1, F_2, F_3\}$ (where F_1, F_2, and F_3 are mutually exclusive and collectively exhaustive). Then we can say that *X* interacts with the partition $\{F_1, F_2, F_3\}$ with respect to *Y* if the probabilistic significance of X for Y, there being three kinds, is different in the presence of each of the three factors of the partition.

It was perhaps natural to pick *three*-way partitions, since we have discussed three kinds of causal significance (aside from mixed, which we would like to disappear within specifications of interactive factors). But there is no reason to stop at three. Another natural generalization would be to let a factor X interact with partitions of any size. We could say that the different possible probabilistic significances of X for Y are positive, negative, and *neutral at any particular value of the probability of* Y. Then the possibilities are: $Pr(Y/X) > Pr(Y/\sim X)$, $Pr(Y/X) < Pr(Y/\sim X)$, and $Pr(Y/X) = Pr(Y/\sim X) = r$, for any r, $0 < r < 1$. In this sense, there are infinitely many different probabilistic significances that X could have for Y. Now we can say that X interacts with a partition $\{F_1, \ldots, F_n\}$ if, in this broader sense, the probabilistic significance of X for Y is different within each of the F_i's. This too is a natural understanding of interaction, for in this sense of interaction, which of the F_i's X or $\sim X$ combines with surely makes a difference for Y, or for the kind of impact X has on Y.

Here is what seems to be the most general possible understanding of X's interacting with a partition $\{F_1, \ldots, F_n\}$ with respect to Y. For each $i = 1, \ldots, n$, let $p_i = Pr(Y/F_i \ \& \ X)$ and $q_i = Pr(Y/F_i \ \& \sim X)$. Then let us say that X interacts with $\{F_1, \ldots, F_n\}$ with respect to Y if the n pairs $\langle p_i, q_i \rangle$ are all distinct. Again, this is a very natural understanding of probabilistic causal interaction, since again in this sense of interaction, which of the F_i's X or $\sim X$ combines with surely makes a difference for Y, or for the kind of impact X has on Y.

Here the revised theory of probabilistic causality that accommodates this generalized understanding of interaction. It suffices now to characterize the background contexts K_i. Let $F_1, \ldots, F_n$ be just those factors that are, in a population P, either (i) a positive or negative cause of Y independently of X or (ii) a member of a partition with which X interacts with respect to Y. Then the K_i's are those maximal conjunctions of the F_i's that, relative to population P, have nonzero probability both in conjunction with X and in conjunction with $\sim X$.[9]

Now consider the partition of $K_{k,l}$'s, $k = 1, \ldots, n$, $l = 1, \ldots, N - n$, of the solution to the general case of the problem of disjunctive causal factors given in the previous section. Call this $n(N - n)$ membered partition D. It is easy to see that all that is required for X to interact with D with respect to Y on the general understanding of interaction is for the following two conditions to hold:

(i) the $Pr(Y/X_{ik})$'s are all distinct, $k = 1, \ldots, n$ (recall that $Pr(Y/K_{k,l} \& X) = Pr(Y/X_{ik})$, for each k and l),

and

(ii) the $Pr(Y/X_{jl})$'s are all distinct, $l = 1, \ldots, N - n$ (recall that $Pr(Y/K_{k,l} \& \sim X) = Pr(Y/X_{jl})$, for each k and l).[10]

But in Section 3, the X_{ik}'s and X_{jl}'s were defined in terms of the Z_{ik}'s and Z_{jl}'s in just such a way as to guarantee (i) and (ii). We saw in section 2 that it is important to hold fixed factors with which a causal factor interacts. And in the previous section we saw that holding fixed the members of D in the causal background contexts provided a solution to the problem of disjunctive causal factors. So we now see that when the idea of interaction is suitably generalized, the solution to the problem of probabilistic causal interaction provides a solution to the problem of disjunctive causal factors as well.

6. THE CASE OF PROBABILISTIC CONSEQUENTS

In this section, I handle the general problem of disjunctive causal factors described in Section 4 *without* the assumption that there is a determinate X_{jl} that an X would be if not an X and that there is a determinate X_{ik} that a non-X would be if it were an X. This means that all the $K_{k,l}$'s (if X then X_{ik}, and if $\sim X$ then X_{jl}) may be false, and have probability of zero. However, I sall assume that the various individuals that are X would, were they not X, distribute themselves over the X_{jl}'s with different *probabilities* (perhaps different probability distributions for different individuals that are X); and similarly for the non-X's.

As a first approximation to the solution to the general case of the general problem, let r range over probabilities such that $r(X_{i1} \vee \ldots \vee X_{in}) = 1$ and s over probabilities such that $s(X_{j1} \vee \ldots \vee X_{j(N-n)}) = 1$. Then for each such r and s, let $K_{r,s}$ be the factor: If X, then one of $X_{i1}, \ldots, X_{in}$ with probabilities given by r; and if $\sim X$, then one of $X_{j1}, \ldots, X_{j(N-n)}$ with probabilities given by s.

Will treating the $K_{r,s}$'s as causal background contexts provide a formally adequate and principled solution to the general problem —

assuming again that nothing else needs to be held fixed in background contexts (or that all other such factors have already been held fixed in the context of this discussion)? Parelleling the treatment of the deterministic special case in the previous two sections, two questions arise. First, of course, is the question: Does each $K_{r,s}$, determine *single causally significant* probabilities for Y in the presence and in the absence of X? And second, can treating the $K_{r,s}$'s as causal background contexts be motivated independently, e.g., in terms of a generalized understanding of interaction?

The answer to the first question is "yes", given a plausible assumption similar to one used in the special case above, and one other. Consider first the probability of Y in the presence of X, in a background context $K_{r,s}$:

$$Pr(Y/K_{r,s} \& X) = \text{SUM}_{k=1}^{n} Pr(X_{ik}/K_{r,s} \& X)\, Pr(Y/X_{ik} \& K_{r,s} \& X).$$

Of course X_{ik} implies X, so the conjunct X can be eliminated from the conditioning part of the second terms. Also, $K_{r,s}$ can be eliminated there since, by assumption, the X_{ik}'s specify everything causally relevant to Y (this is similar to a move made in section 4 above). So we have:

$$Pr(Y/K_{r,s} \& X) = \text{SUM}_{k=1}^{n} Pr(X_{ik}/K_{r,s} \& X)\, Pr(Y/X_{ik}).$$

Also, given the meaning assigned to $K_{r,s}$, it is natural to identify $Pr(X_{ik}/K_{r,s} \& X)$ with $r(X_{ik})$. So we have:

$$Pr(Y/K_{r,s} \& X) = \text{SUM}_{k=1}^{n} r(X_{ik})\, Pr(Y/X_{ik}).$$

Similarly,

$$Pr(Y/K_{r,s} \& \sim X) = \text{SUM}_{l=1}^{N-n} s(X_{jl})\, Pr(Y/X_{jl}).$$

So it is clear that the $K_{r,s}$'s yield single probabilities for Y in the presence and in the absence of X — indeed, "causally significant" probabilities, since individuals are characterized by an r and an s that are supposed to reflect their causal propensities to distribute themselves over the different kinds of X's and non-X's where these different kinds of X's and non-X's are already assumed to confer genuinely causally significant probabilities on Y.

The question now is whether the $K_{r,s}$'s *should* be treated as causal background contexts (or as factors that should be held fixed in causal background contexts). This cannot be justified in the way this treatment of the $K_{k,l}$'s was justified in the previous section. This is because X does *not* necessarily interact with the partition of $K_{r,s}$'s with respect to

Y, in the general sense of interaction described in Section 5. A $K_{k,l}$ of the deterministic case is the same as some $K_{r,s}$ of the general case in which r and s assign probability 1 to some X_{ik} and to some X_{jl}, respectively. And the distinctness of all the X_{ik}'s and of all the X_{jl}'s guaranteed interaction in the deterministic case. However, multiple distinct values can be *averaged* in many *different* ways so as to yield the *same* average. So the last two expressions displayed above can remain the same in value when calculated in terms of different probabilities r and s. And this just means that X may not interact with the $K_{r,s}$'s with respect to Y (depending on which of the $K_{r,s}$'s get nonzero probability).

However, now that we have the $K_{r,s}$'s to work with, it is easy to construct, from them, a partition with which X *does* interact, with respect to Y, each member of which will still determine single causally significant probabilities for Y in the presence and absence of X. Where p and q are any numbers between 0 and 1, inclusive, let $K_{p,q}$ be the disjunction of all the $K_{r,s}$'s such that $Pr(Y/K_{r,s}\ \&\ X) = p$ and $Pr(Y/K_{r,s}\ \&\ {\sim}X) = q$.

Then the $K_{p,q}$'s are mutually exclusive, collectively exhaustive factors, each of which *determines single causally significant probabilities* for Y given X and Y given ${\sim}X$; and X clearly *interacts with the partition* of $K_{p,q}$'s with respect to Y, in the general sense of interaction explained in the previous section. We have interaction, of course, because if $K_{p,q}$ and $K_{p',q'}$ are distinct, then the pairs $\langle p, q \rangle$ and $\langle p', q' \rangle$ are distinct, so that the two pairs $\langle Pr(Y/K_{p,q}\ \&\ X), Pr(Y/K_{p,q}\ \&\ {\sim}X) \rangle$ $(= \langle p, q \rangle)$ and $\langle Pr(Y/K_{p',q'}\ \&\ X,) Pr(Y/K_{p',q'}\ \&\ {\sim}X) \rangle$ $(= \langle p', q' \rangle)$ will be distinct.[11] And I say the probabilities determined are causally significant because of they way they are determined by different kinds of individuals' propensities (the r's and s's) to be different kinds of X's and non-X's were they different, with respect to X and ${\sim}X$, from the way they are; the values determined are not simply averages depending on how individuals happen to be actually distributed among the different kinds of X's and non-X's.[12]

So we see that in the general case as well, the solution to the problem of interaction — according to which we should hold fixed factors with which the causal factor interacts (or, more generally, factors in a partition that the causal factor interacts with, on the general understanding of interaction developed in the previous section) — also provides a solution to the problem of disjunctive causal factors.

University of Wisconsin-Madison

NOTES

* I thank the John Simon Guggenheim Foundation and the National Science Foundation (grant no. SES-8605440) for financial support during the time this paper was written.

[1] The formulation given here is slightly different in details from Cartwright's. For other versions of the theory, and discussion of some important issues not covered here, see — in addition to Cartwright (1979) — Suppes (1970), Skyrms (1980, 1984), and Eells and Sober (1983), for example; for a review see Skyrms (1987).

[2] It may be worth explaining why this is so. Consistent with the description of the example, lung cancer could be more prevalent among those with the condition than among those without it; it's just that for those with the condition, smoking helps a little. In that case the condition could be causally positive for lung cancer. On the other hand, the condition might itself cause smoking, where the combination of the condition with smoking gives the best possible protection, in which case the condition could be causally negative for lung cancer. Or the truth could be a mixture of these two possibilities, such that the benefits of the condition exactly offset the costs, so that the condition would be neutral for lung cancer. Finally, it could be that among smokers the condition is beneficial and among nonsmokers it is detrimental, so that it has a mixed causal role.

[3] For more details on such difficulties, see my (1987).

[4] This characterization of interaction will be generalized in several ways below.

[5] It is easy to see that the two characterizations of interaction are different in more than just formulation. For example, interaction of X and F for Y is symmetric in X and F on Cartwright's formulation; but X's interacting with F with respect to Y is not symmetric in X and F on the characterization given above.

[6] This point will be more fully explained in sections 4 and 6, in which it is a main theme.

[7] This idea will be more fully elaborated in the next section.

[8] Of course, this inference depends in turn on an appropriate interpretation of probability, an issue I will not address in this paper.

[9] It is perhaps an obvious detail, but perhaps helpful to note, that if X interacts, with respect to Y, with a partition J of some *nonmaximal conjunction* of some the F_i's (in that subpopulation or context), then X will also interact, with respect to Y, with a partition of the *general population* that includes all members of J, but at most one, where the exception is included in a member of the partition of the general population.

[10] Suppose $K_{k,l}$ and $K_{k',l'}$ are two distinct members of D, and compare the pair $\langle Pr(Y/K_{k,l} \,\&\, X), Pr(Y/K_{k,l} \,\&\, {\sim}X)\rangle$ with the pair $\langle Pr(Y/K_{k',l'} \,\&\, X), Pr(Y/K_{k',l'} \,\&\, X)\rangle$. The first pair is the same as $\langle Pr(Y/X_{ik}), Pr(Y/X_{jl})\rangle$, and the second is $\langle Pr(Y/X_{ik'}), Pr(Y/X_{jl'})\rangle$. Since we have two distinct members of D, either $k \neq k'$ or $l \neq l'$. In the first case, the two pairs will differ in their first members, by (i); and in the second case, they will differ in their second members, by (ii).

[11] This technique of disjoining K's could have been used in the previous two sections to handle the deterministic case, thereby avoiding the move from the Z_{ik}'s and Z_{jl}'s to the X_{ik}'s and X_{jl}'s that was made in Section 3. For the deterministic case, we had a choice; but in the general case, since there is more than one way to average distinct

values to arrive at a given value, we are forced to disjoin $K_{r,s}$'s that yield given fixed values for the probabilities of X given Y and X given $\sim Y$.

[12] Note, incidentally, that this formulation gives an alternative way of characterizing positive, negative, neutral, and mixed causal relevance. X is causally positive (negative, neutral) for Y if and only if all the $K_{p,q}$'s that have positive probability are such that $p > q$ ($p < q$, $p = q$); and X is mixed for Y if not positive, negative, or neutral.

Note added in proof: X is causally mixed for Y in P if not causally positive, negative, or neutral. Another oversight: It is easy to envision examples that show we must also hold fixed factors that are, independently of X, causally mixed for Y, in addition to independent factors that are positive, negative, or interactive for Y.

BIBLIOGRAPHY

Cartwright, Nancy (1979), "Causal Laws and Effective Strategies", *Nous* **13**: 419—437.

Dupré, John (1984), "Probabilistic Causality Emancipated". In *Midwest Studies in Philosophy IX: Causation and Causal Theories*, edited by P. A. French, T. E. Uehling, Jr., and H. K. Wettstein. Minneapolis: University of Minnesota Press, pp. 169—175.

Eells, Ellery (1986), "Probabilistic Causal Interaction", *Philosophy of Science* **53**: 52—64.

Eells, Ellery (1987), "Probabilistic Causality: Reply to John Dupré", *Philosophy of Science* **54**: 105—114.

Eells, Ellery and Sober, Elliott (1983), "Probabilistic Causality and the Question of Transitivity", *Philosophy of Science* **50**: 35—57.

Skyrms, Brian (1980), *Causal Necessity*. New Haven: Yale University Press.

Skyrms, Brain (1984), *Pragmatics and Empiricism*. New Haven: Yale University Press.

Skyrms, Brian (1987), "Probability and Causation", *Journal of Econometrics*, forthcoming.

Suppes, Patrick (1970), *A Probabilistic Theory of Causality*. Amsterdam: North-Holland Publishing Company.

ELLIOTT SOBER

THE PRINCIPLE OF THE COMMON CAUSE

In *The Direction of Time*, Hans Reichenbach (1956) stated a principle that he thought helps guide nondeductive inference from an observed correlation to an unobserved cause. He called it *the principle of the common cause.* Wesley Salmon (1975, 1978, 1984) subsequently elaborated and refined this idea.

Reichenbach thought that philosophy had not sufficiently attended to the fact that quantum mechanics had dethroned determinism. His principle was intended to give indeterminism its due; the principle bids one infer the existence of a cause, but the cause is not thought of as part of a sufficient deterministic condition for the effects. It is ironic that as plausible as Reichenbach's principle has seemed when applied to macro-level scientific phenomena and to examples from everyday life, it has been refuted by ideas stemming from quantum mechanics itself. The very idea that Reichenbach wanted to take seriously has come to be seen as his idea's most serious problem.

In the light of this difficulty, Salmon (1984) has modified his endorsement of the principle. It applies, not to all observed correlations, but to correlations not infected with quantum mechanical weirdness. Salmon's fallback position has been that Reichenbach's principle would have been quite satisfactory, if quantum mechanics had not made such a hash of our "classical" picture of causality. This reaction, I think it fair to say, has been the standard one, both for philosophers like Suppes and Zinotti (1976) and van Fraassen (1982) who helped make the case that the principle as originally formulated was too strong and for philosophers like Salmon who had to take this fact into account.

In this paper, I want to describe some consequences of thinking about this principle from an entirely different scientific angle. My interest is to see how the principle fares, not in connection with quantum theory, but with respect to the theory of evolution. Evolutionists attempt to infer the phylogenetic relationships of species from facts about their sameness and difference. We observe that chimps and human beings are similar in a great many respects; we infer from this that there exists a common ancestor of them both, presumably one

James H. Fetzer (ed.) Probability and Causality. 211—228.

from whom these shared traits were inherited as homologies. This very approximate description of how phylogenetic inference proceeds suggests that it should be an instance of the problem that Reichenbach and Salmon's principle was meant to address. Inferring common ancestry is a case of postulating a common cause.[1]

In what follows, I'll begin by saying what the principle of the common cause asserts. I'll then argue that the principle has a very serious defect — one that has nothing to do with quantum mechanics, but with a familiar fact about nondeductive inference that discussions of the principle of the common cause have not taken sufficiently to heart.

Let's begin with one of Reichenbach's simple examples. Suppose we follow a theatre company over a period of years and note when the different actors get gastro-intestinal distress. We observe that each actor gets sick about once every hundred days. We follow the company for long enough to think that this frequency is stable — reflecting something regular in their way of life. So by inferring probabilities from frequencies, we say that each actor has a chance of getting sick on a randomly selected day of 0.01. If each actor's illness were independent of the others, then we would expect that two actors would both get sick about once every 10 000 days. But suppose our observations show that if one actor gets sick, the others usually do as well. This means that the probability that two get sick is greater than the product of the probabilities that each gets sick (1/100 being greater than 1/100 × 1/100). We have here a positive correlation.

The principle of the common cause says that this correlation should be explained by postulating a cause of the two correlates. For example, we might conjecture that the actors always dine together, so that on any given day, all the actors eat tainted food or none of them does. Suppose, for the sake of simplicity, that eating tainted food virtually guarantees gastro-intestinal distress, and also that a person rarely shows the symptom without eating tainted food. Then if we further suppose that the food is tainted about once every hundred days, we will have explained the phenomena: we will have shown why each actor gets sick about once every hundred days and also why the actors' illnesses are so highly correlated.

The postulated common cause has a special probabilistic property; it renders the observed correlates conditionally probabilistically independent. The common cause is said to *screen off* one effect from the other. We can see what this means intuitively by considering the idea of predicting one effect from the other. Given that the two actors' states of health are positively correlated, the sickness or wellness of one on a given day is a very good predictor of the sickness or wellness of the other. However, if we know whether the shared food is tainted or not, this also helps predict whether a given actor is sick or well. The idea that the cause screens off one effect from the other can be stated like this: if you know whether the food is tainted or not, additional knowledge about one actor provides no further help in predicting the other's situation.

The principle of the common cause, then, has two parts. First, it says that observed correlations should be explained by postulating a common cause. Second, it says that the common cause hypothesis should be formulated so that the conjectured cause screens off one effect from the other.

Notice that I have formulated the principle as a piece of epistemology, not ontology. I did not have it say that each pair of correlated observables has a screening off common cause. It seems that this existence claim is thrown in doubt by quantum mechanics. But the bearing of this physical result on the epistemological formulation is less direct. Perhaps it is sound advice to postulate a common cause of this sort, even if we occasionally make a mistake. I suspect that may be why Salmon, van Fraassen, and others have not doubted that the principle of the common cause is sensible, once one leaves the quantum domain.

This doesn't mean that the physical result has absolutely no methodological upshot. It has a very modest one. Suppose we observe a correlation and postulate a common cause that satisfies a few simple "classical" requirements (including, but not limited to, Reichenbach's principle). This combination of observed correlation and theoretical postulate implies a further claim about observables — namely, Bell's (1965) inequality. Apparently, when this prediction is tested in the right sort of particle experiment, the inequality comes out false. So by *modus tolens*, we are forced to reject at least one of our starting assumptions. Since the observed correlation is well attested, our search for a culprit naturally gravitates to the theoretical postulates. Reichenbach's principle may have to be discarded; at least it isn't clear that the principle should be retained.[2]

This means that if you face a situation in which you have certain beliefs about the observables, it might *not* be reasonable for you to postulate a screening-off common cause. But this is a very small challenge to the principle construed methodologically, since, by and large, we do not find ourselves in that epistemic circumstance. We are not stopped short, nor should we be, in the theatre troupe example by thoughts of Bell's inequality.

So much for the principle of the common cause. Now I'll sketch a few facts about the problem of phylogenetic inference. This problem does not take the form of deciding whether two species are related. This is usually assumed at the outset. Rather, the real question concerns propinquity of descent. We want to know, for example, whether human beings are more closely related to chimps than they are to gorillas; that is to say, whether human beings and chimps have an ancestor that is not also an ancestor of gorillas.

It is now very controversial in evolutionary theory how such questions are to be decided. There are several methods in the field; these sometimes yield contradictory results. So a heated debate has ensued, with biologists trying to justify their pet methods and to puncture the pretensions of rival approaches. This controversy has mainly focused on the issue of what the various methods assume about the evolutionary process. The idea that correlation of characters is evidence for closeness of evolutionary relationship is anything but widely accepted; it has its partisans, but it certainly does not represent a consensus, there presently being no such thing.[3]

So without emersing ourselves too deeply in the details of this dispute, we can nevertheless discern a simple surface feature that is not without its significance. Biologists of every stripe concede that correlation is evidence for propinquity of descent only if certain contingent assumptions are made about the evolutionary process. There is dispute about what these are and about whether this or that assumption is plausible. But it goes without saying in the biology that correlation is not a self-evident indicator of common ancestry, but needs to be defended by a substantive argument.

This may, of course, be because the biologists are misguided. Perhaps the principle does have some sort of *a priori* claim on our attention, quantum difficulties not withstanding. But I think that this is a mistake. My main claim against the principle will be that correlations are evidence for a common cause only in the context of a background theory.

We do not need recondite evolutionary ideas to grasp this point. There are many correlations that we recognize in everyday life that do not cry out for common cause explanation. Consider the fact that the sea level in Venice and the cost of bread in Britain have both been on the rise in the past two centuries. Both, let us suppose, have monotonically increased. Imagine that we put this data in the form of a chronological list; for each date, we list the Venetian sea level and the going price of British bread. Because both quantities have increased steadily with time, it is true that higher than average sea levels tend to be associated with higher than average bread prices. The two quantities are very strongly positively correlated.

I take it that we do not feel driven to explain this correlation by postulating a common cause. Rather, we regard Venetian sea levels and British bread prices as both increasing for somewhat isolated endogenous reasons. Local conditions in Venice have increased the sea level and rather different local conditions in Britain have driven up the cost of bread. Here, postulating a common cause is simply not very plausible, given the rest of what we believe (Sober, forthcoming-a).

Let me generalize this point by way of an analogy. Hempel (1945) formulated the raven paradox by asking why observing black ravens confirms the generalization that all ravens are black. He asked that we address this problem by adopting a methodological fiction — we are to imagine ourselves deprived of all empirical background information. In this barren context, we then are to show why observing a positive instance confirms a generalization.

Philosophers like Good (1967), Suppes (1966), Chihara (1981), and Rosenkrantz (1977) responded by showing that there are cases in which background knowledge can make it the case that positive instances disconfirm a generalization. Hempel (1967) replied by saying that these examples were not to the point, since they violated the methodological fiction he had proposed. Good (1968) answered by claiming that if we have no background information at all, then positive instances have no evidential meaning. There is no saying whether they confirm, or are neutral, unless empirical background assumptions are set forth.

The principle of the common cause has an important point of similarity with the idea that positive instances always confirm generalizations. Both, I suggest, are false because they are overstated. Correlations always support a common cause hypothesis no more than positive instances always confirm a generalization.

A fallback position, analogous to the one Hempel adopted by

proposing his methodological fiction, suggests itself. Rather than saying that it is *always* plausible to postulate a common cause, why not propse that it is plausible to do so, *if* one knows nothing at all about the subject matter at hand? My view is that there is no assessment of plausibility, except in the light of background assumptions. If there is anything that deserves to be called a principle of the common cause, it must be stated as a conditional; we must say *when* a common cause explanation is preferable to an explanation in terms of separate causes.

I just described the correlation between Venetian sea levels and British bread prices as better explained by postulating separate causes. However, it might be objected that the two correlates do have a common cause, which we can recognize if we are prepared to talk vaguely enough and to go far enough back in time. For example, perhaps the emergence of European capitalism is a common cause. This brings me to a second issue concerning the Reichenbach/Salmon principle, one which the biological problem helps clarify.

I claimed before that a problem of phylogenetic inference is never formulated by focusing on two species and asking if they are related. Rather, the systematist's problem is always a comparative one. We'd like to know, in our running example, whether human beings and chimps are more closely related to each other than either is to Gorillas.

I suggest that a parallel point applies to causality in general. Suppose that the causal relation between token events is transitive. Suppose further that the universe we inhabit stems from a single origination event. If we hold that every event in our universe, save this first Big Bang, has a cause, then we have here a general structure of which the phylogenetic situation is a special case. It now is trivially true that every pair of events has a common cause.

In this hypothetical circumstance, would it suddenly become meaningless to consider the credentials of a "separate" cause explanation? I think not. But this means that a nontrivial question about common causes must have more content than merely inquiring whether two events have a common cause.

Returning to Venice and Britain, perhaps our question should be stated as follows: does the increase in Venetian sea levels and in British bread prices have a common cause that is not also a cause of French industrialization? Perhaps the answer here is *no*. But if we ask, instead, does the increase in Venetian sea levels and in British bread prices have a common cause that is not also a cause of Samoan migrations, the answer may be *yes*. Varying the third term in the question about common causes changes the problem. By analogy with the phylogenetic case, we can see that this third term plays the role of specifying the level of resolution at which the problem about common causes is posed. When I said before that the Venice/Britain correlation is not plausibly explained by postulating a common cause, I had in mind a particular level of analysis, but there are others.

Earlier, I segmented the principle of the common cause into two parts. First, there is the idea that observed correlations should be explained by postulating a common cause. Second, there is the demand that the common cause be characterized as screening off each of the correlates from the other. My Venice/Britain example was intended to challenge the first of these ideas. It now is time to see whether a postulated common cause should be thought of as inducing the special probabilistic relationship of *screening off*.

I'll begin by outlining a circumstance in which the screening off idea makes perfect sense. Let's imagine that we are considering whether the observed correlation between E_1 and E_2 in the following figure should be explained by postulating a common cause or a separate cause. These are event types, whose frequencies, both singly and in combination, we wish to explain. Figure (1a) describes the common cause and (1b) the separate cause pattern of explanation.

How are we to evaluate the quality of explanation provided? The

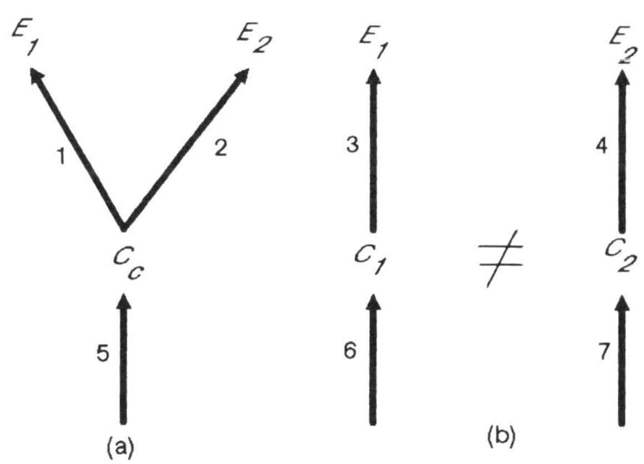

Figure 1.

idea of *likelihood* is intuitively appealing, at least as a start. In the theatre troupe example, it is natural to reason as follows: if the actors share their meals, then it is not very surprising that their sick days are strongly correlated. But if they ate quite separately, then it would be very surprising indeed that their sick days covary so closely. In reasoning in this way, we consider how probable each hypothesis says the observations are. The common cause hypothesis makes the correlation quite probable, whereas the separate cause explanation makes it almost miraculous. We prefer the former explanation because it strains our credulity less.

It is essential to see that in considering how probable the hypotheses make the observations, we are not considering how probable the observations make the hypotheses. Nowhere in the thought just described did we consider how probable it is that the actors eat together. Rather we considered how probable it is that they get sick together, on the supposition that they eat together. The likelihood of a hypothesis is the probability it confers on the observations; at all costs, it must not be confused with the probability the observations confer on the hypothesis.

So in comparing the pattern of explanation displayed in Figure 1, we want to determine what probability the two hypotheses confer on the observations. In the theatre troupe example, we observe that the actors get sick about once every hundred days, but that when one actor gets sick, the other almost always does so as well. Our question, then, is how probable do the explanations represented in Figure 1 say these observations are.

There is no answer to this question, because the two hypotheses are incompletely specified. Each of the branches in the Figure can be thought of as a causal pathway. To calculate the probability of the events at the tips, we must know what probabilities to associate with the branches. These are not stated, so no likelihood comparison can be made.

Let us suppose that our assignment of values to these branches is wholly unconstrained by any prior theoretical commitments. We can simply *make up* values and flesh out the common cause and the separate cause explanations in any way we please.

We will find the *best case scenarios* for the common cause and the separate cause patterns. We will assign probabilities to branches 1, 2, and 5 so as to make the common cause pattern maximally likely. And we will similarly assign probabilities to branches 3, 4, 6, and 7 so as to make the separate cause pattern maximally likely. We then will compare these two best cases. In saying that we choose values so as to maximize the likelihood, we are saying that the hypothesis is set up so as to make the observations as probable as possible.

If we assume that the two separate causes C_1 and C_2 are independent of each other and that all the assigned probabilities must be between 0 and 1 noninclusive, we can obtain a very pleasing result.

We can set up the common cause explanation so that it is more likely than any separate cause explanation. If E_1 and E_2 are correlated, we can assign probabilities in Figure (1a) so that the implied probabilities of the two singleton events and the implied probability of their conjunction perfectly match their observed frequencies. This is the way to maximize the likelihood. If E_1 occurs about once in a hundred times, then the most likely hypothesis will assign that event a probability of 0.01. If E_1 and E_2 almost always cooccur, so that the conjoint event occurs approximately once every 120 times, then the most likely hypothesis will assign that conjoint event a probability of 1/120. However, this matching of postulated probabilities to sample frequencies cannot be achieved in Figure (1b). If we match the singleton probabilities to the singleton frequencies, we will fail to have the conjoint probability match the conjoint frequency. So the best case scenario for the common cause explanation is more likely than the best case scenario for the separate cause explanation.

This kind of reasoning, I think, underlies much of the plausibility of the examples discussed by Reichenbach and Salmon. We *invent* hypotheses of common and separate cause and then compare them. There is nothing especially implausible about the idea that actors should always eat together, or that they should eat separately, so the common cause and the separate cause explanations are not radically different on that count. That is, the *probability* of common meals or separate meals is not an especially telling point of comparison. What remains is the consideration of likelihood. Reichenbach and Salmon were, I think, picking up on the fact that in this special context, the common cause explanation has a virtue that the separate cause explanation cannot claim. That virtue is pinpointed by the concept of likelihood.

However, this is a special case, for at least two reasons. First, we have already seen, in the Venice/Britain example, that there is more to be considered than likelihood. Even though the supposition of a common cause might explain the correlation of sea levels and bread prices, it is immensely implausible to suppose that there is such a common cause. This judgment seems to go beyond the dictates of likelihood. To bring this point to bear on the acting example, merely suppose that we are told that it is enormously improbable that the actors took many of their meals together. Of course, if they had done so, that might be a beautiful explanation of the correlation. But we might nevertheless have to reject this likely story because it is immensely improbable.

The second limitation arises even if we restrict ourselves to the issue of likelihood. The problem is that we filled in the branch probabilities in Figure 1 in any way we pleased. However, suppose we have background information that constrains how these can be specified. Suppose I tell you that when these actors eat together, the food is checked before hand, so that the probability of tainted food getting through to them is one in a billion. This information makes the common cause explanation plunge in likelihood. Even if it were true that the actors take their meals together, it isn't so clear any more that this is a very plausible explanation of the observations. If the probabilities that figure in the separate cause explanation are fleshed out by other kinds of background information, it may even emerge that the separate cause explanation is more likely after all.

So highly specific background information may prevent the common cause hypothesis from being more likely. I now want to show that there can be quite general theoretical reasons for thinking that plausible assignments of probabilities to elaborate the common cause structure of Figure (1a) should *not* induce a screening off relation. Let's consider first the causal set-up depicted in Figure (2a). Here we have a distal cause (C_d) producing a proximal cause (C_p), which then causes the two correlated effects (E_1 and E_2). Let's assume, first, that all probabilities are between 0 and 1 noninclusive and that the proximate cause makes a difference to the probability of each effect. If we further assume that the proximal cause screens off the distal cause from the effects and that the proximal cause screens off the effects from each other, it follows that the distal cause does *not* screen off the effects from each other.[4] This means that if I postulate a common cause of the two effects, and I imagine that the cause is distally related to the effects as just described, then I should *not* formulate my common cause hypothesis so that it involves screening off. But this, we can be sure, will not prevent me

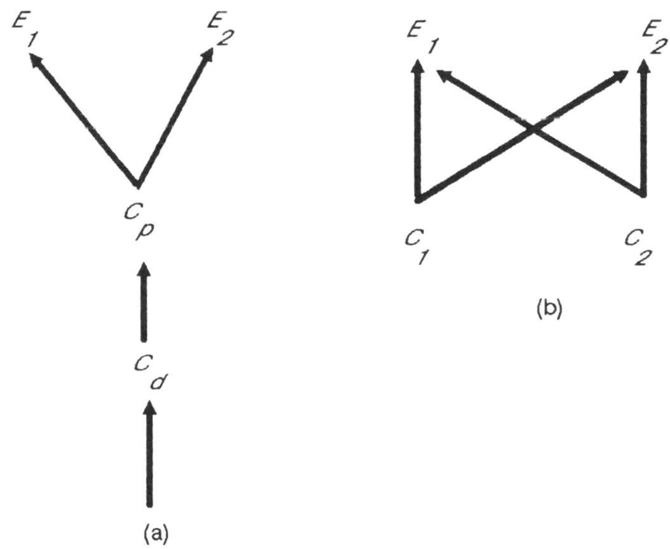

Figure 2.

from testing that hypothesis against other competing hypotheses, and perhaps thinking that it is plausible.

The same lesson arises from the set-up depicted in Figure (2b). Suppose I think that if there is one common cause of the two effects, then there probably is another. If I imagine that these causes contribute to their effects by Mill's rule of composition of causes, then neither cause, taken alone, will screen off one effect from the other.[5] So if I propose to develop a common cause explanation of the correlation, one which I hope will be true, though it may be incomplete, I will not want to formulate my description of, say, C_1, so that it screens off the two effects from each other. But once again, this will not prevent me from finding that the hypothesis that fleshes out the idea that C_1 is a common cause is more plausible than some separate cause explanation I choose to consider.

This point does not go counter to the idea that a totally complete picture of *all* the causal facts should induce a screening off relation. That intuition receives its come-upance from quantum mechanics, not from the kinds of issues explored here. Rather, the thought behind the examples displayed in Figure 2 is that at any stage of inquiry in which we introduce a common cause without the presumption that we are laying out the whole causal story, the demand for screening-off may well be misplaced. For the fact of the matter is that most common causes in the real world do not screen off.

I do not deny that when causal models are developed, it is often natural to formulate them with proximal causes (like C_p in Figure 2a) screening off effects from each other and with proximal causes screening off distal causes from their effects. But this should be regarded as a useful fiction, not as reflecting some fundamental principle about how a common cause must be understood.

It may be objected that my discussion of the examples depicted in Figure 2 does not go counter to the principle of the common cause. After all, that principle says that it is reasonable to think that there exists a screening off common cause; it does not assert that every common cause screens off. I concede that if the principle did no more than recommend this existence claim, then my complaints would not apply. However, I then would argue that the principle has very little to do with the ongoing process of inventing and evaluating causal explanations. However, in the work of Reichenbach and Salmon, the principle does do more; it tells one how to compare the credentials of fleshed out common cause and separate cause explanations. The point of Figure 2 is to show why screening off can count as a weakness rather than a strength in common cause explanation.

So my claim against the principle of the common cause is two-fold. Not only do correlations not always cry out for explanation in terms of a common cause; in addition, common cause explanations are often rendered implausible by formulating them so that they screen off the correlates from each other.

When the principle of the common cause is applied to favorable examples, it seems natural and compelling. When one further sees that it is cast in doubt by quantum mechanics, this merely reinforces the conviction that it reflects some very deep and fundamental fact about how we structure our inferences about the world. However. I have been arguing that both the favorable examples and the physical counterexamples make the principle look more fundamental than it really is. It does not state some ultimate constraint on how we construct causal explanations. Rather, it has a derivative and highly conditional validity.

It is useful to represent the problem of comparing the credentials of a common cause (*CC*) and a separate cause (*SC*) hypothesis as a variety of Bayesian inference. This does not mean that Bayesianism is without its problems. But in a wide range of cases, a Bayesian representation describes in an illuminating way what is relevant to this task of hypothesis evaluation. Bayes' theorem allows us to calculate the posterior probabilities of the two hypotheses, relative to the evidence (E) they are asked to explain, as follows:

$$Pr(CC/E) = Pr(E/CC)Pr(CC)/Pr(E)$$
$$Pr(SC/E) = Pr(E/SC)Pr(SC)/Pr(E).$$

Since the denominators of these two expressions are the same, we find that the common cause explanation is more probable than the separate cause explanation precisely when

$$Pr(E/CC)Pr(CC) > Pr(E/SC)Pr(SC).$$

We might imagine that the evidence E considered here consists in the observed correlation of two or more event types. We have not considered whether there might be other sorts of reasons for preferring a common cause to a separate cause explanation. I believe that there are such, but this point would only affect the principle's completeness, not its truth.

When will a common cause explanation be preferable? I doubt that one can say much about this question in general, save for the inequality stated above. This is as much of a general principle of the common cause as can be had. If the likelihoods are about the same, then a difference in priors will carry the day; and if the priors differ only modestly, a substantial difference in likelihood will be decisive.

An example of the first kind is afforded by our Venice/Britain example: a common cause explanation of the correlation of sea levels and bread prices can be concocted, one which makes the observed correlation as probable as any separate cause explanation could achieve. This explanation may therefore have a high likelihood, but it may be implausible on other grounds. A Bayesian will represent this other sort of defect by assigning the hypothesis a low prior probability.[6]

An example of the second kind is provided by Reichenbach's example of the theatre troupe. Our background information may tell us that there is no substantial difference in prior probability between the hypothesis that the actors usually take their meals together and the hypothesis that they usually eat separately. What makes the former hypothesis overall more plausible in the light of the observations is the way it predicts the covariance of sick days. Here, it is likelihood that is decisive in comparing common cause and separate case explanations.

So what I'm suggesting is that the principle of the common cause is not a *first* principle. It is not an ultimate and irreducible component of our methodology of nondeductive inference. What validity it has in the examples that have grown up around this problem can be described pretty well within a Bayesian framework. And of equal importance is the fact that a Bayesian framework allows us to show why the preference for common cause explanations can sometimes be misplaced. It is specific background information about the empirical subject matter at hand, not first principles about postulating causes, that will settle, in a given case, what the relevant facts about priors and likelihoods are.

The idea that some first principle dictates that postulating common causes is preferable to postulating separate causes has had a long history. Indeed, in the *Principia* Newton gives this idea pride of place as the first of his "Rules of Reasoning in Philosophy":

> *We are to admit no more causes of natural things than such as are both true and sufficient to explain their appearances.* To this purpose the philosophers say that Nature does nothing in vain, and more is in vain when less will serve; for Nature is pleased with simplicity and affects not the pomp of superfluous causes.

The principle of the *common* cause (with emphasis on the penultimate word) deserves to be situated in this tradition: it is a parsimony principle intended to constrain causal explanation.

There is no magical primitive postulate that says that one cause is always better than two. Everything depends on the details of what the one is and what the two are. Nor, I think, is there any principle that says that in a theoretically barren context — one in which the investigator knows nothing about the subject under investigation — that one cause is preferable to two. This I take to be a natural application of what we have learned about induction and the raven paradox to this problem concerning inference to the best explanation.

I mentioned earlier that it is controversial, to say the least, to take correlation as the right way to estimate propinquity of descent. Another controversial idea about phylogenetic inference goes by the name of parsimony. It holds that the best phylogenetic hypothesis is the one that requires the fewest multiple originations. When we find that human beings and chimps are similar, we prefer a genealogical hypothesis that permits these similarities to be derived as homologies from a common ancestor, rather than as having originated independently. The intuition that common causes offer better explanations than separate causes again is very much at work here.

The ideas sketched above about the principle of the common cause apply with equal force to the philosopher's general notion of parsimony and to the specific idea of parsimony that biologists have suggested for use in the context of phylogenetic inference. If parsimony is the right way to infer genealogical relationships, this must be because certain contingent process assumptions are true concerning how evolution works. What is now less than evident, I think, is what those contingent assumptions actually are. But that there must be nontrivial assumptions of this sort is the lesson I draw from this analysis of the strengths and limitations of the Reichenbach/Salmon idea. There is a useful corrective to the formulations we have come to call *the principle of the common cause* and *the principle of parsimony*; it is the idea that *only in the context of a background theory do observations have evidential meaning.*

APPENDIX A

Claim: If (i) a proximate cause (C_p) screens off one effect (E_1) from the other (E_2), (ii) the proximate cause screens off a distal cause (C_d) from each effect, (iii) all probabilities are intermediate, and (iv) the proximate cause makes a difference to the probability of each effect, then the distal cause does not screen off the one effect from the other.

We begin with the following probabilities:

$$P(C_p/C_d) = p$$
$$P(E_1/C_p) = a$$
$$P(E_1/-C_p) = b$$
$$P(E_2/C_p) = x$$
$$P(E_2/-C_p) = y$$

Conditions (i) and (ii) imply that $P(E_1 \ \& \ E_2/C_d) = P(E_1/C_d)P(E_2/C_d)$ precisely when

$$pax + (1-p)by = [pa + (1-p)b]\,[px + (1-p)y].$$

We will disprove this equality by reductio. First, we simplify it to

$$p(1-p)ax + p(1-p)by = p(1-p)ay + p(1-p)bx.$$

Given assumption (iii), this becomes

$$a(x-y)=b(x-y),$$

whose falsehood is guaranteed by (iv), which I interpret to mean $a \neq b$ and $x \neq y$.

APPENDIX B

Claim: Let C_1 and C_2 each be a common cause of the effects E_1 and E_2. If (i) the total state of both causes (i.e., a specification of the presence or absence of each) screens off each effect from the other, (ii) all probabilities are intermediate, and (iii) the probability of each effect is an increasing function of the number of causes that are present, then neither cause taken alone screens off one effect from the other.

First, let

$$\begin{aligned}&P(C_i)=c_i\\&P(E_i/C_1\ \&\ C_2)=w_i\\&P(E_i/C_1\ \&\ -C_2)=x_i\\&P(E_i/-C_1\ \&\ C_2)=y_i\\&P(E_i/-C_1\ \&\ -C_2)=z_i,\end{aligned}$$

for $i = 1, 2$. I will prove the claim for the case in which both causes are present; the other three cases follow the same pattern.

Condition (i) implies that $P(E_1\ \&\ E_2/C_1)=$

$$w_1w_2c_2+x_1x_2(1-c_2),$$

whereas $Pr(E_1/C_1)=$

$$w_1c_2+x_1(1-c_2),$$

and $Pr(E_2/C_1)=$

$$w_2c_2+x_2(1-c_2).$$

C_1 screens off the effects from each other precisely when

$$w_1w_2c_2(1-c_2)+x_1x_2c_2(1-c_2)=w_1x_2c_2(1-c_2)+x_1w_2c_2(1-c_2).$$

Again, we proceed by reductio. Condition (ii) allows this equality to be simplified to

$$w_1(w_2-x_2)=x_1(w_2-x_2).$$

I interpret condition (iii) to mean that $w_i > x_i$, $y_i > z_i$ ($i = 1, 2$), which guarantees that this equality is false. Note that a stronger compositional principle (which might be termed a Millian principle of composition of causes), wherein $w_i - x_i = y_i - z_i > 0$ ($i = 1, 2$), is not necessary, though it does suffice.

University of Wisconsin, Madison

NOTES

[1] A fuller treatment of the bearing of phylogenetic inference on the principle of the common cause is provided in Sober (forthcoming-b), Chapter 3.

[2] I owe this exposition of the logic of Bell's (1965) argument to van Fraassen (1982).

[3] For some indication of the diversity of opinion on the proper methodology of phylogenetic inference, see the essays in Duncan and Stuessy (1984).

[4] See the proof in Appendix A.

[5] See the proof in Appendix B.

[6] Critics of Bayesianism — for example, Edwards' (1971) espousal of likelihood as a sufficient analytical tool — will look for some other representational devices.

REFERENCES

Bell, J. (1965): On the Einstein Podolsky Rosen Paradox. *Physics* **1**: 196—200.

Chihara, C. (1981): Quine and the Confirmational Paradoxes. In P. French, H. Wettstein, and T. Uehling (eds.), *Midwest Studies in Philosophy* **6**. Minneapolis: University of Minnesota Press. 425—52.

Duncan, T. and Stuessy, T. (1984): *Cladistics: Perspectives on the Reconstruction of Evolutionary History*. New York: Columbia University Press.

Edwards, A. (1972) *Likelihood*. Cambridge: Cambridge University Press.

Good, I. J. (1967): The White Shoe is a Red Herring. *British Journal for the Philosophy of Science* **17**: 322.

Good, I. J. (1968): The White Shoe qua Herring is Pink. *British Journal for the Philosophy of Science* **19**: 156—57.

Hempel. C. (1945): Studies in the Logic of Confirmation. In *Aspects of Scientific Explanation and Other Essays*. New York: Free Press. 1965.

Reichenbach, H. (1956): *The Direction of Time*. University of California Press.

Rosenkrantz, R. (1977): *Inference, Method, and Decision*. Dordrecht: D. Reidel.

Salmon, W. (1975): Theoretical Explanation. In S. Korner (ed.). *Explanation*, pp. 118—45. Oxford: Blackwell.

Salmon, W. (1978): Why ask "Why?" *Proceedings and Addresses of the American Philosophical Association* **51**, pp. 683—705.

Salmon, W. (1984): *Scientific Explanation and the Causal Structure of the World*. Princeton: Princeton University Press.

Sober, E. (forthcoming-a): Explanation and Causation: A Review of Wesley Salmon's *Scientific Explanation and the Causal Structure of the World. British Journal for the Philosophy of Science.*

Sober, E. (forthcoming-b): *Reconstructing the Past: Parsimony, Phylogeny, and Inference.* Cambridge: Bradford/MIT Press.

Suppes, P. (1966): A Bayesian Approach to the Paradoxes of Confirmation. In J. Hintikka and P. Suppes (ed.), *Aspects of Inductive Logic.* Amsterdam: North-Holland Publishing Co. 1978—207.

Suppes, P. and Zinotti, M. (1976): On the Determinism of Hidden Variable Theories with Strict Correlation and Conditional Statistical Independence of Observables. In P. Suppes, *Logic and Probability in Quantum Mechanics.* Dordrecht: D. Reidel.

van Fraassen, B. (1982): The Charybdis of Realism: Epistemological Implications of Bell's Inequality. *Synthese* **52**: 25—38.

D. H. MELLOR

ON RAISING THE CHANCES OF EFFECTS

1. INTRODUCTION

There is no doubt that effects need probabilities. Deterministic causes — sufficient and/or necessary conditions — give effects probability 1 with their causes and/or probability 0 without them. And the indeterministic causation of Salmon (1984) and others has probabilities of effects built into its very foundations.

Doubt and debate enter with the questions: what probabilities must effects have, and of what kind? The kind, I agree with Salmon, is what Carnap called 'statistical': objective physical probability, which for brevity I call 'chance'. Causal probabilities cannot be merely subjective or inductive: short-circuits cause fires neither by making people expect them nor by providing evidence for them (effects can be as good evidence for their causes as their causes are for them).

Salmon and I differ on what chance is: he favours a frequency account of it, I favour a propensity one (Salmon 1979; Mellor 1982). But we also differ on what chances effects must have: I and others say they must be greater with their causes than (in the circumstances) they would be without them; whereas Salmon denies this. That is the dispute I wish to settle here. And since it can be settled in the same way whatever chances are, that question can be waived for the time being.

What can't be waived is the question of what the right way to settle the present dispute is. Invoking causal intuitions case by case isn't the way, because they too are unsettled. Take Salmon's (1984, p. 200) excited atom that decays improbably from an excited energy level to a ground state via an improbable intermediate level (1). He thinks it is obviously in the ground state *because* it was previously in level 1, whereas I think it is just as obviously in the ground state *despite* that fact, not because of it. (The low *chance* of the atom's second decay may be caused — deterministically — by its first decay; but that's another matter.)

In the end, no doubt, the view should prevail that makes most sense of most plausible cases of causation. But plausibility depends on more

James H. Fetzer (ed.) Probability and Causality. 229—239.

than case-by-case intuitions about causation itself. To vary the old (and much overrated) adage: don't ask for the use, ask for the *point* of the use. Ask, in other words, what we mean to imply when we call a situation causal. That, I believe, is how we should settle our dispute: consider how causation's connotations depend on the chances of effects, and let that settle our disputed cases.

2. THE CONNOTATIONS OF CAUSATION

Causation's main connotations are the following:

(1) *Temporal*: Causes precede their effects.
(2) *Evidential*: Causes and effects are evidence for each other.
(3) *Explanatory*: Causes explain their effects.
(4) *Means-end*: If an effect is an end, its causes are means to it.

These connotations clearly don't entail determinism: (1) causes can precede their effects without being either sufficient or necessary for them; (2) evidence for something need not raise its probability from 0, or to 1; (3) explanations need not be deductive, nor need the falsity of an explanans entail the falsity of its explanandum; and (4) a means to an end can be worth taking even if it is neither an infallible nor the only possible means.

And what makes paradigm cases of indeterministic causation plausible is precisely that all these connotations apply to them. When we say that an atom was caused to split by being bombarded, we imply that the bombardment preceded the splitting, was evidence for it (and *vice versa*), explains it, and would be a means to it. Similarly, when we say that someone's smoking caused his or her cancer, we imply that the smoking preceded the cancer, was evidence for it (and *vice versa*), explains it, and would have been a means to that (admittedly perverse) end.

Theories of causation and of the other concepts involved should therefore respect these connotations, and say why causation has them. Theories of causation and time should between them say why causes precede their effects, and similarly with our theories of causation and evidence, of causation and explanation, and of causation and what means to ends are. And even though causation's connotations do not entail determinism, they may still set some bounds to the chances of effects, which theories of it should respect. But do they? And if so, which ones do so, and what bounds do they set?

Salmon's theory will not say, since it bases causation on 'the two basic concepts of *propagation* and *production*' (Salmon 1984, p. 139). But in his examples (e.g., 'the electrical discharge produces a fire') 'produces' is just a synonym for 'causes'; while 'propagates' just adds to causation's temporal connotation the truism that the cause-effect relation is not intransitive. And these concepts will not serve to settle the question at issue.

Most other theories base causation on its evidential or explanatory connotations. Hume in effect defines causation as the basis of non-demonstrative inference (Hume 1777, Section 4, Part 1). Hempel in effect defines it as providing a certain kind of explanation (Hempel 1965, Section 2.1). I think they are both wrong, for reasons I have given elsewhere (Mellor 1987a): I think the theory of causation should be based on its means-end connotation. Causation is best conceived of as the feature of the world that gives ends means. And this conception, as we shall see, does set bounds to the chances of effects.

There is however an obvious objection to basing causation on its means-end connotation: namely, that to be a means to an end just *is* to be one of its causes — that is all 'a means' means. And if that were so, the means-end connotation would indeed tell us little about causation. In particular, constraints on the chances of ends would have to follow from independent constraints on the chances of effects, not *vice versa*. But in fact the objection is mistaken. We can say enough about means to limit the chances of ends without invoking causation at all. And the means-end connotation will then limit the chances of effects accordingly.

The trick is to use a principle of non-causal decision theory (Jeffrey 1983) to say what it takes to be a means to an end. The principle is the *expected utility principle*: act to maximise expected utility. Only here (*pace* Jeffrey and others) the expected utility must be objective: in particular, it must incorporate chances, not merely inductive or subjective probabilities (Mellor 1983). And then the principle yields the independently plausible condition that an end's chance with a given means must in the circumstances exceed its chance without that means.

I have given the argument for this elsewhere (1987, 1987a), but it may be summarised as follows. Call the end (a desirable fact) *E*, and the prospective means (a realisable fact) *C*. The basic idea is that *C* will be a means to *E* only when expected utility prescribes it (i.e. when *C*'s expected utility exceeds that of $\sim C$). Of course this won't be true in general, because C may have some intrinsic value or disvalue of its own. But it will be true if that is not the case, i.e. if the C and $\sim C$ rows in the relevant utility matrix are the same:

	E	$\sim E$
C	u	u'
$\sim C$	u	u'

Now suppose that E's chances with and without C would in the circumstances be:

	E	$\sim E$
C	p	$1-p$
$\sim C$	p'	$1-p'$

Then the relevant expected utilities $EU(C)$ and $EU(\sim C)$ of C and $\sim C$ will be:

$$EU(C) = up + u'(1-p);$$
$$EU(\sim C) = up' + u'(1-p');$$

so that $EU(C) > EU(\sim C)$ if and only if

$$(u-u')(p-p') > 0.$$

But to say that E is the end is to say that u exceeds u', so that expected utility prescribes C just when

$$p > p',$$

i.e. just when the chance of the end would be greater in the circumstances if C were brought about than if it were not. Unless that is so, bringing C about will be no way to bring the end about: C will be no means to it.

3. CAUSES AS MEANS TO ENDS

The means-end connotation now imposes this condition on the cause-effect relation: an effect's chance with a cause must be greater than it would be in the circumstances without that cause. For otherwise, effects could be ends without their causes being means to them.

In deriving this condition, I do not of course assume that effects *are* ends, nor that their causes could actually be brought about, or are only worth bringing about for the sake of their effects. None of this can be true, if only because effects in turn cause other effects. And none of it is entailed by the means-end connotation, which does not say that effects must actually be ends, nor that their causes must be realisable, and lack all value of their own. All it says is that *if*, while C is related to E as cause to effect, E *were* an end, C would *ipso facto* be a means to it: i.e. that if in those circumstances C had no intrinsic value or disvalue, and could be brought about directly when E could not, the expected utility principle would prescribe bringing C about.

In short, the means-end connotation is conditional. It does not require a causal world to be full of ends and means. It can perfectly well allow causation in a world without either value or agents, and hence without either ends or means. In such a world, as in ours, causation could still exist as what, if only there were ends, would give them means.

I must emphasise also that using the expected utility principle to define causation's means-end connotation does not require us to endorse all that principle's prescriptions. We needn't for example ignore, as it does, the distinction between action and inaction. Suppose a patient in great and incurable pain so much wants to die (E) that a doctor should let him die even though he shouldn't kill him. Whether a means C to the end E should be adopted will then depend on whether it takes positive action to bring it about. We can admit that, while still insisting that for C to *be* a means to E, it must be such that under the conditions stated above the expected utility principle would prescribe it.

Similarly, we can let expected utility give way to the 'maximin' principle (Luce and Raiffa 1957, p. 278) in some cases. Some possibilities are arguably too bad to be worth running any risk of for any end, however valuable. If so, perhaps C should never be brought about when $C \& \sim E$'s utility falls below some minimum value, even if $EU(C)$ exceeds $EU(\sim C)$. What this threshold value (if any) is, below which expected utility should give way to maximin, is a moot point. But fortunately we needn't settle it. For the maximin principle does not deny that C *is* a means to E, merely that C should be brought about just because its expected utility exceeds $\sim C$'s.

Conflicts between expected utility and the 'dominance' principle (Jeffrey 1983, p. 8) are more serious. If in each column of the C/E utility matrix, the utility in the $\sim C$ row exceeds that in the C row, dominance will prescribe $\sim C$, on the grounds that whether E is a fact or not, it will be better if *C* isn't. Like maximin, dominance can contradict expected utility in such cases because it never lets *E*'s value make *C* worth bringing about, however much *C* raises *E*'s chance and thus *C*'s expected utility. But the case for dominance, unlike that for maximin, seems to imply that *C* is not merely unjustified by *E*, but is not a means to it. So if expected utility is to define the means-end relation, it must outrank dominance when the two principles conflict.

And so it does when the expected utilities incorporate chances, i.e. statistical as opposed to inductive or subjective probabilities. For then the situation is one of real *risk*, as opposed to mere *uncertainty* (Luce and Raiffa 1957, p. 13), and all parties agree that expected utility outranks dominance then. Doubt arises only when expected utilities are based on inductive or subjective probabilities. And I confess I share the doubt: I doubt if an unpleasant medicine is worth taking just to increase the evidence (inductive probability), or the degree of my belief (subjective probability), that I will recover from an illness (Mellor 1983). But the point need not be argued here. No one takes inductive evidence for a proposition to be *ipso facto* a *means* of making it true. Nor, when subjectivists advocate wishful thinking, do they do so as a *means* of attaining the wished-for end. The authority or otherwise of inductive or subjective expected utility is therefore irrelevant to the means-end relation. Basing that relation on the expected utility principle only requires the principle to outrank dominance when it uses chances; and no one will deny that it does that.

4. CAUSAL DISPOSITIONS

In short, the means-end connotation makes causes raise their effects' chances — by which of course I don't mean that they cause their effects' chances to increase. That condition would be both circular and ineffective, since it both invokes causation and leaves it deterministic (merely replacing the real effects with their chances). What I mean is what I said in section 3: an effect's chance with a cause must be greater than it would be in the circumstances without it.

But the phrase 'in the circumstances' raises well-known problems, which I cannot really tackle here (see Mellor 1987a) but should say something about. First, it at least means 'given the other causes', which may look circular, but isn't. It simply requires each cause *C* of an effect *E* to raise *E*'s chance above what it would be without *C* but with all

E's other causes, i.e. with everything else that (*inter alia*) meets this condition.

We need something like this to allow for alternative causes that would occur if *C* didn't, and would make *E*'s chance as great or greater than *C* makes it. And what really makes this hard to allow for is that our condition makes causation depend on what *E*'s chance would be without *C*, and that too may depend on *C*. But we want the causal relation to depend on the *actual* value of what *E*'s chance would be without *C* — not on what that value would be if *C*'s absence would make a difference to it. The latter is what the phrase 'in the circumstances' is supposed to rule out. The question is what, if it is to do so, must 'the circumstances' be?

Basically, the causal circumstances we need are dispositions of objects and fields. To see why, consider first a deterministic disposition like solubility (in water). The fact that an object *a* is soluble is what makes putting it in water cause it to dissolve. It does so because while *a*'s chance of dissolving when not in water is 0, its being soluble means (roughly) that its chance of dissolving if it were put in water would be 1.

But this definition of solubility is not quite right. Suppose *a* is not only soluble but valuable, so that we take care to keep it dry. We would only put it in water if it were *in*soluble, so that in fact if *a* were put in water, its chance of dissolving would be 0 — which contradicts our definition of solubility.

Fortunately, however, this true subjunctive is easily falsified by adding to its antecedent the condition that *a* be soluble. If *a* were put in water while it was soluble, its chance of dissolving would be 1, not 0. So the right definition of solubility (*D*) is of a property *S* such that *a*'s chance of dissolving if put in water would be 1 if and only if it were then *S*. But then *a*'s being *S* is the very circumstance in which our condition on causation applies to this case.

(*D*) again may look viciously circular, but it really isn't. (If it were a quantified material conditional, instead of a subjunctive, it would be what Carnap (1936, sections 5—6) called a reduction sentence.) But nor is (*D*) an eliminative definition, since it refers to *S*. We must still say what *S* is; presumably some arrangement of *a*'s molecules that makes them easily separated by water molecules. But that arrangement will no doubt depend on *a*'s chemical composition, so that *S* will vary from one chemical substance to another. So to say that *a* is soluble is not to ascribe a specific property to it, but to make an existential claim:

a has *some* property *S* such that if and only if *a* were put in water while being *S* would its chance of dissolving be 1. But then whatever *S* is, it will supply all 'the circumstances' we need, since by satisfying (*D*) it will automatically comprise all the other causes of *a*'s dissolving: e.g. the chemical composition that makes *a*'s molecular structure satisfy (*D*).

And as for deterministic causes, so for indeterministic ones. The relevant circumstances are tacitly defined by the matrix of chances given in section 2. Suppose *E* is my recovery from an illness, *C* my taking some medicine. The matrix does not just state what my chances of recovery would be were I to take, or not to take, the medicine. It tacitly credits me with metabolic properties such that, while I have them, my chances of recovery would be as it states. The matrix of course does not say what these properties are, any more than (*D*) says what *S* is. Their identity is an empirical matter, to be settled by medical research. But whatever the properties may be, my having them is the very circumstance in which my taking the medicine would raise the chance of my recovery and thereby cause it.

In short, causation needs dispositions to make true the subjunctives implied by the condition that causes raise their effects' chances. To identify causes and effects is to give a dispositional specification that some properties of the objects or fields involved must meet; and to identify the causal relation is to find the properties that meet that specification. These of course are tasks for science, and are clearly inter-dependent: finding the property that meets a dispositional specification shows how (and hence that) it is met; while failure to find it casts doubt on the disposition, and hence on the causation that entails it. Thus on the one hand, finding the nuclear structure that makes an atom which has it more likely to split when bombarded shows how bombarding such an atom can cause it to split. And on the other, until we identify the properties of cigarette smoke that make those who inhale it more likely to get cancer, the tobacco industry can still dispute the causal connection between smoking and cancer.

5. THE EFFICACY OF CAUSES

Causation must of course do more than raise its effects' chances, and not all its other features will follow as this one does from its giving ends means. One that does not is the fact that causes are contiguous to their immediate effects — or rather, given the denseness of time, that they are linked to their effects by dense 'ropes of causation', i.e. by the processes that Salmon (1984, Chapter 5) takes to be causation's basic entities. But if the means-end connotation does not explain the denseness of causation, nor does it deny it: the two conditions are perfectly compatible.

But the means-end connotation does explain other aspects of causation. In particular, it explains its temporal connotation. For if causes raise their effects' chances, causal loops, and hence simultaneous and backward causation, are impossible (Dummett 1964; Mellor 1981, Chapter 10). The condition that makes causes means when their effects are ends makes even the most underdetermined effects come later than their causes. The means-end connotation not only lets causes precede their effects: it makes them do so.

Another thing the means-end connotation does is give sense to the idea that causal efficacy comes by degrees — which shows incidentally that our condition (that causes raise their effects' chances) is too weak: to be a means to an end *E*, *C* must raise *E*'s chance more than an infinitesimal amount. The reason is that not all means are equally useful, and the more *C* raises *E*'s chance (from p' without *C* to p with it) the more useful it is. For when *C* costs something (i.e. lowers the utility of both *E* and $\sim E$), the greater $p - p'$ is, the less valuable *E* needs to be before the expected utility principle will justify incurring that cost. While as $p - p'$ gets less, the less use *C* becomes: i.e. the more valuable *E* must be to justify employing *C* at any cost, however small.

In short, the usefulness of means comes by degrees, of which $p - p'$ is the natural measure; and a means must be *some* use to be a means at all. So a means must raise its end's chances by more than some minimum amount. The amount will no doubt vary from case to case, and even then be hard to specify. But the amount doesn't matter, any more than it matters how hot something must be to be hot. A thing's temperature tells us all we need to know about how hot it is; and $p - p'$ similarly tells us all we need to know about how useful *C* is as a means to *E*.

Nor is causation's means-end connotation the only one that comes by degrees. The temporal connotation doesn't (temporal order doesn't come by degrees), but the others do. The evidence a cause *C* and its effect *E* provide for each other clearly increases the more *C* raises *E*'s chance. And so does the extent to which *C* explains *E*, for reasons I have given elsewhere (Mellor 1976).

Jeffrey (1969), Fetzer (1981, Chapter 5), Salmon (1984) and others dispute this. They deny that to explain *E*, *C* must raise its chance (or make it high), the higher the better. Their objections do not persuade me, but I shall not argue the point here. (Except to deny that everything, however improbable, must be explainable (*pace* Fetzer 1981, p. 134); and to press the 'pivotal principle' which Salmon needs 'the temerity to . . . reject' (1984, p. 113), that what explains *E* could not also explain $\sim E$: e.g. if my smoking explains my getting cancer, it could not also explain my not getting it. Causes that raise the chances of the effects they explain automatically satisfy this principle. Salmon's and Fetzer's causes need not.) For Salmon *et al.* agree that causes explain their effects by giving them chances. What they deny is not the explanatory connotation, merely the idea that it comes by degrees. And though I think it does, that is not essential to the idea that causation does. That idea gets sense enough from its means-end connotation. Its entailing degrees of evidential and explanatory strength is a real but dispensable bonus.

The efficacy, $p - p'$, of the causal relation between *C* and *E* is thus primarily a measure of *C*'s usefulness as a means of bringing *E* about; but also of how much evidence *C* and *E* provide for each other, and — I maintain — of how far *C* explains *E*. So the most effective causes are those that provide the most useful means, the best evidence and the best explanations for their effects. But those are the traditional deterministic causes, which raise their effects chances' from 0 to 1 — and this is no doubt why causes have traditionally been required to determine their effects. Now we know they need not, because though indeterminism progressively weakens causation's connotations, it does not destroy them all at once. But it does destroy them in the end, when causes no longer raise their effects' chances. And because determinism gives causation all its connotations in their highest degree, deterministic causation is still the best. So I confess to having the 'lingering desire for Laplacian determinism, or if worse comes to worst, as close an approximation thereto as possible' of which Salmon (1984, p. 113) accuses me. And so will anyone who understands the implications that make causation worth ascribing in the first place.

Cambridge University

REFERENCES

R. Carnap (1936), Testability and Meaning. In H. Feigl and M. Brodbeck, editors (1953), *Readings in the Philosophy of Science*, New York, pp. 47—92.

M. Dummett (1964), Bringing About the Past. In *Truth and Other Enigmas* (1978), London, pp. 333—50.

C. G. Hempel (1965), Aspects of Scientific Explanation. In *Aspects of Scientific Explanation*, New York, pp. 331—496.

D. Hume (1977), An Enquiry Concerning Human Understanding. In *Enquiries Concerning the Human Understanding and Concerning the Principles of Morals*, edited by L. A. Selby-Bigge (1902), Oxford, pp. 5—165.

J. H. Fetzer (1981), *Scientific Knowledge: Causation, Explanation and Corroboration*, Dordrecht: D. Reidel.

R. C. Jeffrey (1969), Statistical Explanation vs. Statistical Inference. In N. Rescher, editor, *Essays in Honor of Carl G. Hempel*, Dordrecht: D. Reidel pp. 104—113.

R. C. Jeffrey (1983), *The Logic of Decision* (2nd edition), Chicago: University of Chicago Press.

R. D. Luce and H. Raiffa (1957), *Games and Decisions*, New York: John Wiley and Sons.

D. H. Mellor (1976), Probable Explanation, *Australasian Journal of Philosophy* **54**, 231—41.

D. H. Mellor (1981), *Real Time*, Cambridge: Cambridge University Press.

D. H. Mellor (1982), Chance and Degrees of Belief. In R. McLaughlin, editor, *What? Where?, When? Why?*, Dordrecht: D. Reidel, pp. 49—68.

D. H. Mellor (1983), Objective Decision Making, *Social Theory and Practice* **9**, 239—309.

D. H. Mellor (1987), Fixed Past, Unfixed Future. In B. Taylor, editor, *Contributions to Philosophy: Michael Dummett*, Dordrecht: Nijhoff, pp. 166—86.

D. H. Mellor (1987a) Indeterministic Causation, *African Philosophical Inquiry* (forthcoming).

W. C. Salmon (1979), Propensities: A Discussion-Review, *Erkenntnis* **14**, 183—216.

W. C. Salmon (1984), *Scientific Explanation and the Causal Structure of the World*, Princeton: Princeton University Press.

RICHARD JEFFREY

HOW TO PROBABILIZE A NEWCOMB PROBLEM

1. PROBABILIZING

"Dogmatism" (in a non-invidious sense) is an ancient term for the view that judgment is and ought to be a matter of assertion and denial, of belief (Greek, "dogma") and disbelief. Probabilism is a relatively recent view, urged in the mid-17th century, that judgment isn't or shouldn't generally be a matter of believing but of — what to call it? — probabilizing. Will it rain today? I don't know; I'm not in a position to assert or deny it. But still I may have an action-guiding probabilistic judgment, e.g., my odds on rain might be 7 : 3, my probability for rain might be 70%.

100% probability corresponds to infinite odds, 100 : 0. That's a reason for thinking in terms of odds, i.e., to remember how momentous it may be to assign a hypothesis probability 1.[1] It means you'd stake your all on its truth, if it's the sort of hypothesis you can stake things on. To assign probability 1 to rain is to think it advantageous to bet your life on it in exchange for any petty benefit. We forget that because of the ambient dogmatism in which we ordinarily speak and think. We imagine that we'd assign probability 1 to whatever we'd simply state as true, but of course it's not so. If you ask me the time, I'll glance at my watch and tell you, even though I don't utterly rule out the possibility that I've misread it, or that the thing has gone haywire. When I say "Ten past two" I'm dogmatizing, not probabilizing. (The corresponding probabilistic judgment might be a density with 90% of its mass concentrated within 5 minutes of ten past two — not that there's any uniform way of giving probabilistic equivalents of dogmatic judgments.)

Experience with a variety of probability distributions can help you find a distribution that adequately represents your current state of mind. But often you're dissatisfied with that state of mind, and think it would be pointless to work at probabilizing it. Maybe option A_i would be distinctly preferable to the others iff you thought C_i far the most probable cell of a partition $\{C_1, \ldots, C_n\}$, but you doubt that any cell would stand out in that way if you were to probabilize your current

James H. Fetzer (ed.) Probability and Causality. 241—251.

thought. Then you may do well to seek new experience that's likely to put you in a state of mind which, probabilized, would clearly settle the matter for you. Effort spent in probabilizing your current state of mind is better spent on the forthcoming one.

This is not Carnap's (1950) vision of a Laplacean intelligence that integrates successive experiences by conditioning on appropriate observation sentences. I take it that only a few of the relevant features of our experience are usefully characterized by sentences, and that when they are, the sentences need not have probabilities that are as close to 1 as makes no odds. Nor is it Carnap's probabilistic modification of that vision (see Jeffrey 1975), or Field's (1978) revision, which sees experience as setting input parameters that serve to update probabilities. Rather, probabilizing is here seen as an art that calls on subject matter dependent skills that need learning and polishing. The artifact is a probability distribution or a partial characterization of one as meeting certain constraints. "How to do it" books can help even if they don't enable the novice to do it simply by following directions.

A fine point that might be treated in such a book after the basics have been attended to is the following version of the prisoners' dilemma, construed as a Newcomb problem (Lewis, 1979).

2. THE SIMILAR PRISONERS' DILEMMA

Recall the story. If one prisoner confesses and the other doesn't, the confessor goes free and the other serves a 10 year prison sentence. If both confess, each serves 9 years. If neither does, each serves only 1 year. Now here's the twist.

The prisoners (Alma and the Count, say) aren't all that different; her decision, A or $-A$ (Alma confesses or not), wouldn't be totally useless as a predictor of his, C or $-C$ (the Count confesses or not). Alma sees their similarity as like the similarity of two urns of similar but unknown composition: each is filled with tickets marked "do" or "don't", and Alma thinks that the unknown fractions of tickets marked "do" in her urn and the Count's are nearly the same. In fact, let's suppose, she thinks they're exactly the same. Then in effect, Alma thinks that she and the Count decide by drawing (with replacement) from the same urn. And she thinks that the Count thinks that, too.

There's nothing bizarre in this view of the matter. The composition of the notional urn is meant to represent their common final chance of confessing, as determined by their common tough attitudes and similar ways of thinking about their common predicament, i.e., egotistical attention to one's own narrowly conceived interest.

What is Alma's judgmental probability distribution for the unknown fraction p of tickets marked "do" among all the tickets in the urn? Perhaps initially it is the uniform distribution over the unit interval. On this assumption Alma's probabilities for the hypotheses A and AC are[2]

$$P_0(A) = \int_0^1 p\,dp = 1/2$$

$$P_0(AC) = \int_0^1 p^2\,dp = 1/3$$

whence can be calculated her conditional probabilities

$$P_0(C \mid A) = P_0(-C \mid -A) = 2/3$$

and her unconditional probabilities for the four conjunctions of her options with the Count's,

$$P_0(AC) = P_0(-A-C) = 1/3, \qquad P_0(A-C) = P_0(C-A) = 1/6,$$

as in Table I(a).

TABLE I

(a) Probabilities P_0

	C	$-C$	
A	1/3	1/6	1/2
$-A$	1/6	1/3	1/2

(b) Desirabilities D_0

	C	$-C$	
A	1	10	4
$-A$	0	9	6

3. AN UNSATISFACTORY EVIDENTIARY SOLUTION

If we suppose that Alma's desirabilities for the 4 conjunctions $A-C$, $-A-C$, AC, $-AC$ are 10, 9, 1, 0 as in Table I(b) (i.e., 10 minus the number of years she serves) then her current desirabilities for A and

$-A$ are 4 and 6: she now prefers confessing. And so does he, for his attitudes are just like hers (i.e., after "A" and "C" are interchanged throughout Table I). The term "desirability" is used as in Jeffrey (1983), so that $D_0(A)$ and $D_0(-A)$ are Alma's expectations of utility conditionally on the hypotheses A and $-A$ that she does and doesn't confess:

$$D_0(A) = [1P_0(AC) + 10P_0(A-C)]/P_0(A) = 4$$
$$D_0(-A) = [0P_0(-AC) + 9P_0(-A-C)]/P_0(-A) = 6$$

Then if choiceworthiness goes by desirability as in evidentiary decision theory (Jeffrey, 1965), it would seem that Alma had better not confess.

But despite her preference for A's falsity, Alma should make A true if she thinks that the significance of her choice for the Count's is purely diagnostic: if she thinks that her choice has no influence on his. Why? Because she sees that on either hypothesis about his choice she's better off by 1 unit of desirability if she makes A true than if she makes A false. And she thinks that the Count has the same view (with rôles reversed), and that neither sees the taking of this unit advantage as having any adverse effect on the other's choice. Then both expect to confess; both expect an outcome in the upper left-hand corner of Table I(b), i.e., the nearly worst outcome, even though both prefer the nearly best outcome in the lower right-hand corner, which they'd have if neither confessed: both prefer truth of $-A-C$ to truth of AC. But alas! Making $-A-C$ true is an option for neither, even though making $-A$ true is an option for one, and making $-C$ true is an option for the other.

That's their dilemma, so-called: seeing what's better for both, they wisely choose what's worse. (Of course, "they" names no agent, wise or foolish. That's why, if each real agent chooses wisely, the result will be foreseeably worse than what the opposite choices would have yielded.)

The choices are wise, given the agents' appreciation of their interests. It's a case where it doesn't pay to be narrowly self-seeking. (That does happen.) But although each can see that their shared selfishness is self-defeating, that realization is of no use to them in the present problem, for their utility functions represent character-based dispositions that can't be trimmed to individual decision problems. Indeed, it's her appreciation of their common character that determines Alma's conditional probabilities, and, so, her preference for remaining silent even as she ruefully ratifies her foreseen decision to confess, and forsees the Count's decision, and her 9 years in jail.

4. CHOICEWORTHINESS AS FINAL DESIRABILITY

Does this mean that choiceworthiness doesn't necessarily go by desirability? That's not yet clear, for the probability matrix of Table I(a) shows Alma far from decision; her odds $P_0(A):P_0(-A)$ between confession and silence are even. The question "Does choiceworthiness go by desirability?" must concern final desirability, just before decision. If she does finally decide to confess then the final desirabilities of A and of $-A$ will be computed from the desirabilities in Table I(b) via some successor of Table I(a) that assigns A a probability near 1.

What successor? Given that the row sums are to be near 1 and 0, that question is answered approximately once we know the ratios of entries in each row, or, equivalently, once we know Alma's final conditional probabilities for the Count's choosing just as she does. Initially those conditional probabilities were both 2/3. If they are still 2/3 when she chooses, then choiceworthiness splits cleanly away from desirability.

But why would Alma keep those conditional probabilities through her deliberation? It's thing she'd do if she used her conditional expected utilities of options at the start of deliberation as measures of their choiceworthiness. But if she did that she'd remain silent, or, anyway, she'd think it foolish to confess, and by hypothesis, that's not so. Then we must look closer to see choiceworthiness and desirability part company.

The clearest cases where choiceworthiness dependably goes by initial desirability are those where probabilities conditionally on options remain fixed throughout deliberation, as Alma's probabilities for A and $-A$ move toward the extremes. For contrast, let's now see how those conditional probabilities vary in the present example when Alma moves from her initial distribution P_0 in Table I(a), derived from a uniform judgmental distribution for the chance p that she'll confess (= the chance that the Count will confess), to distributions P_n that push more and more of her judgmental probability onto A as her judgmental distribution for p squeezes right in the unit interval, toward 1.

In particular, suppose that Alma's judgment about p at the end of her deliberation will be characterized by a normalized density function f_n proportional to p^n, for some sizeable n:

$$f_n(p) = (n+1)p^n$$

We have supposed that her initial state has that form with $n = 0$, i.e., the flat distribution $f_0(p) = p^0$, which is identically 1.[3]

In general, her final density f_n for p determines Alma's final probabilities for her confessing and for the Count's confessing as her f_n-expectation of the random variable p

$$P_n(A) = P_n(C) = \int_0^1 pf_n(p)dp = (n+1)/(n+2)$$

and similarly determines her final probability for their both confessing as

$$P_n(AC) = \int_0^1 p^2 f_n(p)dp = (n+1)/(n+3)$$

TABLE II

(a) Chances

	C	$-C$
A	p^2	$p(1-p)$
$-A$	$(1-p)p$	$(1-p)^2$

(b) Unconditional Probabilities P_n

	C	$-C$
A	$(n+1)/(n+3)$	$(n+1)/(n+2)(n+3)$
$-A$	$(n+1)/(n+2)(n+3)$	$2/(n+2)(n+3)$

By the elementary probability calculus her other unconditional final probabilities, for $A-C$, $-AC$ and $-A-C$, are as in Table II(b), and her conditional final probabilities for the Count's acts given hers are

$$P_n(C \mid A) = (n+2)/(n+3), \qquad P_n(-C \mid A) = 1/(n+3)$$
$$P_n(C \mid -\mathrm{A}) = (n+1)/(n+3), \qquad P_n(-C \mid -A) = 2/(n+3)$$

Thus f_n determines Alma's final desirabilities for her options as

$$D_n(A) = P_n(C \mid A) + 10P_n(-C \mid A) = (n+12)/(n+3)$$
$$D_n(-A) = 0 + 9P_n(-C \mid -A) = 18/(n+3)$$

The conceit is that in this particular context Alma's deciding to make A true is a matter of her entering into a probabilistic state corresponding to f_n for some large n – say, $n = 97$, so that she's about 99% sure that she'll confess. (She allows for the possibility that something may prevent her carrying out her decision.) Thus $P_n(A) = 98/99$, $P_n(C|A) = 99\%$, $P_n(-C|-A) = 2\%$, and, after all, confession is the option she prefers: at $n = 97$, $D_n(A) = 1.12$ while $D_n(-A) = 0.18$. Note that

$$D_n(A) - D_n(-A) = (n-6)/(n+3),$$

which is positive for n greater than 6, i.e., for $P_n(A) > 87.5\%$, and approaches 1 as a limit as n increases without bound.

What if Alma chose not to confess, e.g., ending her deliberation in a judgmental state corresponding to a density

$$g_n(p) = (n+1)(1-p)^n$$

for some sizeable n? The density g_n would determine probability and desirability functions Q and E in place of the functions P and D that f_n determined, but the rejected act of confessing would still have the larger desirability when Alma's probability of confessing is greater than 87.5%. Proof: integration yields

$$Q_n(-A) = Q_n(-C) = (n+1)/(n+2),$$
$$Q_n(-A-C) = (n+1)/(n+3),$$

so that by the elementary probability calculus the unconditional Q_n probabilities are as in Table II(b) but with entries for AC and $-A-C$ exchanged; conditional probabilities are then

$$Q_n(C|A) = 2/(n+3), \qquad Q_n(-C|A) = (n+1)/(n+3),$$
$$Q_n(C|-A) = 1/(n+3), \qquad Q_n(-C|-A) = (n+2)/(n+3);$$

and final desirabilities of Alma's confessing and not would be

$$E_n(A) = 2/(n+3) + 10(n+1)/(n+3) = (10n+12)/(n+3)$$
$$E_n(-A) = 0 + 9(n+2)/(n+3) = (9n+18)/(n+3)$$

respectively. Then for $n = 97$, when her probability for confessing would only be $Q_n(A) = 1/99$, her desirabilities for truth and falsity of A would be 9.82 and 8.91, respectively.

Alma's final strength of preference for truth over falsity of A is the same, whether desirability is measured by D or by E:[4]

$$D_7(A) - D_7(-A) = E_7(A) - E_7(-A) = (n-6)/(n+3)$$

Then if Alma's final density for p is g_n with $n > 6$, so that $Q_n(A)$ is near 0 and she expects not to confess, she expects to act against her final preferences as indicated by her final desirabilities. In the light of her final probabilities, making A false is not a *preferential* choice for her.

5. CONCLUSION

Alma's is a "Newcomb" problem because she regards her possible acts as mere signs of conditions that she would promote or prevent if she could. Mere signs: she thinks that if the truth about chances were known, her acts and the Count's would be seen as independent, e.g., the probability that they'd both confess would be seen as the product of their separate probabilities of confessing. As we have modelled the problem, those separate probabilities are both p, i.e., a random variable for which various possible densities were canvassed, all of the beta form, proportional to $p^a(1-p)^b$, with $a = 0$ or $b = 0$, and $a + b = n$. When $b = 0$ we wrote the density as $f_n(p) = (n+1)p^n$, and when $a = 0$ we wrote it as $g_n(p) = (n+1)(l-p)^n$.

Newcomb problems require two-level probability models. At the lower level of Alma's problem is the unknown chance p that she will confess. At the upper level are her possible judgmental probabilities (P_n, Q_n) for A, AC, etc., which are determined by her possible judgmental densities (f_n, g_n) for p.

The two layers are needed because it's basically p that she needs to make up her mind about. An acceptable "final" density for p will determine her preferential choice, whether to confess or remain silent, in such a way that her probability for A is near 0 (choice of silence) or 1 (choice of confession) depending on whether her final desirability for A is less or greater than that for $-A$.

For small n (i.e., $n < 6$) both of those densities are flat enough to make Alma see a strong positive correlation between A and C, i.e., a correlation strong enough to make confessing the less desirable act. But with such n, $P_n(A)$ and $Q_n(A)$ are further than 12.5% from the extremes, 0 and 100%, so that unless Alma has strong doubts about her ability to carry out her decision it is implausible to think that she has yet chosen.

For $n > 6$ both of those densities are extreme enough to make A

and C look nearly enough independent so that confessing is the more desirable act; and $P_n(A)$ and $Q_n(A)$ can be extreme enough to make it plausible that she has chosen (e.g., within 1% of the extremes for n = 97). But as she and we can see, the choice will be preferential only if it is a choice to confess.

Note that if Alma doesn't confess, her not confessing needn't be an act of hers; indeed, it may be someone else's doing, as, e.g., if she is killed to secure her silence. Nor is it clear that if she confesses, that must be an act of hers, as, e.g., if she confesses while unconscious under the influence of sodium pentothal.

Can Alma really be said to *choose* silence when she remains silent while regarding confession as preferable? We have seen that to be a technical possibility; but perhaps Socrates was right in holding that one can't choose what one sees as the worse option. If he was wrong, choice can be counterpreferential; if he was right, then *a fortiori* silence cannot be a preferential choice for Alma.

In either case, the evidentiary decision theory of the first (1965) edition of *The Logic of Decision* seems to be satisfactory when the Newcomb problem is probabilized on two levels as here — once we see that it's final desirabilities (final preferences) that must rule.[5]

In chapter 1 of that book I supposed that Alma's points of view as they would be if she chose to make A or $-A$ true could be approximated by conditioning her initial probability function P_0 on A or $-A$. To make the approximation more realistic one need only replace conditioning on A or $-A$ by a less extreme change of probabilities from $P_0(A)$ and $P_0(-A)$ to values closer to the extremes, and use the kinematical scheme floated in chapter 11 of the book to extend those changes so as to get new probabilities for AC, $A - C$, etc. So I thought. Now I know better.

Indeed that device works in ordinary decision problems, where the change in probabilities of A and $-A$ is not accompanied by any change in conditional probabilities given A or $-A$. But Newcomb problems are not ordinary in that way. In Newcomb problems the initial judgmental relevance of acts (A, $-A$) to conditions (C, $-C$) derives from uncertainty about chances — an uncertainty that is effectively nullified by choice.

In Newcomb problems choice is represented by an extreme distribution of some random variable, which determines final probabilities near 0 and 1 for all options. Conditionally on these options, hypotheses may have different initial and final probabilities, e.g., if Alma expects to confess,

$$P_0(C \mid A) = 2/3, \qquad P_7(C \mid A) = 9/10,$$
$$P_0(C \mid -A) = 1/3, \qquad P_7(C \mid -A) = 8/10,$$

and initial relevance may be greater than final relevance:

$$P_0(C \mid A) - P_0(C \mid -A) = 1/3, \qquad P_7(C \mid A) - P_7(C \mid -A) = 1/10.$$

If Alma expects not to confess, some the corresponding figures are different; Q_0 is the same function as P_0, but

$$Q_7(C \mid A) = 2/10, \qquad Q_7(C \mid -A) = 1/10.$$

Yet the relevances are the same; $Q_7(C \mid A) - Q_7(C \mid -A) = 1/10$.

Alma can see all that in advance; and she can see in advance that although silence is initially preferable, confessing is preferable if she's (say) 90% sure of what she'll do, either way:

$$P_7(A) = Q_7(-A) = 9/10,$$
$$D_7(A) - D_7(-A) = E_7(A) - E_7(-A) = 1/10$$

So she can see that confessing is her only preferential choice.

In an adequate probability model of the prisoners' dilemma "for clones", choiceworthiness does go by conditional expected utility, i.e., relative to the agent's final judgmental state, no matter whether that state represents a preferential choice to confess or a perverse, counter-preferential choice to remain silent.

Princeton University

NOTES

[1] By using log odds instead of odds we could treat the lower extreme of the probability scale symmetrically with the upper end: as p ranges from 0 to 1/2 to 1 and $p/(1-p)$ ranges from 0 to 1 to $+\infty$, $\log[p/(1-p)]$ ranges from $-\infty$ to 0 to $+\infty$.

[2] This P_0 is Carnap's (1950, appendix) function m^* for a language with two individuals (Alma, the Count) and one primitive property (confesses).

[3] Formally, f_n is the posterior value of Alma's f_0 density conditionally on n of her clones having confessed, in similar prisoners' dilemmas: $f_n(p) = f_0(p \mid A_1 \ldots A_n)$. Therefore $P_n(H) = P_0(H \mid A_1 \ldots A_n)$, where P_0 is c^* for a language with individuals Alma, the Count, and no end of virtual individuals Alma_i and Count_i $(i = 1, 2, \ldots)$.

[4] More generally, with 9 and 10 in Table I(b) replaced by d and $d + i$ $(d > 1, i >$

0)), the final desirability of A exceeds that of $-A$ iff $n > (d - i - 2)/i$, no matter whether final desirabilities are determined by D_n or by E_n.

[5] The present treatment has the virtues of the flawed heroic cure, "ratificationism", that was floated in sec. 1.7 of the second edition (Jeffrey, 1983), but escapes the difficulties over positively correlated failures of Alma and the Count to carry out their decisions that were noted there (i.e., van Fraassen's counterexample). This is not to say that the present treatment is superior to that of "causal" decision theory of Gibbard and Harper, Lewis, Skyrms, Sober, and others. (For references see Jeffrey, 1983 sec. 1.8.)

REFERENCES

Carnap, Rudolf. (1950, 1962) *Logical Foundations of Probability*. University of Chicago Press.

Field, Hartry. (1978) 'A Note on Jeffrey Conditionalization.' *Philosophy of Science* **45**, 361—367.

Jeffrey, Richard. (1975) 'Carnap's Empiricism.' *Induction, Probability, and Confirmation*, ed. Grover Maxwell and Robert M. Anderson, Jr. University of Minnesota Press.

Jeffrey, Richard. (1965) *The Logic of Decision*. McGraw-Hill. Second edition, revised: University of Chicago Press, 1983.

Lewis, David. (1979) 'Prisoners' Dilemma is a Newcomb Problem.' *Philosophy and Public Affairs* **8**, 235—240.

PAUL HUMPHREYS

NON-NIETZSCHEAN DECISION MAKING*

INTRODUCTION

Two ways have been suggested to protect evidential decision theory[1] from the kinds of counterexamples which have given rise to causal decision theories.[2] One, which I shall call the 'Cartesian defence', relies on using the agent's privileged access to his own reasons for acting in order to render the counterexamples innocuous, and has been presented in a number of papers by Ellery Eells.[3] The other, known as the 'ratifiability defence', has been described by Richard Jeffrey,[4] and consists in ratifying that a preferred act will be preferred even when that act is actually chosen — i.e. having decided that an act is optimal, there is no other act which would then be preferred, conditional on that decision having been made. Both of these defences require that restrictions be placed on the set of situations within which it is appropriate to use decision theory. Ingenious and inventive as these defences are, the required restrictions are, I believe, unjustifiable as normative criteria for rationality. Furthermore, if implemented, they would so restrict the application of decision theory that its interest as a guide to life would be almost completely erased.

THE PROBLEM

Potential conflicts between the recommendations of dominance principles and the recommendations of evidential decision theory were first brought to notice by Newcomb's Problem.[5] A more credible although conjectural example involves the Fisher smoking hypothesis.[6] Here is another example, structurally similar to the smoking case, but with the advantage of being true.

You have some reason to believe that you may be suffering from Klinefelter's syndrome, but you are not sure. The Klinefelter syndrome is one in which a male has an extra X chromosome in an XXY configuration. Possession of the extra X chromosome is associated with an increased chance of socially maladjusted behaviour, including crimi-

James H. Fetzer (ed.) Probability and Causality. 253–268.

nal activity, and it is also associated with a twenty-fold increase in the chance of male breast cancer later in life. At the moment, as many of us do from time to time, you are considering whether to engage in some mildly anti-social activity which will result in monetary gain with little chance of being caught (perhaps you are considering becoming a lawyer). You realize that performing the anti-social act will be a sign of having the extra chromosome, and hence, of an increased risk of cancer later in life. You also know all the above information, have taken a course in decision theory, and are able to deliberate about many things on which Klinefelter's syndrome has no influence. One last thing. Like any sensible person, you know that deciding to perform the anti-social act will not guarantee that you will actually do it, because, for example, you might suffer a loss of nerve. But your theory of how Klinefelter's syndrome works is this: in addition to the influence which your decisions have on your acts, the syndrome tends to produce anti-social acts directly. That is, when it comes to misdemeanours, Klinefelter victims are simply more likely to do them, all other things being equal, than are men with the normal XY chromosome structure. What should you do? And is it even rational to deliberate in this situation?

I say that it is, and that you should go ahead and do it. Both the Cartesian and ratifiability defences say that such cases lie outside the domain of decision theory, and give no guidance as to what to do.

THE ISSUE

The situation just described is similar to the Fisher smoking case, with one important difference. The influence of the syndrome contributes directly to the performance of the act, independently of the decision to do it. The causal structure of the situation is thus:

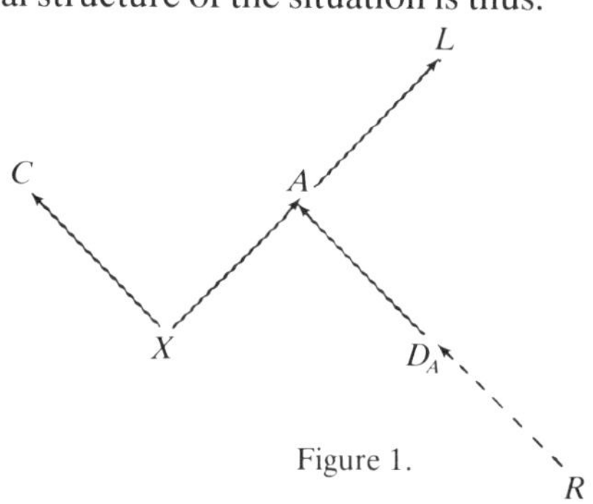

Figure 1.

It is this feature which renders ineffective the two methods which have been suggested to rescue evidential decision theory from Newcomb-style counterexamples.

Suppose that we were to analyze the decision process in the original evidential way, with one stipulation: the agent believes that all the connections in the structure are probabilistic in form. Then this situation can be represented by the following desirability and probability matrices:

TABLE I

	X	$\overline{X}$
$\overline{CRIME}$	0	b
$CRIME$	a	$a+b$

Desirability Matrix

$P(X/A) = p$; $P(C/X) = q$; $P(L/A) = s$.
$P(X/\overline{A}) = p'$; $P(C/\overline{X}) = q'$; $P(L/\overline{A}) = s'$.
Use Jeffrey Mixing Rule: $V(A) = \Sigma_{i=1}^{2} P(S_i/A)\, \Sigma_{j=1}^{4} P(O_j/S_i \,\&\, A) V(O_j \,\&S_i \,\&\, A)$
where
$[S_i]_{i=1}^{2} = [X, \overline{X}]$ $\quad [O_j]_{j=1}^{4} = [L \,\&\, C, \overline{L} \,\&\, C, L \,\&\, \overline{C}, \overline{L} \,\&\, \overline{C}]$
and V is the evidential (expected) utility, EU.
Here X = presence of extra X chromosome; A = committing crime; C = getting breast cancer; L = reward of loot from crime.

It is then easy to show that:

$$EU(\text{No Crime}) - EU(\text{Crime}) > 0 \text{ iff } (p - p') > a(s - s')/b(q - q').$$

Because committing the crime is evidentially relevant to possession of the extra chromosome i.e. $(p - p') > 0$, then assuming that the act is positively relevant to getting the loot and that the syndrome is positively relevant to getting cancer, if the desirability of not getting cancer outweighs the value of the loot sufficiently, which it presumably does, then evidential theory entails that the agent should refrain from committing the crime. But the agent already has the extra chromosome, or he does not, and nothing the agent can do can affect this state of affairs.

So the agent might as well go ahead and commit the crime, as in either chromosomal state he will certainly gain the incremental value of the loot. This, of course, is what dominance suggests, and so we have the standard conflict between the straightforward Bayesian recommendation and dominance in this common cause case.

The problem here is clear enough and widely acknowledged. The subjective probabilities capture the fact that the symptomatic act is evidence for ('news of') the undesirable outcome,[7] but they do not capture the lack of causal influence of the first on the second. Because a central advantage claimed for Jeffrey's theory is that there is no requirement that acts be independent of states (thus avoiding the need for choosing special partitions for states or outcomes) advocates of this theory are reluctant to give it up. Hence the need for the defences.

THE CARTESIAN DEFENCE

This kind of defence has been presented in a number of papers by Ellery Eells (see Chapters Six and Seven of his (1982), also his (1981), (1984a), (1984b), (1985), (1986)). The argument in its full form is sophisticated, detailed, and carefully argued. However, the skeletal form which I shall give here contains all of its basic features, and none of the aspects which I have omitted affect the points which will be made later. The first assumption is one concerning the specific causal form of Newcomb-type examples:

Assumption 1. "The way in which a common cause causes a rational person to perform a symptomatic act is by causing him to have such beliefs and desires that a rational evaluation of the available acts in light of these beliefs and desires leads to the conclusion that the symptomatic act is the best act. And I shall assume that our agent believes this hypothesis about how the common cause causes the symptomatic act." (Eells (1982), p. 152).

If we let R constitute the agent's beliefs and desires, which we shall identify with his reasons for performing an act, and let D_A be the decision to perform an act A, then the causal picture which emerges from Assumption 1 is, in the Klinefelter case:

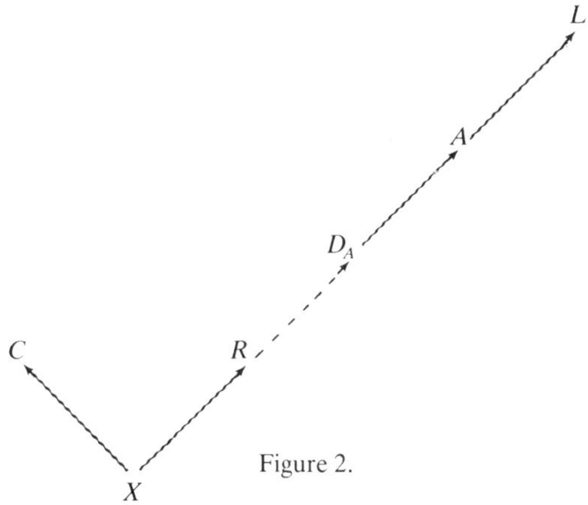

Figure 2.

Assumption 2. For the agent, $P(R) = 1$. That is, he has full knowledge of his own reasons for acting.

Assumption 3. The agent is such that $P(A \text{ iff } D_A) = 1$. That is, the agent performs an act just in case he decides that the act is rationally optimal for him.

Assumption 4. The causal chains in the model are Markov.
Conclusion. For the agent, $P(X/A) = P(X/-A)$.

Once the evidential connection between the common cause and the act has been broken, then there will be no evidential relation between the act and the undesirable outcome, and so evidential decision theory will not be open to this kind of counterexample. The argument for the conclusion is quite straightforward, given the assumptions. From A1 together with A4, we have that $P(D/R \,\&\, X) = P(D/R \,\&\, -X)$. Because the reasons produce the decision directly, given those reasons, it is irrelevant how they arose, whether from the chromosomal factor or its absence. Then, using A2 and a principle of eliminability of certain knowledge, we have $P(D/X) = P(D/-X)$. Using A3 and a substitution principle for belief equivalents, we have $P(A/X) = P(A/-X)$. Finally, using the symmetry of stochastic independence we have $P(X/A) = P(X/-A)$.

THE RATIFIABILITY DEFENCE

In regular decision theory, probabilities are conditioned on the actual performance of the act, rather than on the decision to do it. Richard Jeffrey's ratifiability approach to Newcomb problems requires that the probabilities be conditioned on both. We then have the definition: An action A is ratifiable just in case $EU(A/D_A) > EU(B/D_A)$ for every option B. Rational acts are ratifiable acts.

The idea here is that one should not, having chosen (although not yet performed) an act as optimal, immediately recognize that, given that choice, the act was not optimal. That is, optimal acts should be optimal after one has chosen them. Jeffrey's primary concern is with Prisoner's Dilemma situations, and here the belief is that once you have made your decision, what the other actually does should be irrelevant, because the similarity between you and him rests on rationality grounds and not performance grounds. In this case, your decision screens off your act from his act and since it is his act being evidentially relevant to your act which produces problems for evidential theory, screening it off alleviates that problem. Consider a case of Prisoner's Dilemma which has the desirability matrix:[8]

TABLE II

		Agent B		
		CONFESS	$\overline{\textit{CONFESS}}$	
Agent A	$\overline{\textit{CONFESS}}$	0	b	$b > a$
	CONFESS	a	$a + b$	

Desirabilities (Standard Prisoner's Dilemma)

Elementary calculations show that EU(not confessing) $-$ EU(confessing) > 0 iff P(other confesses/agent confesses) $-$ P(other confesses/agent does not confess) $> a/b$, and hence evidential theory (without ratifiability) directs the agent not to confess when the relevance difference is large enough (a/b can of course, be quite small). If we assume that the assumption that agent B is similar to agent A entails that (at least) my confessing would be evidence for his confessing, then evidential theory without either ratifiability or a Cartesian defence gives what seems to most to be incorrect advice.

However, if we condition the probabilities on the decision to do *A*, as well as on *A* itself, we have: *P*(*B* confesses/*A* decides to confess & *A* confesses) = *P*(*B* confesses/*A* decides to confess & *A* doesn't confess). Then the problematical evidential relations are removed, and ratifiability tells you to pick the dominant act, i.e. to confess.

WHAT IS WRONG WITH THE DEFENCES

Both defences of evidential theory just described require a restriction to be placed on the domain of rational decision making. In particular, they both insist that to the extent that factors outside the agent's control influence the act directly, to that extent the situation is outside the realm of rational decision making. Illustrative quotes here are: "If he takes the performance itself to be directly promoted by the presence of [the common cause], there is no question of preferential choice: the performance is compulsive." (Jeffrey, (1983), p. 25). "Although causal decision theory may give correct answers even if the decision maker does believe that the correlation is, in part, enforced by a factor that sometimes causes the irrational act, this poses little threat to evidential decision theory. In this kind of case, the causal theory fares better than the evidential theory to the extent that the decision situation is not one in which the agent should find it appropriate to apply standards of rational decision in the first place" (Eells and Sober (1986), p. 241) Because in the example with which we began, the performance of the criminal act *is* directly promoted by the presence of the syndrome, and the X factor, when present, biases the action toward fulfillment of the decision to commit the crime, both the ratifiability and the Cartesian defences claim that there is something irrational about that situation. What are we to make of these assertions that such situations are "outside the realm of rational decision making"?

First, some clarification of what is at issue. There are three separate questions which could be asked of such situations: (1) In these problematical situations, is it rational even to deliberate? (2) If the deliberations take place, are they rational or irrational? (3) If the deliberations are rational, are the consequent acts rational or non-rational (i.e. irrational or arational) at least in part? The second question is the easiest to answer. We are dealing with situations of the form of Figure 1, and not of this form:

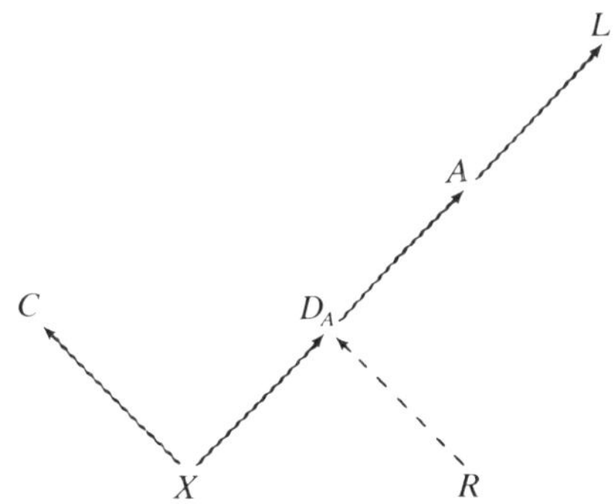

Figure 3.

where the non-deliberative influences act directly on the decision itself. Such situations clearly exist. For example, my decision to go to a party may be partly due to deliberation and partly due to the effects of alcohol. In those kinds of case, the decision process is not entirely rational. In the situation with which we are concerned, however, the ratiocinations themselves take place entirely independently of other influences. Hence they can be fully rational according to Bayesian standards. So the trouble does not lie with a negative answer to question two. With regard to question three, although there is a tendency to talk of acts as rational or irrational, it is the decisions or choices on which these acts are based which are rational, arational or irrational. The acts themselves may be optimal or non-optimal for an agent, and performed as a result of a process of rational deliberation or not, as the case may be. Saying that an "irrational act" was performed is ambiguous: it may mean that the act was performed without preliminaries counting as rational deliberation, or that an act was performed which was different from that chosen as optimal by a process of rational choice. This latter case may reduce to the former if, having deliberated, the results of that deliberation are discarded and no serious effort is made to carry out the chosen act. But where such conscientious efforts are made, it is improper to label the consequent act as irrational simply because for reasons (causes) outside the agent's control the optimal act turns out not to be performed. Indeed, to avoid being misled, we should classify acts as most preferable or not, optimal or not, and reserve rationality talk for decisions and decision procedures.

It is also worth noting that how an act is carried out is irrelevant to the issue of its rationality. That is, the process leading from a rational choice to an act does not ordinarily introduce non-rationality into the act. For a physically helpless agent could act out his decisions through purely mechanical means, or, by the use of a mad but totally dutiful servant, and the acts would still be rational if the decisions were. It is why the act is performed that counts, and here we have the real issue. Note first that we do not lose the ability to explain an individual's actions when there are multiple causal influences on an act, as there are in Figure 1. For as has been amply demonstrated (see e.g. Humphreys (1981), (1983), (1987), Salmon (1984)) satisfactory explanations in terms of multiple probabilistic causal influences are clearly possible. (I assume here that decisions can be causes of acts. Bayesians seem to allow this — if not, everything that follows should be construed within a causally dispositional account of mental states, as, for example, detailed in Stalnaker (1984).) Is then, the act partly non-rational because its explanation consists in part of factors having non-rational origins? In order to answer this question, we must separate the use of decision theory as an explanatory theory of the actions of rational agents from its use as a guide to rational choice. We can grant that, to the extent that an explanation of an agent's act involves influences other than the agent's rational decision alone, to that extent the explanation involves more than just decision theory. But from this it does not follow that for the agent involved in making an informed choice that this decision situation falls outside the realm of rational decision theory. For what the agent who needs to make a decision has to know is this: should I or should I not employ the apparatus of rational decision theory in deciding what the optimal act for me would be? The answer to this question seems to me to be clear: as long as you have some influence on what comes about, even though that may be through the medium of acts over which you do not have full control, you should carry out the rational analysis of what would be the optimal act for you, and then do your best to implement it. This position needs both direct and indirect support. Let me begin with another example.

A MODIFIED PRISONER'S DILEMMA

Dissidents *A* and *B* are being held in solitary confinement in the state prison. The state torturer will be along to visit them soon, but because of the press of business, he'll spend just five minutes trying to extract a confession from each. Members of their underground group have been allowed one (simultaneous) visit to them, and they have brought the same offer to each: if you don't confess and the other does, you'll get ($a + b$) units of reward when released, and he'll get nothing. If you both confess, you'll get b units of reward for having undergone the torture, and if neither of you confesses, you'll get a units. The group has made this offer because it seems to them that the result will be that both A and B will decide that the optimal thing to do is to not confess. A and B realize this additional fact: If I don't confess during the five minutes, I'll have endured $-c$ units of unpleasantness; if I do confess after t minutes, I'll have endured $-x$ units.

We let $c = a$, and assume that $b > x$, and hence from Table III we see that not confessing dominates confessing.[9] So you should decide that not confessing is the optimal act for you. The torturer then arrives and despite A's decision that not confessing is optimal for him, and his firm resolve to carry out that act, he breaks down after a while and confesses. Is this situation outside the realm of rational decision theory? It is difficult for me to see why it should be. The agent has factored into the decision the effects of the external influence on him. Should the agent refuse to use decision theory here? Should he be "partly rational", whatever that means? Neither of those options seems sensible. Not confessing is the best thing to do, it's just that he isn't omnipotent in this case.

Both defences of evidential theory described above allow (and in fact require) that decisions do not determine acts.[10] How can this happen?

TABLE III

		Agent B		
		$\overline{\mathit{CONFESS}}$	*CONFESS*	
Agent A	$\overline{\mathit{CONFESS}}$	$a - c = 0$	$a + b - c = b$	$a = c$
	CONFESS	$-x$	$b - x$	$b > x > 0$

Desirabilities (Modified Prisoner's Dilemma)

In the above example, the act of confessing was due to the direct influence of an external cause and not through weakness of will. My own view is that any agent who can withstand torture for even a few minutes is not suffering from weakness of the will, but if that counts as weakness of the will for you, no matter. For as long as the non-ratiocinative cause is not filtered through rational decisions or reasons, neither defence is available. And whatever goes on when victims are tortured, their breaking down and confessing need not happen by means of anything that counts as the dynamics of deliberation. The agent can be saying to himself right up to the time he confesses that not confessing is best, and that he has decided not to confess. Then he confesses in spite of himself. At that point he may say that what he did was not fully rational, but that is for explanatory purposes. It does not mean that his original deliberations were inappropriate or irrational, or that his reasons will not figure in the explanation, or that he did not know that this might happen when he did his calculations.

To keep things simple, the external cause in this example was not made a common cause of an outcome with large disutility, and one need not even consider those aspects of the situation peculiar to Prisoner's Dilemma. The important point is that brute, non-rational influences on an act do not, and should not, always place the situation outside the realm of rational decision making.

I find this example sufficiently transparent to convince me of the incorrectness of limiting decision theory in the way suggested. It is worthwhile, however, to look at the options which are open to someone who wishes to employ the restriction on decision theory suggested by the evidential defences, because the implausibility of those options is a second reason to avoid placing such restrictions on the theory.

BASIC ACTIONS, ATTEMPTS, AND MEGALOMANIA

To begin with extremes, it is clearly pointless to deliberate about what to do with matters which are known to be determined or precluded by factors entirely outside one's control. However, as I emphasized at the beginning of the paper, we are dealing here with properly probabilistic situations. What is of concern for us is the situation where I rationally decide that *A* is the optimal thing to do, and honestly attempt to do it, but counteracting causes result in (although they do not necessitate) the opposite act's being carried out. To assess this situation requires due consideration being given to the probabilistic nature of the influences. We can agree that, trivially, if having decided that A is the optimal act, but $-A$ is actually performed, whether as a result of weakness of the will or counteracting influences (or both), then the optimal thing was not done. Does this entail that what was done was irrational to some degree? No, because it is in the very nature of such probabilistic situations that different outcomes sometimes occur with exactly the same set of initial conditions. Thus an act can be non-optimal for the agent without the process which led to it thereby being irrational.

There is, though, a residual source of worry coming from the direction of the extreme case. The less we feel we will contribute to the performance of the act by our deliberations, the more we tend to feel that there is something peculiar about deliberating in the circumstances. That is why we never call outcomes rational or irrational. Such considerations push us in an obvious direction. Perhaps we should make a strict separation between acts and outcomes,[11] include in the former only those acts over which the agent believes he has full control, and insist that because the theory is properly applicable only to ideally rational agents and acts, we must reject any situation or agent whose features do not meet this strict demarcation criterion. Any evidential theorist who takes this route will find himself in an uncomfortable trilemma. Either he will have to restrict his decision theory to basic actions, or deliberations will be restricted to attempts to act, or the agents involved will have to be a special kind of "ideally rational" agent. Let me consider these options in turn.

The first way to ensure that there are no external influences on the act is to make the acts basic in form. But this way is clearly unpromising, for this would be achieved at the price of divorcing decision theory from any kind of ordinary application. I decide to turn left at the stop-light, and not just to make a leftward motion with my arms. Assassins, unfortunately, decide to kill people even though, fortunately, they often fail. They do not deliberate about whether to move their index finger.

The interest of decision theory has always been twofold. It appears to provide us with a systematic theory of a particular kind of rationality and also to serve as the basis of a normative theory of action having widespread applications in economics, statistics, business, warfare and so on. This first way out would require a loss of almost all of the applications of this normative theory. Philosophically, this would be no great loss but if that is what is required to hold on to evidential theory, we should be clear about the severely constrained domain. Furthermore, the division between acts and outcomes could no longer be arbitrarily drawn.

The second option, that of restricting deliberations to decisions about attempts to act, seems equally unpromising.[12] Firstly, many of our rational deliberations and decisions are clearly not of that form. Schubert did not decide to try to write his 6th (Unfinished) Symphony, he decided to write it and was thwarted by circumstances outside his control, just as the Allies did not decide to try to invade Europe in 1944: despite formidable obstacles, they decided to invade. It is true that the more we feel that circumstances are outside our control, the more we speak of deciding to try to act rather than of acting, and that such attempts are fully under our control. Even granting this point, this second option is unavailable to the ratifiability defence, for as Jeffrey has noted, "The notion of ratifiability is applicable only where, during deliberation, the agent finds it conceivable that he will not manage to perform the act he finally decides to perform, but will find himself performing one of the other available acts instead." (Jeffrey [1983], p. 18) If this were not assumed, some of the probabilities would be ill-defined, being conditioned (in the prisoner's dilemma case) on such propositions as "decide to try to confess and don't try to confess". If the propositions were about such attempts (over which we do have full control) and the agent did not even make the attempt, the deliberation would be merely idle speculation. Also, as Eells has noted, the Cartesian defence cannot use this rejoinder either, because as soon as a decision has been made to (try to) perform an act, all other acts will have zero probability, and the rationale for performing the optimal act will be unavailable as soon as the decision has been made.

These considerations suggest that where decisions to act do not guarantee that the action will be successfully carried out, the preference ordering among acts must be calculated by weighing the expected utilities of acts by the probabilities that a decision to perform them will result in their being performed, in order to deal with cases where optimal acts are extremely difficult to carry out but suboptimal acts are easy (see e.g. the Moscow case below).[13]

Could, then, the evidential theory be saved as an idealization of rational activity? I hardly think so, for the idealization needed involves a requirement of personal omnipotence. Even disregarding the fact that immunity of one's actions from counteracting forces is descriptively hopeless, such an idealization of personal power is not an idealization involving rationality. Unlike an epistemological assumption, such as the Cartesian defence's assertion that we have perfect knowledge of our own reasons for acting, assumptions involving the presence or absence of causal influences outside our control cannot be rationality assumptions, at least not without a great deal more argument than has been given. As I mentioned above, evidential theorists have been willing to weaken the connection between decisions and actions, so that the former do not determine the latter, but only when the other influences do not affect the act directly. But a view which held that it was ideally rational to believe that the only way a decision could fail to result in an action was through a lack of resolve on the part of the agent would seem to be itself a sign of an irrational individual. It bears the marks of a kind of megalomania, of an agent who believes that he is immune from the effects of an uncooperative world. In particular, if the only situations in which ideally rational agents would trouble to deliberate would be those in which they believed that they had full control over their acts, those rational agents must hold that there is a connection between rationality and degrees of (believed) omnipotence. That is, because all this is formulated in terms of the agent's beliefs, if he believes that he has complete control over things which to most of us seem clearly outside his realm of influence, he is, according to the criterion used by evidential Bayesians, acting rationally. Indeed, the more you tend toward megalomania, the more you are able to rationally deliberate about, according to this view; a position which not only seems downright unfair to us more modest types, but appears to recommend some serious revisions in modern psychiatric practice. For if an agent is convinced that by setting out for Moscow he can singlehandedly capture it, then deliberating about what to do (and presumably deciding to capture Moscow) is rational (for him), according to the evidential view. Speaking for myself, I would have him incarcerated.

University of Virginia

NOTES

* It is a great pleasure to contribute an essay to a volume honouring Wes Salmon. For many years his work in explanation, probability, and causality has served as the

inspiration for much of my own work in those areas and his influence on the literature as a whole has been widespread and beneficial. He has also had a longstanding interest in Bayesian reasoning. This last subject is not quite as far from the first three as one might think, for it is exactly the kind of probabilistic causal relations underpinning explanation and objective probabilities which turn out to be required for a satisfactory decision theory. I do not know if he would agree with what follows but even if he does not, it is irrelevant, because Wes is one of those rare people whose interest lies in getting things right rather than insisting that he *has* got it right. Would that we were all as careful and honest.

Previous versions of this paper were read at the Center for Philosophy of Science, University of Pittsburgh; the Pacific Division of the American Philosophical Association; and Virginia Commonwealth University. I am grateful to Brad Armendt, Ellery Eells, John Heil, Mark Overvold, and Nicholas Rescher for helpful discussions on issues connected with this paper. It should not be assumed, of course, that they concur with the conclusions of this paper.

[1] By 'evidential decision theory' I mean the theory presented in Jeffrey (1983).

[2] As variously given in Gibbard and Harper (1978), Skyrms (1980), Lewis (1981) and others. A representative collection of papers in evidential and causal decision theory can be found in Campbell and Sowden (1985), Section III.

[3] See Eells (1981), (1982), (1984a), (1984b), (1985), and Eells and Sober (1986).

[4] In Jeffrey (1983) Ch 1.7, and (less explicitly) in Jeffrey (1981), section VI.

[5] See Campbell and Sowden (1985), Section III. I shall avoid discussing the Newcomb problem, because its causal structure is so unclear. Those who relish fantasies can construct their own parallels.

[6] Discussed in e.g. Jeffrey (1981).

[7] The differences between causal and evidential relations are also erased in the construal of preferences given by Jeffrey as preferences between news items. "To say that A is ranked higher than B means that the agent would welcome the news that A is true more than he would the news that B is true . . ." (Jeffrey (1983), p. 82). But also ". . . there is no effective difference between asking whether he prefers A to B as a news item or as an act, for he makes the news." (ibid., p. 84).

[8] I have represented the game theoretic situation as a decision problem for agent A, in order to bring out the structual similarity with the previous example.

[9] For simplicity I have let x have the same value for both agents here. The decision theoretic structure of the situation remains the same if symmetry is broken by x being different for each, perhaps because one resists longer than the other. Letting $c = a$ is reasonable — the group exactly compensates the agents for five minutes of torture when they both resist. How reasonable is it to set $b > x$? It might be said that if an agent confesses, then x must have been greater than c, and so we should need $b > a$ which seems to reward an individual for confessing, contrary to the group's needs. In fact this odd reward system still produces the group's desired result, that not confessing is optimal for each. Furthermore, the claim that confessing entails that x must have been greater than c requires backtracking conditionals, where the state of the world conforms to what the agent does. In contrast, I assume that the world goes on as it does, the torture has whatever disutility it has, and that longer is worse.

[10] Jeffrey (1983), p. 18; Eells and Sober (1986) p. 235ff.

[11] A separation deliberately avoided within evidential decision theory.
[12] Jeffrey discusses this possibility (not in relation to the present problem, however) in (1983), pp. 83—84 and concludes "An act is then a proposition which is within the agent's power to make true if he pleases."
[13] The need for some modification along these lines was suggested by an anonymous commentator on a previous version of this paper, to whom I am indebted for this point.

REFERENCES

Campbell, R. and Sowden, L. (eds) (1985): *Paradoxes of Rationality and Cooperation*, Vancouver, University of British Columbia Press.

Eells, Ellery (1981): 'Causality, Utility, and Decision' *Synthese* **48** 295—329.

Eells, Ellery (1982): *Rational Decision and Causality*, Cambridge University Press, Cambridge.

Eells, Ellery (1984a): 'Newcomb's Many Solutions' *Theory and Decision* **16** 59—105.

Eells, Ellery (1984b): 'Metatickles and the Dynamics of Deliberation' *Theory and Decision* **17** 71—95.

Eells, Ellery (1985): 'Causal Decision Theory' in *PSA 1984 Volume 2*, P. Asquith and P. Kitcher (eds.) Philosophy of Science Association, East Lansing.

Eells, E. and Sober, E. (1986): "Common Causes and Decision Theory," *Phil. Sci.* **53**, 223—245.

Gibbard, A. and Harper, W. (1978): 'Counterfactuals and Two Kinds of Expected Utility' in *Foundations and Applications of Decision Theory, Vol 1*, C. A. Hooker, J. J. Leach, and E. F. McClennen (eds). Dordrecht, D. Reidel; 125—62.

Humphreys, Paul (1981): 'Aleatory Explanations' *Synthese* **48** 225—232.

Humphreys, Paul (1983): 'Aleatory Explanations Expanded' in *PSA 1982 Volume 2*, P. Asquith and T. Nickles (eds.), Philosophy of Science Association, East Lansing, 208—233.

Humphreys, Paul (1987): 'Scientific Explanation: The Causes, Some of the Cause, and Nothing but the Causes' in *Minnesota Studies in the Philosophy of Science, Volume XII*. P. Kitcher and W. Salmon (eds), University of Minnesota Press, Minneapolis.

Jeffrey, Richard (1981): 'The Logic of Decision Defended' *Synthese* **48** 473—492.

Jeffrey, Richard (1983): *The Logic of Decision* (2nd Edition), University of Chicago Press, Chicago.

Lewis, David (1981): 'Causal Decision Theory', *Australasian Journal of Philosophy* **59**, 5—30. Reprinted with postscript in his *Philosophical Papers, Volume 2*, Oxford University Press, Oxford, 1986.

Salmon, Wesley (1984): *Scientific Explanation and the Causal Structure of the World*, Princeton University Press, Princeton.

Skyrms, Brian (1980): *Causal Necessity*, Yale University Press, New Haven.

Stalnaker, Robert (1984): *Inquiry*, MIT Press, Cambridge.

EPILOGUE

WESLEY C. SALMON

PUBLICATIONS: AN ANNOTATED BIBLIOGRAPHY

BOOKS

Logic. Englewood Cliffs, NJ: Prentice-Hall, 1963.

This book grew out of a small multilithed pamphlet, "Logic Reference Notes," that I produced for introductory philosophy students (in courses other than logic) at Brown University. It was intended to serve as a handbook, incorporating parts of logic that would have useful applications in areas outside of logic. An enterprising publisher's representative saw its potential for expansion into a small book. In the first and subsequent editions two aims were paramount: first, to emphasize the applications of logic, both deductive and inductive; and, second, to present elementary material accurately — that is, in ways that would require only supplementation, not correction, in order to proceed to more advanced levels of logic.

Spanish translation. Mexico: Union Tipografica Editorial Hispano Americano, 1965.
Japanese translation. Japan: Baifu Kan, 1967.
Italian translation. Italy: Societa editrice il Mulino, 1969.
Portuguese translation. Brazil: Zahar Editores, 1969.

The Foundations of Scientific Inference. Pittsburgh: University of Pittsburgh Press, 1967.

In the winter and spring quarters of 1962—63, I held an appointment as Visiting Research Professor at the Minnesota Center for the Philosophy of Science. The major part of my effort was devoted to the study of Carnap's then most recent work on probability and induction. Herbert Feigl, Director of the Center, arranged a week's escape from the rigors of winter in Minneapolis, during which he, Grover Maxwell, and I visited Santa Monica, California, and held daily discussions with

James H. Fetzer (ed.) Probability and Causality. 271—336.

Carnap (under the orange trees in his yard) on his concept of logical probability. These discussions were of inestimable value. While in Minnesota I also had considerable contact with the distinguished statistician I. Richard Savage (brother of the late Leonard J. Savage), from whom I received an excellect introduction to Bayesian statistics and personal probability. My previous training with Reichenbach had emphasized the frequency interpretation.

Coincidentally, during a three year period commencing while I was in Minnesota, I was invited to give five public lectures on probability and induction at the Pittsburgh Center for Philosophy of Science. These lectures, which drew heavily on the work at Minnesota, provided the basis for a monographic essay, "The Foundations of Scientific Inference" (1966). In this essay I attempted to survey the major issues regarding probability and inductive inference. The first three chapters are devoted to the problem of the justification of induction, and the next two to the various interpretations of probability. These five chapters consist mainly of critical discussions of the most influential viewpoints that were around at that time.

The next chapter presents my best effort to provide a pragmatic justification of the rule of induction by enumeration as a means of establishing values of probabilities (construed as limiting frequencies). Prior to my visit to Minnesota I had believed that I possessed an adequate justification of that sort (see "On Vindicating Induction", 1963, and "Inductive Inference," 1963). Richard Savage (in conversation) first pointed out one flaw in the argument, and Ian Hacking exhibited another ("Salmon's Vindication of Induction," *Journal of Philosophy* **LXII** (May, 1965)). At the time I thought it would not be difficult to repair the damage, but that turned out to be quite wrong. The final chapter advances an approach to scientific confirmation that could appropriately be characterized as *objective Bayesianism*.

This essay was published, along with a small addendum, as a separate book in 1976. Although, after twenty years, it is somewhat dated in various respects, I think it still constitutes a sound introductory survey of these topics. Moreover, I believe that objective Bayesianism is still worthy of serious consideration as a theory of the logic of scientific inference. From time to time I wonder whether it may be possible to overcome the objections of Savage and Hacking and to carry through a successful pragmatic justification of induction by enumeration. At present I have no firm opinion about this matter.

Zeno's Paradoxes. Indianapolis: Bobbs-Merrill, 1970. [Editor]

Zeno's paradoxes — which have, of course, challenged philosophers and mathematicians for two and a half millenia — enjoyed a period of particularly active discussion during the 1950s and early 1960s. It seemed that a small volume on this topic would be a useful addition to the Prentice-Hall *Contemporary Perspectives in Philosophy Series* of which I was a co-editor. Because I was enormously impressed by Adolf Grünbaum's treatment of them (in several articles and his doctoral dissertation) I approached him with the proposal that he edit such a volume. He declined, but urged me to do so, promising help with the project.

One of the items I wanted to include was an elegant brief article by Grünbaum, "Modern Science and Refutation of the Paradoxes of Zeno." He granted permission, but said that he felt that it was rather incomplete and needed supplementation. In particular, it did not include any discussion of the "infinity machines" that figured prominently in the then recent discussions. In the end, he contributed an additional essay, "Modern Science and Zeno's Paradoxes of Motion," which runs to 50 printed pages. My anthology grew beyond the space limitations of the *Contemporary Perspectives Series*, a series that was being phased out by the publishers anyhow. After a rather long delay it was finally published, quite appropriately, in the *Library of Liberal Arts* (Bobbs-Merrill).

Mainly as a result of his additional work on the paradoxes of motion, Grünbaum published a book, *Modern Science and Zeno's Paradoxes*, which contains, among other things, a definitive treatment of the infinity machines. It pleases me to think that our interactions played a noticeable role in stimulating that work. Zeno's paradoxes have a truly remarkable capacity for eliciting new and profound issues. For example, a brand new Zeno-type paradox was apparently discovered for the first time less than 20 years ago (see "A Zenoesque Problem" (1971) and *Space, Time, and Motion: A Philosophical Introduction*, pp. 48–52). Still more recently, at least two authors, Brian Skyrms and Michael White, have used non-standard analysis in dealing, respectively, with the paradox of plurality and the paradox of the flying arrow.

My anthology has been out of print for a number of years, but a second edition is now in the works with the University Press of America. I continue to find Zeno's paradoxes intriguing and stimulating. Bertrand Russell's at-at theory of motion, by means of which he resolved Zeno's paradox of the flying arrow, played a key role in my work on causality in *Scientific Explanation and the Causal Structure of the World.*

Statistical Explanation and Statistical Relevance. Pittsburgh: University of Pittsburgh Press, 1971. [With contributions by Richard C. Jeffrey and James G. Greeno.]

At the 1963 meeting of the American Association for the Advancement of Science, Section L (History and Philosophy of Science), I presented a paper, "The Status of Prior Probabilities in Statistical Explanation" in which I criticized Carl G. Hempel's then new theory of statistical explanation. The main point — although it was very obscurely expressed — was that statistical relevance, not high probability, is the key idea in statistical explanation. Henry E. Kyburg, Jr., commented on it. This paper was published in 1965, along with Kyburg's critique and my response. In his discussion, Kyburg showed that the same kind of relevance objection I leveled at Hempel's theory of statistical explanation can also be brought against his deductive-nomological model of explanation. Kyburg's discussion also contained his own extremely brief proposal for a characterization of explanation, along with the remark that it is easy to say what scientific explanation is. I found his proposal rather patently unsatisfactory, and pointed out its shortcomings in my reply.

But if Hempel's models do not succeed, and if Kyburg's notion will not work, what is scientific explanation? I tried to answer that question in a paper, "Deductive and Inductive Explanation," that was presented at a 1965 workshop at the Pittsburgh Center for Philosophy of Science. This paper was never published under that title, but a completely revised and greatly expanded version appeared under the title "Statistical Explanation" (1970). Again, the University of Pittsburgh Press proposed publishing this essay as a separate paperback.

Between the writing of "Statistical Explanation" and the time at which *Statistical Explanation and Statistical Relevance* was prepared for publication, I became acquainted with the essays by Richard C. Jeffrey and James G. Greeno that were included in the volume. On the basis of these developments, I was able to add an introduction in which the statistical-relevance (S—R) model was presented by name and succinctly placed in sharp contrast to the Hempelian models.

Although I no longer hold the S—R model to be adequate, I do think its development was (for me, at least) an indispensable step along the way to what I now consider the much more satisfactory treatment of scientific explanation offered in *Scientific Explanation and the Causal Structure of the World.*

Logic, 2nd ed. Englewood Cliffs, NJ: Prentice-Hall, 1973.

This edition is substantially enlarged by the addition of truth tables, Venn diagrams, some material on binary relations, and the use/mention distinction.

Japanese translation. Japan: Baifu Kan, 1975.
German translation. Stuttgart: Philipp Reclam, 1983.

Space, Time and Motion: A Philosophical Introduction. Encino, CA: Dickenson Publishing Co., 1975.

My interest in philosophical problems of space and time dates back to my days as a graduate student taking Hans Reichenbach's courses. Over a period of many years I gave public lectures on a variety topics in this general area. The first chapter of this book was a lecture on "Philosophy and Geometry" presented in a mathematics course at Indiana university (conceived and organized by students) called *Math and Everything Else.* Professors from a great variety of fields gave lectures on the relationship between some part of mathematics and something important in their fields. I spoke on the epistemological importance of the discovery and application of non-Euclidean geometries.

Chapter 2, "A Contemporary Look at Zeno's Paradoxes," has essentially the content of a lecture I delivered in the Pittsburgh Philosophy of Science Series. A funny thing happened on the way to that lecture. The Philosophy of Science Center sent the title to a university publicity office to make up an announcement and press release. Someone in that office evidently looked in the *Encyclopedia of Philosophy* to find out who Zeno was. Since Zeno of Citium (the founder of stoicism) comes alphabetically before Zeno of Elea (the one with the paradoxes), the write-up on my lecture gave a brief sketch of the life and works of the founder of stoicism. This provided me with a snappy introduction for the the lecture: how ironic — given Zeno of Elea's polemic against plurality — that there should be more than one Zeno!

Since the fourth chapter deals with some technical aspects of special relativity, it was necessary to provide a basic introduction to that theory. Chapter 3, "A Trip on Einstein's Train," does just that, and it does so in the simplest and most intuitive way I have ever seen. In an introductory physics course for non-physical-science students (the physics-for-poets sort of thing) that I taught jointly with physicists and historians for a number of years at the University of Arizona we used this chapter (instead of the presentation given in the textbook for the course) with quite reasonable success. Unlike some popularizations of special relativity, it is scientifically sound; in contrast to many sound introductions, it is intuitive and easy.

Chapter 4, "Clocks and Simultaneity in Special Relativity," was my contribution to a workshop held at Ohio State University; it was also presented at physics colloquia at Indiana University and the University of Arizona. It deals with three main topics. First is the so-called 'twin paradox' — which is not really a paradox at all, but rather an effect that falls into the category of *strange, but true* — namely, that a traveling twin who takes a long ride on a space ship at very high velocity will actually be younger than his/her stay-at-home sibling upon returning home. Shortly before the chapter was written, this differential aging effect had been demonstrated experimentally by taking atomic clocks on around-the-world trips on regularly scheduled jet airplanes.

The second topic is the so-called "clock paradox." If we take the traveling twin's space ship as the stationary frame of reference it would appear that the twin who remains on earth should be younger when next they meet. However, to say that the earth-bound twin is both older and younger than his/her sibling would be paradoxical. Many discussions of this paradox resolve it by appealing to general relativity, but following an approach initiated, though not completed, by Grünbaum, I show how to resolve it entirely within special relativity. The resolution consists in demonstrating that, on the basis of the principles of special relativity, the twin who takes the ride in the space ship is unequivocally singled out as the younger of the two when the ride is over (and the amount of the age difference can be calculated uniquely). This is important. The paradox arose in the special theory in the first place, so if we are worried about the consistency of *that* theory, that is the context within which is must be resolved.

The third topic is the much-debated issue of conventionality of simultaneity. In the special theory of relativity Einstein had established the *relativity of simultaneity* — the fact that two events that are simultaneous with respect to one inertial frame of reference will not be simultaneous with respect to other inertial reference frames in motion relative to the first. But Einstein seems to have suggested that a convention is involved in establishing relations of simultaneity within any given intertial reference frame — i.e., synchronizing the clocks located at different places in that frame. Reichenbach articulated that thesis explicitly and argued for it in detail. Others, especially Brian Ellis and Peter Bowman, argued against the conventionality thesis. Chapter 4 reviews the arguments, pro and con, and concludes that the conventionality thesis is correct.

When the manuscript for this book was completed, I made a copy for Adolf Grünbaum (to whom it is dedicated) and had it bound for presentation to him. In place of the genuine title page another was included bearing the title "Everything You Ever Wanted to Know about Grünbaum's Philosophy of Space and Time* (*But Were Afraid to Ask)." I think he was amused. His wife, Thelma, said, "Surely not *everything*," to which I had, in all honesty, to agree.

Hans Reichenbach: Logical Empiricist. Dordrecht: D. Reidel Publishing Co., 1979. [Editor]

When Reichenbach died in 1953 a volume on Logical Empiricism, which would have dealt with both Carnap and Reichenbach, was in preparation for Paul A. Schilpp's *Library of Living Philosophers.* As a result of his untimely death it became *The Philosophy of Rudolf Carnap.* This development aroused in me a strong feeling that the philosophical world deserved a comprehensive volume on Reichenbach's philosophy, comparable, insofar as possible, to those in *The Library of Living Philosophers.* Although no one could write Reichenbach's intellectual autobiography, and no one could furnish his replies to his critics, it was feasible to bring together a collection of critical essays that would exhibit the scope, importance, and unity of his contributions to 20th century philosophy. For a variety of reasons — not all of them good, by any means — I did not get around to undertaking this task for more than twenty years.

Two events furnished the needed impetus. The most important was the decision by Friedr. Vieweg & Sohn (Wiesbaden) to bring out a nine-volume edition of Reichenbach's collected works (*Hans Reichenbach, Gesammelte Werke*, edited by Andreas Kamlah and Maria Reichenbach), and their invitation to me to write a comprehensive introduction. The entire collection was to be published in German. Those of Reichenbach's works that were originally published in German were republished in their original form. Those that were originally published in other languages — of which there were many, mainly in English — were translated into German for inclusion in this collection. (To my complete astonishment I learned that Reichenbach's main epistemological treatise, *Experience and Prediction*, had never previously been published in German — this shows the degree to which his work had been neglected in his native land.) My introduction was translated into German by his widow, Maria Reichenbach.

At more or less the same time, Jaakko Hintikka (editor-in-chief of *Synthese*) and I were discussing the possibility bringing out one or more issues of *Synthese* devoted to Reichenbach's work. When I agreed to write the introduction for the Vieweg collection I received permission to publish my English original in *Synthese*, and D. Riedel (publishers of *Synthese*) agreed to bring out an expanded collection of essays on Reichenbach as a volume in the *Synthese Library*. This is what, in fact, transpired.

I have never tried to make a secret of my intellectual debt to Reichenbach, and for a long time during the beginning of my career I was regarded by many (not without justification) as a slavish follower. The publication of *Hans Reichenbach: Logical Empiricist* must have been the fulfilment of a filial duty. It is not that I now want to deny the debt — to do so would be absurd — but somehow I now feel that the debt has been repaid to the best of my ability. I still believe, without diminution, that Reichenbach was an extraordinarily important contributor to philosophy of science during the first half of the 20th century and beyond (his posthumous work, *The Direction of Time*, was published in 1956). I tried to say why in "The Philosophy of Hans Reichenbach," my introduction to this volume. With both candor and humility I can say that I consider this a book of lasting importance —

not because of its editor, but because of the subject, and its other contributors as well.

Space, Time, and Motion: A Philosophical Introduction, 2nd ed. Minneapolis: University of Minnesota Press, 1981.

The first edition of this book was not a commercial success; it soon went out of print. It was, it seems, a book that many wanted to own (if they could get a free copy) but few wanted to prescribe as a text — a textbook publisher's nightmare. Not long thereafter the publishing company closed up shop and its president decided to go into a different kind of business. I do not accept full responsibility for these latter developments.

On the strength of an extremely favorable published review, I approached the University of Minnesota Press concerning a second edition. They agreed. The major change is the addition of an extensive annotated bibliography. This edition, too, is now out of print. It is also out of date. An up-to-date introduction to space, time, and motion would require treatment of a good deal of recent work by John Earman, Michael Friedman, David Malament, and Larry Sklar, among others.

Logic, 3rd ed. Englewood Cliffs, NJ: Prentice-Hall, 1984.

This edition is expanded over the previous one by the addition of substantial material on Mill's methods, controlled experiments, and causal reasoning.

Japanese translation. Tokyo: Baifukan Co., Ltd., 1987.

Scientific Explanation and the Causal Structure of the World, Princeton, NJ: Princeton University Press, 1984.

When I wrote the main essay in *Statistical Explanation and Statistical Relevance* I was clearly aware of the obvious fact that causality plays a central role in scientific explanation. At that time I held out some hope that causal concepts could be adequately explicated entirely on the basis of statistical relations. During the next few years that hope vanished. At the same time, I felt a pressing need to come to terms with theoretical explanation. My first attempts to deal with this topic made the indispensability of causality more vivid than ever. This focus on causality made me acutely aware of the need to face squarely Hume's celebrated critique of causality.

As we all know, Hume raised a deep and recalcitrant problem concerning causal connections. Somewhere along the line it occurred to me that the world is full of causal processes, and that these might provide the long-sought causal link. A key to the nature of this kind of connection lies in the distinction between causal processes (e.g., light rays or moving material particles) and pseudo-processes (e.g., shadows). Reichenbach had offered the ability to transmit a mark as a criterion to distinguish causal processes from pseudo-processes. This is an extraordinarily fertile idea.

Mark transmission involves two causal concepts, a *causal interaction* that imposes a mark and *causal propagation* whereby the mark is transmitted from one spacetime locale to another. One can say, very roughly, that a causal interaction occurs when two or more processes intersect, and each is modified at the intersection in a way that persists beyond the intersection. A mark is transmitted from point A to point B when, having been introduced in an intersection at A, it appears at each intervening stage between A and B *without any additional interactions.* This is the crux of the at-at theory of causal transmission that was inspired by Russell's at-at theory of motion. An adequate understanding of causal processes and causal interactions provides, I believe, a suitable basis for a theory of causal explanation, which is what I tried to present in this book.

My first serious involvement with scientific explanation was in the context of statistical explanation, and it concerned the role of statistical relevance relations. At the time I thought that these relevance relations had intrinsic explanatory import. After reflecting on the role of causality in explanation, I came to the conclusion that statistical relevance relations have no direct explanatory import; rather, their significance lies in the evidence they provide regarding causal relations. The explanatory import lies in the causal relations rather than in the statistical relations per se. As a result, the S-R model, which had formerly been taken as a full-fledged model of scientific explanation, is transformed, in the current theory, into *the S-R basis* — i.e., a foundation upon which bona fide scientific explanations can be erected.

The emphasis on causality does not, however, involve an implicit assumption of determinism. Causal relations occur, I believe, in indeterministic contexts — there is some sort of probabilistic causality in the world. Nevertheless, I do not think it is possible to explicate these causal concepts in terms of statistical relationships alone. Appeal must also be made to causal processes and interactions.

This point may be illustrated in connection with theoretical explanation. The fundamental problem of theoretical explanation concerns the existence of unobservable entities. Using the debates about the reality of molecules and atoms around the turn of the century as an actual historical example, I argue that there is a cogent argument for the existence of such unobservables. Scientists found it compelling and so should we. It is a common cause argument. It appeals to what Reichenbach called a *conjunctive fork*. The conjunctive fork is an extremely useful explanatory device; it is often used in science and in everyday life. It is defined in terms of statistical relations alone. Examples show, however, that not all conjunctive forks constitute common causes; connecting causal processes have to be taken into account as well.

The causal account I am advocating stands in sharp contrast to the deductive-nomological (D—N) and inductive-statistical (I—S) accounts so ably elaborated and defended by Hempel. Using a distinction first introduced, I believe, by Alberto Coffa, we can say that Hempel advocates an *epistemic conception* of scientific explanation, while I am trying to defend an *ontic conception*. Hempel's conception is epistemic because it takes an explanation to be an argument to the effect that the event-to-be-explained *was to be expected* by virtue of certain explanatory facts. The causal concept is ontic because it construes an explanation as *a causal nexus* of some sort *that exists in the physical world*. An event is explained by fitting it into such a causal pattern.

D. H. Mellor's 1976 article, "Probable Explanation" (*Australasian Journal of Philosophy*) brought into clear focus (for me) the fact that there is a third major conception of scientific explanation — *the modal conception* — that has often been advanced (e.g., by G. H. von Wright). In its most primitive form it takes an explanation to be a demonstration that the event-to-be-explained *had to happen.* Mellor extends this conception to the indeterministic context by appealing to degrees of necessitation.

A good deal of attention is devoted in this book to the elaboration and discussion of these three basic conceptions of scientific explanation. I claim that all three can be found in Aristotle, and that they persist in the history of scientific thought right down to the present time. I also suggest that the distinction among them doesn't amount to much in the deterministic context of classical physics, but that they diverge sharply when we consider seriously explanation in indeterministic contexts. Close examination shows, by the way, that each of the major conceptions comes in more than one version. The epistemic conception, for example, comes in an *erotetic version* (e.g., Bas van Fraassen) and an *information theoretic version* (e.g., James Greeno), as well as the more familiar *inferential version* of Hempel and many other authors. Recognition of these different conceptions and their various versions is part of the indispensable process called by Carnap *clarifying the explicandum.* If this task is neglected our explication is apt to go astray, for we may be misconceiving the very concept we are trying to explicate.

The epistemic conception, in its inferential version, stood for a long time as *the received view.* The ontic conception, in the causal/mechanical version I defend, it radically different. In the closing pages I try to show how profoundly different they are. It takes nothing less than a sort of gestalt shift to switch from the one to the other. Yet I believe the effort needs to be made, for the formerly received view has been seriously undermined in recent years.

In the end, it must be emphasized, I am not attempting to provide a model or set of models to which every bona fide scientific explanation must conform. Unlike Hempel and Opphenheim, I am not discussing *the logic* of scientific explanation. I doubt that any such logic exists. What constitutes a satisfactory pattern for scientific explanation depends crucially upon the kind of world in which *in fact* we live, and perhaps upon the domain of science with which we are concerned. In particular, although I regard causal explanation as quite pervasive, I do not see how it can be extended into the microcosm — the quantum mechanical domain. I do not have any theory of quantum mechanical explanation, and I don't think anyone else does either. To explicate the nature of quantum mechanical explanation is, I believe, a premier challenge to 20th century philosophy of science. Perhaps the reason for this lack is that we do not yet have an adequate understanding of the physics of the microcosm.

I have often been complimented over the years on the clarity of my writing, and I am naturally pleased by such flattery. It is even more pleasing that some people have found this book interesting. I am painfully aware of many of its shortcomings, but if it stimulates others to deal more adequately with these problems, I will be happy indeed. Elliott Sober, for example, has severely criticized my treatment of the common cause principle, and on good grounds. He has also provided a vastly improved account. That sort of thing is deeply satisfying.

The Limitations of Deductivism. Berkeley & Los Angeles: University of California Press, forthcoming, [Co-editor, with Adolf Grünbaum]

This volume is dedicated to the memory of our beloved friend and esteemed colleague J. Alberto Coffa. It contains, basically, the proceedings of a workshop held at the Pittsburgh Center for Philosophy of Science in 1980. Several of the papers are significantly revised: my paper, "Deductivism Visited and Revisited," was composed expressly for this volume (my workshop paper having been rendered superfluous by the publication of *Scientific Explanation and the Causal Structure of the World*). This paper makes extensive use of Coffa's notion of *deductive chauvinism*, and I offer a characterization of a *deductive chauvinist pig*. Philosophers would profit, I think, from a wider acquaintance with these concepts. I shall also provide an Introduction for this volume. The other contributers are Ronald N. Giere, Carl G. Hempel, Henry E. Kyburg, Jr., and Frederick Suppe.

Scientific Explanation. Minneapolis: University of Minnesota Press, forthcoming. [Co-editor with Philip Kitcher]

This volume is the result of the first quarter — which was devoted to scientific explanation — of an academic year Institute funded by the National Endowment for the Humanities on the question, Is a New

Consensus Emerging in Philosophy of Science? It was held at the Minnesota Center for Philosophy of Science in 1985–86. I am contributing a monographic essay, "Four Decades of Scientific Explanation," which will trace the development of philosophical discussions and controversies regarding the nature of scientific explanation from the epoch-making Hempel-Oppenheim "Studies in the Logic of Explanation" (1948) to the present. I shall also contribute a detailed response to reviews and criticisms of *Scientific Explanation and the Causal Structure of the World* that have appeared to date.

Philip Kitcher is contributing a monographic essay, "Explanatory Unification and the Causal Structure of the World," in which he develops his version of the theory of explanation as unification. Although the basic idea has been around for a long time, it was first systematically elaborated by Michael Friedman in 1974. Kitcher effectively undermined Friedman's version, and then offered a novel twist to the basic idea. In the present essay he provides a detailed articulation and defense of that approach. It contrasts sharply in many respects with my position.

The other contributors to this volume are Nancy Cartwright, Paul Humphreys, David Papineau, Peter Railton, Merrilee Salmon, Matti Sintonen, and James Woodward.

Zeno's Paradoxes, 2nd edition. Lanham, MD: University Press of America, forthcoming.

This will be a photo-reproduction of the first edition, which has been out of print for a number of years. I feel that the papers in this volume constitute a resource that should be readily available to students and scholars.

DISSERTATIONS

"Whitehead's Conception of Freedom," M.A., University of Chicago, 1947. A relic, best forgotten, of the days when I was totally committed to Alfred North Whitehead's metaphysics.

"John Venn's Theory of Induction," Ph.D., University of California at Los Angeles, 1950.

As I was casting around in the history of inductive logic for a suitable dissertation topic, I ran across John Venn's *The Logic of Chance*. The originality, clarity, cogency, and beauty of his treatment of the limiting frequency interpretation of probability appealed enormously to me. Although some earlier writers — e.g., Aristotle and Robert Leslie Ellis — might qualify in a pinch as precursors, in my opinion Venn fully deserves the title of the first real frequentist. Time has not decreased my admiration for his book on probability in the least. It constitutes an epoch-making work in the fullest sense of the term.

Venn's work on induction, *Empirical Logic*, is not memorable. Curiously, Venn was unable to discern any connection between induction and probability at all, and he criticized many other authors for confusing them. He even rejected the Bernoulli theorem on the ground that it illegitimately combined probability and induction.

Written entirely in 11 days, this thesis (which runs to well over 200 pages) may hold the record for the most speedily composed Ph.D. dissertation in philosophy. The quality of the work accurately reflects the haste with which it was put together.

ARTICLES

"A Modern Analysis of the Design Argument," *Research Studies of the State College of Washington* **XIX**, 4 (Dec., 1951), pp. 207—220.

During my first year of full-time teaching (UCLA, 1950—51) I offered a course in Philosophy of Religion in which great emphasis was placed on the traditional arguments for the existence of God, especially the design argument. My study of Hume's *Dialogues Concerning Natural Religion* convinced me that his argument could fruitfully be analyzed in terms of Bayes's theorem. It is my first example of the utility of the Bayesian point of view. Although this early article contained many defects, I still think the basic idea is right. A drastically revised and much improved discussion appears in "Religion and Science: A New Look at Hume's *Dialogues*," (1978).

"The Frequency Interpretation and Antecedent Probabilities," *Philosophical Studies* **IV**, 3 (April, 1953), pp. 44—48.

This short note, which should never have been published in this way, was a fragment of a much larger article on a frequentist approach to the probabilities of scientific hypotheses via Bayes's theorem. It was emasculated to conform to editorial demands that should never have been honored. Nevertheless, it represents a first stab at developing an *objective Bayesian approach* to scientific confirmation. The gist of the whole article was eventually included in chapter VII of *The Foundations of Scientific Inference*.

"The Uniformity of Nature," *Philosophy and Phenomenological Research* **XIV**, 1 (Sept., 1953), pp. 39—48.

This article was extracted from my doctoral dissertation. The main thrust was that a principle of uniformity of nature is neither necessary nor sufficient for a justification of induction. Before publication, this paper was presented at a meeting of the American Philosophical Association, Pacific Division. It got me my first real teaching position — at the State College of Washington in Pullman — after I held a one-year revolving instructorship at UCLA.

"The Short Run," *Philosophy of Science* **XXII**, 3 (July, 1955), pp. 214—221.

The problem of the short run was poignantly stated by Peirce in terms of having one's eternal fate determined by a single draw of a card from a deck having 25 red cards and one black card or from a deck containing one red card and 25 black cards. Drawing a red card results in eternal felicity; a black card results in eternal torment. The subject must choose from which deck to make the selection. The make-up of the decks are known, and so are the probabilities. What, Peirce asks, makes it more rational to draw from the deck with more red cards, knowing that a draw from that deck may yield a black card, while a draw from the other deck may yield a red card. He concludes that one can be rational *only* by altruistically taking into account the interest of the entire community; one who is entirely selfish cannot be rational. Recall that one of his best known works is entitled *Love, Chance, and Logic*.

I saw the problem as one of applying knowledge of a long run frequency in a single instance — or any finite number from a given sequence, for that matter. It is entirely different from the problem of the single case. The problem of the single case is one of ascertaining what probability value to assign to a particular outcome; it depends on the choice of an appropriate reference class. In the short run problem we may assume that the pertinent probability is known. The problem of the single case may arise for a given individual in connection with many different sequences; indeed, each of us deals with single cases many times, day in and day out. Peirce's problem stands in sharp contrast. Here *everything* hinges on one draw of a card.

In this article I attempted to offer a pragmatic justification of a short run rule that was analogous to Reichenbach's pragmatic justification of induction. As it turned out, my short run justification had all of the shortcomings of Reichenbach's pragmatic justification. See "The Predictive Inference" (1957). I offer a different, and I think better, solution in "Dynamic Rationality," this volume.

Reply to critic:

"Reply to Pettijohn," *Philosophy of Science* **XXIII**, 2 (Apr., 1956), pp. 150—152.

"Regular Rules of Induction," *Philosophical Review* **LXV**, 3 (July, 1956), pp. 385—388.

This brief article is mostly for fun. In one of his discussions of induction Max Black had introduced — as an example of a rule no sensible person would adopt — a counterinductive rule. He remarked however, that anyone who could show why it is a bad rule would have mastered the most basic problems in the philosophy of induction. This note easily shows how use of that rule can lead to outright contradiction. One does not need to know much about induction at all to see what is wrong with it.

There is a much more serious aspect, however. The principles used to rule out Black's counterinductive rule (which I later called *normalizing conditions*) turn out to be extremely useful in dealing with Reichenbach's class of asymptotic rules. See "Vindication of Induction" (1961), "On Vindicating Induction" (1963), and "Inductive Inferece" (1963).

"Should We Attempt to Justify Induction?" *Philosophical Studies* **VIII**, 3 (Apr.,1957), pp. 33—48.

In the early 1950s Max Black published a series of papers in which he severely criticized Reichenbach's pragmatic justification of induction, advocated a sort of ordinary language dissolution of the problem of justifying induction, and proposed a way for inductive arguments to be self-supporting. Sometime around 1955 Herbert Feigl wrote to me saying that he was quite strongly impressed by Black's arguments. I was shocked. I had thought that the defects of those arguments were patent. When I replied in that vein to Feigl, he offered me an entire issue of *Philosophical Studies* in which to make my reply. It was an extraordinarily generous act on Feigl's part; I learned later that it was motivated in part by the fact that my mentor, Reichenbach, had died a short time earlier, and that, as a very junior philosopher, I needed support. The professional benefits that resulted from this publication were large.

Stated more generally, the purpose of this article is to examine the most important arguments, current at the time, designed to show that a justification of induction is impossible, unnecessary or both. In the article I do three things. (1) I argue that Black's criticisms of Reichenbach's justification of induction are unfounded. (2) I attack the ordinary language dissolution of the problem of justification of induction — i.e., the effort to dismiss it as a pseudo-problem. This approach was suggested by Black and spelled out in much greater detail by P. F. Strawson. In this part of the argument I make essential use of Feigl's crucial distinction between two types of justification — validation and vindication. I claim that the ordinary language dissolution trades on an ambiguity between the two. (3) I maintain that Black's inductive justification of induction will not work. Black had stated that his self-supporting arguments are not circular in the standard sense because they do not contain their conclusions among their premises. I claim that they suffer from what might be called *rule circularity*, and show that a strictly parallel self-supporting argument can be constructed for the counterinductive rule. In an effort to clinch the case for those who might remain unconvinced by my philosophical argument, I composed a parody (previously unpublished) of the popular song, "That Old Black Magic," that goes as follows:

That old Black logic has me in its spell —
That old Black logic that you wield so well —
Induction's problems we will circumvent
With self-supporting types of argument.
Assured there isn't any fallacy,
And that it's free from circularity,
'Round and 'round I go —
But no *petitio* —
In a spin,
Loving the spin I'm in,
Using that old Black logic
By Max.

The article concludes that attempts to justify induction are neither futile nor surperfluous.

Reprinted (in part or in toto):

"Should We Attempt to Justify Induction?" in John V. Canfield and Franklin Donnell, eds., *Readings in the Theory of Knowledge* (New York: Appleton-Century-Crofts, 1964), pp. 363—367.

"Should We Attempt to Justify Induction?" in Ernest Nagel and Richard B. Brandt, eds., *Meaning and Knowledge* (New York: Harcourt, Brace & World, 1965), pp. 365—369.

"Should We Attempt to Justify Induction?" Bobbs-Merrill Reprint, Phil 184, 1969.

"Should We Attempt to Justify Induction?" in H. Feigl, W. Sellars, and K. Lehrer, eds., *New Readings in Philosophical Analysis* (New York: Appleton-Century-Crofts, 1972), pp. 500—510.

"Should We Attempt to Justify Induction?" in Alex Michalos, ed., *Philosophical Problems of Science and Technology* (Boston: Allyn & Bacon, 1974), pp. 357—373.

"The Predictive Inference," *Philosophy of Science* **XXIV**, 2 (Apr., 1957), pp. 180—190.

A predictive inference (as defined by Carnap) is an inference from an observed sample of a population to a finite unobserved portion of the same population. In this paper I deal with cases in which the population consists of an infinite ordered sequence of entities. Such inferences might be conceived as proceeding in two steps: an inference from the observed sample to the limiting frequency, and an inference from the limiting frequency to the unobserved 'short run.' It did not occur to me until much later that this way of construing the predictive inference is doomed from the beginning because of the failure of transitivity of the relation of inductive support (see "Consistency, Transitivity, and Inductive Support" (1965)).

This paper shows that, even when the normalizing conditions are imposed, Reichenbach's class of asymptotic rules contains members that license the inference or posit of *any* value between zero and one (endpoints included) as the limiting frequency of a sequence on the basis of *any possible* observed frequency in a sample of *any* size. The remaining problem — one to which I returned many times — is how to eliminate this arbitrariness by narrowing the class of acceptable asymptotic rules or providing a rationale for selecting a uniquely satisfactory rule.

The paper also shows that the pragmatic justification I had offered for a short run rule (see "The Short Run" (1955)) suffers from the same defect as Reichenbach's pragmatic justification of his rule of induction.

" 'Exists' as a Predicate," *Philosophical Review* **LXVI**, 4 (Oct., 1957), pp. 535—542. [with George Nakhnikian].

One day during 1955—56, while Nakhnikian was visiting Brown University, he asked me to explain an argument that had been used by A. J. Ayer, C. D. Broad, and John Wisdom, in connection with the ontological argument for the existence of God, to show that a logical contradiction results from treating existence as a property (or "exists" as a predicate). Nakhnikian was lecturing on the ontological argument in an introductory class. After relatively brief consideration we realized that the argument was logically fallacious.

We then showed how to set up a consistent first order monadic predicate calculus, with a predicate "E" of existence as a logical constant, and an axiom "(x)Ex" in addition to the standard ones. We proved as a theorem "All Gods exist" — i.e., "if x is a God then x exists" or "there exist no non-existent Gods." Inasmuch as the aim of the traditional ontological argument is to demonstrate the statement, "Some Gods (at least one) exist," we argued that "exists" could consistently be treated as a predicate without, as a consequence, validating the ontological argument.

As Nicholas Rescher pointed out, modal realists will not find our approach acceptable, but that consequence does not really disturb me very much.

Spanish translation:

"Existe' como Predicado," Revista "Universidad De San Carlos" **XLI**, pp. 145—155.

"Psychoanalytic Theory and Evidence," in Sidney Hook, ed., *Psychoanalysis, Scientific Method, and Philosophy* (New York: New York University Press, 1959), pp. 252—267.

This essay investigates the relationship between observable behavior and the mechanisms postulated by psychoanalytic theory. The principle of psychic determinism is taken as a point of departure. I propose that it be weakened and reinterpreted as a principle about evidence. It then states, in part, that behavior that cannot be explained solely on the basis of constitutional principles constitutes indirect inductive evidence for conscious or unconscious psychic mechanisms. I defend this principle against well-known trivializing arguments on the ground that *additional evidence* pertaining to these mechanisms is always available. I consider in some detail Freud's discussion of 'counterwish dreams' in order to illustrate the nontrivial character of psychoanalytic explanations.

In this article I adopt the position of scientific realism, a standpoint I did not seriously attempt to defend until much later in "Why Ask, 'Why?'?" (1978), *Scientific Explanation and the Causal Structure of the World*, chap. 8 (1984), and "Empiricism: The Key Question" (1985).

Reprinted:

"Psychoanalytic Theory & Evidence," in Richard Wollheim, ed., *Freud: A Collection of Critical Essays* (Garden City, NY: Anchor/Doubleday, 1974), pp. 271—284.

"Psychoanalytic Theory and Evidence," in Md. Mujeeb ur-Rahman, ed., *The Freudian Paradigm* (Chicago: Nelson-Hall, 1977), pp. 187—200.

"Barker's Theory of the Absolute," *Philosophical Studies* **X**, 4 (June, 1959), pp. 50—53.

Stephen Barker offered a 'theory' involving The Absolute as a counterexample to Carnap's 1956 criterion of cognitive significance. This brief paper analyzes Barker's 'theory.' It shows that, after the correction of some minor technical flaws, this 'theory' does satisfy Carnap's requirements, but it fails as a counterexample to Carnap's criterion because it has empirical content. The result is established by deducing patently observational consequences from the 'theory.' The derivation provides a slightly nontrivial example of the use of symbolic logic in philosophy.
Reply to critic:
"Empirical Statements about the Absolute," *Mind* **LXXVI**, 303 (July, 1967), pp. 430— 431.

"Vindication of Induction," in Herbert Feigl and Grover Maxwell, eds., *Current Issues in the Philosophy of Science* (New York: Holt, Rinehart, and Winston, 1961), pp. 245—256.

As I was reflecting on Carnap's theory of degree of confirmation, it occurred to me that the relationship between his inductive logic and the descriptive language within which it was applied was undesirably intimate. Although internally any such logical system with its accompanying language fulfills an equivalence condition — according to which, if h and h' are logically equivalent hypotheses and e and e' are logically equivalent evidence sentences, then $c(h, e) = c(h', e')$ — the same equivalence condition does not hold if we bridge two different descriptive languages. We can, for example, use Carnap's then-favored confirmation function c^* with two different descriptive languages **L** and **L**$'$ and prove in the metalanguage that there are logically equivalent hypotheses h and h' and logically equivalent evidence sentences e and e' in the unprimed and primed languages respectively for which $c^*(h, e) \neq c^*(h', e')$. In *Logical Foundations of Probability* Carnap blocked this problem by imposing the requirement of descriptive completeness. This unappetizing requirement demands that we have a language that is sufficiently rich to enable us to formulate all of the descriptive sentences that will ever occur in any science before we have made a single inductive inference. Carnap was quite aware of the undesirability of this requirement, and in his later work sought to avoid it.

In this paper I decided to face the issue head on. In order to do so, I formulated *the criterion of linguistic invariance*, which amounts to a metalinguistic imposition of the foregoing equivalence condition across different descriptive languages. And it applies to inductive rules of inference as well as confirmation functions. I argued that this criterion, in conjunction with the *normalizing conditions* (see "Regular Rules of Induction" (1956)) sufficed to rule out all asymptotic rules except for Reichenbach's rule of induction, which is analogous to Carnap's straight rule. Included among the inductive methods ruled out by this argument are all of the methods in Carnap's continuum of inductive methods that embody a logical factor — that is, all but the straight rule.

Moreover, I used the same argument to pick out a unique short run rule. Consequently, it seemed to me that this paper provided a full answer to the problems I had raised in "The Predictive Inference" (1957). In 1963 I learned that this claim to have picked out unique rules by means of the normalizing conditions and the criterion of linguistic invariance was mistaken (see the remarks under *The Foundations of Scientific Inference*). Nevertheless, even though the criterion of linguistic invariance does not do everything I had hoped, I still regard it as a sound principle. For example, its bearing upon Carnap's continuum of inductive methods is quite illuminating.

Reply to critic:

"Rejoinder to Barker," *ibid.*, pp. 260—262.

"On Vindicating Induction," in Henry E. Kyburg, Jr., and Ernest Nagel, eds., *Induction: Some Current Issues* (Middletown, Conn.: Wesleyan University Press, 1963), pp. 27—41.

"Vindication of Induction" was presented orally at a meeting of the American Association for the Advancement of Science in 1959. As the commentator on my paper, Stephen Barker raised the question of whether the rule of induction by enumeration satisfies the criterion of linguistic invariance. Goodman's notorious grue-bleen paradox provided the foundation for his doubt.

"On Vindicating Induction" addresses this issue. A precise formulation of the paradox is given, and a resolution is offered. The key to the resolution lies in the fact (acknowledged explicity by Goodman) that not all grue things match one another in the way that all green things do, and similarly for bleen things and blue things. Goodman attached little importance to these facts about matching, but I think they are crucial. My point is reinforced in a (previously unpublished) song — a parody of the old English folksong "Greensleeves" — entitled "Bleensleeves" — that goes as follows:

Alas, my glove, what have you done,
That you clash with me so outrageously?
Whilst you had matched me so long
I delighted in your proximity.
Bleensleeves is what I am,
Bleen ever my constant hue,
Bleensleeves I shall always be —
Oh, why did you have to turn grue?

Returning to the argument, the result, I claim, is that, whereas "green" and "blue" are ostensively definable predicates, "grue" and "bleen" are not. I am indebted to Stephan Körner for convincing me of the importance of ostensive definability. I argue that the primitive descriptive predicates of our scientific language should be ostensively definable. By restricting primitive induction to statements involving only primitive predicates, I maintain, we can secure the linguistic invariance of the rule of induction by enumeration. The same basic argument as was employed in "Vindication of Induction" is used to show that induction by enumeration is the only rule that satisfies the criterion of linguistic invariance and the normalizing conditions. At the time this paper was presented to a conference on induction at Wesleyan University I thought I had in hand a satisfactory vindication of primitive induction. As I pointed out above (in the remarks on *The Foundations of Scientific Inference*) such optimism was quite unjustified.

Nevertheless, I am still inclined to believe that the basic idea behind my resolution of the Goodman pardox is sound. Goodman demonstrated conclusively, I believe, that no syntactical resolution of the paradox is feasible. In his appeal to entrenchment of predicates he offers a pragmatic resolution. My attempted resolution is semantic. The statement of my resolution has some serious defects, but I do not believe they reflect defects in the argument. A restatement of the resolution has been on my agenda for many years, but somehow I have not gotten around to doing it. Perhaps I should get busy now, for we have only 13 years left to decide whether we should infer that 21st century emeralds are green or grue.

Reply to critic:

"Reply to Black," *ibid.*, pp. 49—54.

Reprinted:

"On Vindicating Induction," *Philosophy of Science* **XXX**, 3 (July, 1963), pp. 252—261.

"On Vindicating Induction," in Sidney Luckenbach, ed., *Probabilities, Problems, and Paradoxes* (Encino, CA: Dickenson Publishing Co., 1972), pp. 200—212.

"Inductive Inference," in Bernard H. Baumrin, ed., *Philosophy of Science: The Delaware Seminar* vol. II (New York: John Wiley & Sons, 1963), pp. 341—370.

In my ill-founded euphoria about having an adequate justification of induction, I was eager to popularize it. This introductory-level essay surveys the issues discussed more technically in "Regular Rules of Induction" (1956), "Vindication of Induction" (1961), and "On vindicating Inducation" (1963). It sets out the problem of justification of induction, and argues for the importance of dealing with it. Two approaches — linguistic dissolution and pragmatic vindication — are discussed. Dissolutionist approaches are judged unsatisfactory. Reichenbach's pragmatic justification, supplemented by the normalizing conditions and the criterion of linguistic invariance, is offered as a sound approach. To illustrate concretely how the argument goes, half a dozen distinct inductive rules are introduced. Reichenbach's convergence criterion, the normalizing conditions, and the criterion of linguistic invariance are applied to them. Even though the claim to have singled out and justified a unique inductive rule is incorrect, the examination of the six different rules has genuine heuristic value, I believe.

Reprinted:

"Inductive Inference," in Baruch A. Brody, ed., *Readings in the Philosophy of Science* (Englewood Cliffs, NJ: Prentice-Hall, Inc., 1970), pp. 597—617.

"The Pragmatic Justification of Induction," in Richard Swinburne, ed., *The Justification of Induction* (London: Oxford University Press, 1974), pp. 85—97.

Spanish translation:

"La justificacion pragmatica de la induccion," in Richard Swinburne, ed., *La justificacion del razonamento inductivo* (Madrid: Alianza Editorial, 1976), pp. 105—118.

"What Happens in the Long Run?" *Philosophical Review* **LXXIV**, 3 (July, 1965), pp. 373—378.

In many discussions of probability the famous aphorism of Lord Keynes is quoted: "In the long run we'll all be dead." Similarly, reference may be made to Carnap's remark: "Unfortunately, we do not live to the ripe old age of denumerable infinity." There is often a plaintive suggestion that if we could experience the long run, our use of probability would make more sense.

This brief paper examines the question of what happens in the long run. Suppose that a game of heads and tails, involving infinitely many tosses of a coin with a probability of one-half for heads, could be played through to completion. Or suppose some 'infinity machine' (of the type discussed in the literature on Zeno's paradoxes) could serve as an equivalent probabilistic device, so that the infinitude of plays could be completed in a finite amount of time. How would a player, who bet one dollar on heads on each toss, fare (assuming that unlimited credit is available)? It is shown that, given only the constraint imposed by the probability value of one-half, the player may win an infinite amount, lose an infinite amount, experience gains and losses that fluctuate within finite bounds, or experience winnings and losses that fluctuate without any finite bounds.

If a randomness assumption is added, it is shown that the last of these alternatives obtains, for the function representing the player's fortune executes a one-dimensional symmetric random walk. Under no circumstances can a player *come out even in the long run* in this *fair game.*

A player who has a finite fortune or a finite line of credit may win, may come out even, may suffer a noncatastrophic loss (may retain some part of his or her fortune), or may loss everything *in the short run.* In the long run, a player with a finite fortune must lose everything. Perhaps it is not such a bad thing not to live forever.

"The Concept of Inductive Evidence," *American Philosophical Quarterly* **II**, 4 (Oct., 1965), pp. 1—6.

There is one particular type of argument for the dissolution of the problem of justification of induction that has been widely used by many important authors. It goes roughly as follows: To be rational is to fashion one's beliefs according to the evidence; inductive evidence is one basic type of evidence; thus, there is no question as to whether it is rational to employ induction. The problem of justification thereby evaporates, or is trivialized into the question of whether it is rational to be rational.

This paper attempts to undermine the foregoing dissolution argument. The critique rests upon the fact that, to have a viable concept of evidence, we must choose the rules we are going to employ from among a wide variety of possible rules. The problem of justifying induction is the problem of justifying such choice. The paper criticizes Carnap's view that such choice must be based upon 'inductive intuition.'

Reply to critics:

"Rejoinder to Barker and Kyburg," *ibid.*, pp. 13—16.

Reprinted:

"Symposium on Inductive Evidence" [with S. Barker and H. Kyburgl, Bobbs-Merrill Reprint, Phil 240, 1969.

"The Concept of Inductive Evidence," in Richard Swinburne, ed., *The Justification of Induction* (London: Oxford University Press, 1974), pp. 48—57.

"Rejoinder to Barker and Kyburg," *ibid.*, pp. 66—73.

Spanish translation:

"El concepto de evidencia inductiva," in Richard Swinburne, ed., *La justificacion del razonamiento inductivo* (Madrid: Alianza Editorial, 1976), pp. 61—71.

"Replica a Barker y Kyburg," *ibid.*, pp. 85—92.

"The Status of Prior Probabilities in Statistical Explanation," *Philosophy of Science* **XXXIII**, 2 (Apr., 1965), pp. 137—146.

This paper was presented orally at a meeting of the American Association for the Advancement of Science in 1963, shortly after the publication of Carl G. Hempel's "Deductive-Nomological vs. Statistical Explanation" (1962). This paper by Hempel is truly epoch-making, for it is, to the best of my knowledge, the first detailed and systematic attempt by any philosopher to explicate the concept of statistical explanation. My paper offers a critique of Hempel's theory. Hempel provided an improved account in "Aspects of Scientific Explanation" (1965), but my argument applies equally to the later version as well.

In this article I focus on fundamental differences between deductive and inductive arguments to show that, in the inductive case, prior probabilities should not be ignored. The upshot is that we must focus, not on the degree of probability of the explanandum relative to the explanans, but rather on the difference between the prior probability and the posterior probability. In this paper I argued that *positive statistical relevance* rather than *high probability* is the key to statistical explanation. In subsequent papers, especially "Statistical Explanation" (1970), I argued that statistical relevance, *positive or negative*, has explanatory import. Still more recently, in *Scientific Explanation and the Causal Structure of the World* (1984), I maintained that *causal relevance* is the relation that embodies explanatory import, and that statistical relevance relations are important to explanation only as they provide evidence for causal relations. In any case, it seems to me, high probability is not at all important with respect to explanation.

This paper is my first writing on scientific explanation. Although it is quite obscure, it is the precursor to later work in which I elaborated the statistical-relevance (S-R) model of scientific explanation.

To my great delight, Henry E. Kyburg, Jr., who was the official commentator on my paper at the AAAS meeting, provided an example to show that precisely the same sort of relevance objection I raised against Hempel's model of statistical explanation also can be brought against the deductive-nomological model as well.

Reply to critic:

"Reply to Kyburg," *ibid.*, 152—154.

"Consistency, Transitivity, and Inductive Support," *Ratio* **VII**, 2 (Dec., 1965), pp. 164—169.

This article investigates certain intuitively appealing analogies between

inductive support and logical entailment, exhibiting the dangers of the 'almost deduction' view of induction. It also examines certain modes of combining inductive and deductive arguments. It is shown that inductive support is not transitive, that deductive inference from inductive conclusions is acceptable, but that inductive inference from deductive conclusions is prohibited. It is also shown that inductive reasoning must satisfy an equivalence condition that is closely related to the criterion of linguistic invariance.
German translation:
"Widerspruchsfreiheit, Transitivität und induktive Bekraftigung," *Ratio* **VII**, 2 (Dec., 1965), pp. 152—157.

"Use, Mention, and Linguistic Invariance," *Philosophical Studies* **XVII**, 1—2 (Jan./Feb., 1966), pp. 13—18.

This article presents a rationale for invoking the criterion of linguistic invariance as a principle for selecting suitable rules of inductive inference. According to this criterion, an inductive rule is unacceptable if it sanctions the sanctions the drawing of two mutually inconsistent conclusions from two logically equivalent (consistent) sets of premises. A number of well-known methods violate this condition. The criterion has been attacked on the ground that, in certain circumstances, the way in which the data are formulated may affect the outcome of an experiment. The criterion is defended against such attacks by appealing to the use/mention distinction.

"Verifiability and Logic," in P. K. Feyerabend and Grover Maxwell, eds., *Mind, Matter, and Method* (Minneapolis: University of Minnesota Press, 1966), pp. 347—376.

This essay evaluates reasons for the general abandonment of verifiability criteria of cognitive meaningfulness. Church's critique of the formulation in the second edition of A. J. Ayer's *Language, Truth, and Logic* is analyzed. It is seen that the difficulty lies in an inadequate characterization of inductive confirmation; Church's critique is patently irrelevant to the verifiability criterion per se. Furthermore, difficulties cited by Hempel, involving combinations of verifiable and unverifiable sentences, can be handled by means of a suitable criterion of *sameness* of cognitive meaning. Finally, it is shown that, if verifiability is understood in terms of *physical* possibility of verification, physical considerations may impose constraints upon the structure of our logic. There are some reasons for believing that this situation actually obtains in the realm of quantum mechanics.

Today (1987) I am not quite sure where I stand on the issue of meaning criteria. As I have understood them, verifiability (or confirmability or testability) criteria were always regarded as criteria of *cognitive* meaning. There was no intention of denying that other sorts of meaning exist. At present I am not strongly tempted to argue for any such general criterion of cognitive meaning, but I would be inclined to express total lack of interest in allegedly factual statements for which it is impossible in principle to obtain any empirical evidence, positive or negative.

I do have considerable sympathy with Karl R. Popper's attempt to establish a criterion of demarcation between science and nonscience. He offers falsifiability as the criterion of demarcation. While I agree that it is often useful to think in terms of the possibility of falsifying a putatively scientific statement, I would be inclined to broaden the criterion to allow the possibility of either positive or negative empirical evidence to qualify a statement as scientific. Scientists, by and large, it seems to me, adopt some such attitude; frequently they use the term "operational" to express that view. In spite of the wildly speculative character of much current physical theorizing, I think that few scientists take seriously theories that are totally immune in principle to empirical testing.

Reprinted:

"Verifiability and Logic," in Malcolm L. Diamond & Thomas V. Litzenburg, Jr., eds., *The Logic of God* (Indianapolis: Bobbs-Merrill, 1975), pp. 456—480.

"The Foundations of Scientific Inference," in Robert G. Colodny, ed., *Mind and Cosmos* (Pittsburgh: University of Pittsburgh Press, 1966), pp. 135-275.

See the comments under *The Foundations of Scientific Inference*.

Reprinted:

Wesley C. Salmon, *The Foundations of Scientific Inference* (Pittsburgh: University of Pittsburgh Press, 1967). [Addendum, April 1967, added]

"The Problem of Induction," in John Perry and Michael Bratman, eds., *Introduction to Philosophy* (New York: Oxford University Press, 1986), pp. 265—286. [Contains pp. 5—27, 40—43, 48—56, 132—136 from *The Foundations of Scientific Inference*.]

"Carnap's Inductive Logic," *Journal of Philosophy* **LXIV**, 21 (Nov. 9, 1967), pp. 725—739.

This was the lead paper in an American Philosophical Association symposium on The Philosophy of Rudolf Carnap. My paper focuses on Carnap's contributions to inductive logic, and it expresses both appreca-tion and criticism. Even those who, like me, disagree fundamentally with Carnap's theory of logical probability must admit that his work has brought the precision of modern logic and semantics to this area, making it possible for us to discuss these issues in a much more rigorous fashion than was previously possible. And his investigations have revealed many problems that were previously unrecognized.

The criticisms fall under four headings. (1) I maintain that the notion of *partial entailment*, though initially appealing, cannot provide any underpinning for a theory of inductive logic. This point is elaborated in much greater detail in "Partial Entailment as a Basis for Inductive Logic" (1969). (2) Fundamental problems attach to Carnap's doctrine that degree of confirmation statements, where true, are analytic. There is serious question as to how such probability statements can function as a guide of life. (3) In a number of places I have criticized Carnap's systems of inductive logic found in *Logical Foundations of Probability* and *The Continuum of Inductive Methods* for violating the criterion of linguistic invariance. In subsequent work Carnap attached his logical measures to models rather than to sentences or predicates, thus avoiding the problem of linguistic variance. In this paper I suggest that we need another invariance principle — a principle of statistical invari-ance — that applies to models or properties, rather than to linguistic entities. (4) Questions are raised about Carnap's changing attitute toward the problem of justification of induction. I suggest that his appeal to "inductive intuition" does not resolve the difficulty.
Italian translation:

"La logic induttiva di Carnap," in Alberto Meotti, ed., *L'induzione e L'ordine Dell'universo* (Milano: Edizioni de Comunita, 1978), pp. 191—192. [Translation of small extract.]

"The Justification of Inductive Rules of Inference," in Imre Lakatos, ed., *The Problem of Inductive Logic* (Amsterdam: North-Holland Publishing Co., 1968), pp. 24—43.

This paper was presented orally at a conference at Bedford College, London, England, in 1965, attended by Carnap and Popper, among many others. In it I argued: (1) contra ordinary language dissolutionists, that inductive rules of inference stand in need of *justification*; (2) contra Popper, that science requires *inductive* rules among its rules of inference, and that Popper's deductivism is a disguised attempt to justify inductive methods; and (3) contra Carnap, that analytic degree of confirmation statements do not provide an adequate 'guide of life,' and hence, that we need bona fide inductive *rules of inference*. Inasmuch as I realized that the vindication I had offered in previous papers was not cogent, I made no claim that an adequate pragmatic vindication of induction had been supplied, but I still held out hope for such a result.

Ian Hacking was the official commentator on this occasion, and he directed his comments entirely at my efforts to provide a Reichenbach type of pragmatic vindication of induction. Although I answered many of his objections, I had to acknowledge the cogency of his argument (given in an appendix to his comments) that to justify the rule of induction by enumeration it is necessary and sufficient to justify three principles: additivity, invariance, and symmetry. I believe that additivity is rather easily justified; that invariance, though it is a bit harder, can be justified; but that symmetry is much more recalcitrant. These issues are discussed in some detail in my lengthy reply.

Hacking's critique was, in one sense, devastating. I have never subsequently published any further attempt to provide the sort of vindication I had been working on until that time. Papers published subsequently on probability or confirmation have skirted that issue. Indeed, about that time I turned most of my attention to scientific explanation. Nevertheless, I have not found any of the attempts to dissolve the problem of induction convincing, and I have not concluded that it is impossible in principle to provide any sort of vindication. As I maintain in "Unfinished Business: The Problem of Induction" (1978), it is a problem we still must face.

Reply to critics:
"Reply," *ibid.*, pp. 74—97.
Spanish translation:
"La justificacion de las reglas inductivas de inferencia," in Luis O. Gomez and Roberto Torretti, eds., *Problems de la Filosofia* (Editorial Universitaria, Universidad de Puerto Rico, 1974), pp. 373—385.

"Inquiries into the Foundations of Science," in David L. Arm, ed., *Vistas in Science* (Albuquerque: University of New Mexico Press, 1968), pp. 1—24.

Directed primarily toward nonphilosophers, this essay attempts to present and clarify basic logical issues concerning the testing and confirmation of scientific hypotheses. The role of plausibility arguments serves as a point of departure for comparison between the standard hypothetico-deductive schema and an approach using Bayes's theorem as the basic schema. The nature of the prior probabilities is discussed from the standpoint of three different interpretations of probability: as a priori probabilities in the logical interpretation, as subjective entities in the personalistic interpretation, and as objective probabilities in the frequency interpretation. The significance of the washing out or swamping of prior probabilities in the presence of increasing amounts of evidence from tests is emphasized.

This paper is an attempt to present objective Bayesianism to a broad audience consisting mainly of scientists. It is a popularization of some of the central material in chapter VII of *The Foundations of Scientific Inference*. I believe it is basically sound, but in my recent thinking about the Bayesian approach I have come to see that there are serious problems in interpreting the likelihoods that occur in Bayes's theorem.
Reprinted:
"Inquiries into the Foundations of Science," in Sidney Luckenbach, ed., *Probabilities, Problems, and Paradoxes* (Encino, CA: Dickenson Publishing Co., 1972), pp. 139—58.

"Introduction: The Context of These Essays," *Philosophy of Science* **XXXVI**, 1 (Mar., 1969), pp. 1—4. [With Adolf Grünbaum]

This is the Introduction to "A Panel Discussion of Simultaneity by Slow

Clock Transport in the Special and General Theories of Relativity." The following paper is my contribution to the panel; the other contributors are Adolf Grünbaum, Allen I. Janus, and Bas van Fraassen.

"The Conventionality of Simultaneity," *Philosophy of Science* **XXXVI**, 1 (Mar., 1969), pp. 44–63.

After describing a new method of synchronizing spatially separated clocks by means of clock transport, this paper discusses the philosophical import of the existence of such methods, including those of Brian Ellis and Peter Bowman and of Bridgman, with special reference to the Ellis-Bowman claim that "the thesis of the conventionality of distant simultaneity . . . is either trivialized or refuted." I argue that the physical facts do not support this philosophical conclusion, and that a substantial part of their argument against Reichenbach, in particular, is misdirected. Finally, I suggest that Ellis and Bowman employ seriously unclear notions of triviality and "good physical reasons" that tend to obscure rather than clarify the basic philosophical issues. An objective *criterion of non-triviality of conventions* is advanced.

The issues raised in this paper are also treated in chapter 4 of *Space, Time, and Motion: A Philosophical Introduction* and in "The Philosophical Significance of the One-Way Speed of Light" (1977).

"Partial Entailment as a Basis for Inductive Logic," in Nicholas Rescher, ed., *Essays in Honor of Carl G. Hempel* (Dordrecht: D. Reidel Publishing Co., 1969), pp. 47–82.

This paper examines a fundamental thesis of the logical theory of probability — namely, that inductive logic is based upon a concept of partial entailment in much the same way as deductive logic rests upon a relation of full logical entailment. The notions of full entailment, partial entailment, and complete independence are examined in the context of propositional logic. This context yields a plausible relation of partial deductive entailment, but this relation is of no use whatever in constructing an inductive logic. It is admitted that the theory of logical probability, after it has been constructed (say, by defining a particular confirmation function) can be said to furnish a relation of partial entailment, but an appeal to partial entailment provides no guidance whatever in selecting the appropriate confirmation function. Hence, the notion of partial entailment cannot serve as a foundation for logical probability. Its heuristic use for this purpose has been positively misleading. The question must then be raised: what legitimate a priori basis can logical probability or inductive logic rest upon? The only satisfactory answer, I think, is that there can be no such a priori basis.

"Statistical Explanation," in Robert G. Colodny, ed., *The Nature and Function of Scientific Theories* (Pittsburgh: University of Pittsburgh Press, 1970), pp. 173–232.

See the comments under *Statistical Explanation and Statistical Relevance*.
Reprinted:

Wesley C. Salmon, et al., *Statistical Explanation and Statistical Relevance* (Pittsburgh: University of Pittsburgh Press, 1971), pp. 29–87.

"Bayes's Theorem and the History of Science," in Roger H. Stuewer, ed., *Minnesota Studies in the Philosophy of Science*, vol. V (Minneapolis: University of Minnesota Press, 1970), pp. 68–86.

This essay embodies an attempt to clarify the relationships between history of science and philosophy of science. First, the concepts, *context of discovery* and *context of justification*, are clarified. Next, the simple hypothetico-deductive account of confirmation is shown to be inadequate; a Bayesian approach is judged to be more satisfactory. Appeal to Bayes's theorem reveals that plausibility considerations have a legitimate place in the context of justification. Finally, it is suggested that historians who view scientific method in hypothetico-deductive terms run a great risk of making erroneous judgments regarding rational vs. nonrational components of science — i.e., of confusing the contexts of discovery and justification.

In 1983 I had the honor of participating, along with Thomas Kuhn and Hempel, in an American Philosophical Association symposium on The Philosophy of Carl G. Hempel. My contribution on that occasion was "Carl G. Hempel on the Rationality of Science" (1983). In that paper I attempted to build a bridge between Kuhn and Hempel by means of Bayes's theorem. This more recent paper picks up on the point made "Bayes's Theorem and the History of Science." I am currently working on a rather large paper, "Rationality and Objectivity in Science," that will discuss this Bayesian bridge-building enterprise in a more complete and detailed fashion. I am pleased that Kuhn has, in conversation, taken a positive attitude toward this venture.

"Determinism and Indeterminism in Modern Science," in Joel Feinberg, ed., *Reason and Responsibility*, 2nd and subsequent eds. (Encino, CA: Dickenson Publishing Co., 1971).

This essay, designed for introductory philosophy students, attempts to clarify the meanings of determinism and indeterminism, and to assess their relations to free will. Different forms of determinism, mechanistic and fatalistic, are distinguished, along with the corresponding forms of indeterminism. Evidence for Laplacian determinism in classical physics, and in the biological and behavioral sciences, is presented. Evidence for indeterminism in contemporary physics is discussed. Such concepts as law of nature, cause, and statistical correlation are explicated. Various models of scientific explanation, appropriate to deterministic and indeterministic frameworks, are presented and compared.

"Logic," *Encyclopedia Americana*, 1972, vol. **17**, pp. 673—687.

This article provides a general survey of logic, primarily deductive, presupposing no previous knowledge of the subject. Basic concepts — e.g., argument, form, validity, deduction, induction — are explained. The historical development of logic is sketched. The main aspects of ancient syllogistic logic and Stoic propositional logic are given. Elements of modern symbolic logic — including truth tables, quantifiers, and relations — are presented. First order logic is outlined. Higher order logic and set theory, including Russell's paradox, are briefly treated. Metatheory — including the distinction of syntax/semantics/pragmatics — is introduced, and the liar paradox is discussed. Some important metatheorems are mentioned.

"Confirmation," *Scientific American* **228**, 5 (May, 1973), pp. 75—83.

This article is a sort of 'budget of paradoxes' concerning the confirmation of scientific hypotheses. Hempel's raven paradox and Goodman's grue-bleen paradoxes are mentioned very briefly. More emphasis is placed upon examples that undermine the crude hypothetico-deductive method and show the failure of transitivity of inductive support. The main emphasis is upon puzzles arising from a failure to appreciate the distinction between incremental confirmation (increasing the probability) and absolute confirmation (rendering highly probable) — for example, a counterintuitive situation in which two pieces of evidence each (incrementally) confirm a hypothesis, but in conjunction refute it. Duhem's problem arises with a vengeance in an example in which negative evidence deductively refutes the conjunction of a hypothesis being tested and an auxiliary hypothesis, but nevertheless (incrementally) confirms each of them individually.

It is great fun to have an article published with extremely clever illustrations in living color in the *Scientific American*. The technical details were worked out much more fully in "Confirmation and Relevance" (1975).

Reply to critic:

"Reply to Bradley Efron," *Scientific American* **229**, 3 (Sept., 1973), pp. 8—10.

"Memory and Perception in *Human Knowledge*," in George Nakhnikian, ed., *Bertrand Russell's Philosophy* (London: Gerald Duckworth & Co., 1974), pp. 139—167.

This article and the one that follows were written in celebration of the Russell centennial. This essay assesses the comparative epistemic merits of memory and perception as sources of basic empirical data. The main thesis is that they enjoy equal status. Noting Russell's remark that the general reliability of memory is an "independent postulate," but noting also that Russell did not include any such postulate in his ultimate list, I argue that no such postulate is needed for memory or perception. This paper deals with various forms of the *immediacy objection* — the claim that perceptual data are to be preferred to memory data on grounds of evidential immediacy. All forms of this objection are analyzed and rejected.

"Russell on Scientific Inference *or* Will the Real Deductivist Please Stand Up?" *ibid.*, pp. 183—208.

This paper was presented orally at a conference honoring Russell on the centennial of his birth. It analyzes and evaluates Russell's theory of nondemonstrative inference as stated in *Human Knowledge, Its Scope and Limits*, Russell's last major epistemological treatise. Careful examination reveals that, although Russell often referred to nondemonstrative inference, he did not admit any form of nondemonstrative inference. Nondemonstrative inferences are, for him, enthymemes; instead of seeking rules of nondemonstrative inference he seeks suitable premises to add to transform the enthymemes into valid deductions. His logic is strictly deductive.

Russell's views are put into historical perspective by comparison with those of Carnap, Popper, and Reichenbach. Striking similarities are found between Russell and Carnap; they provide complementary approaches to the same basic problems in the logic of science.

"An Encounter with David Hume," in Joel Feinberg, ed., *Reason and Responsibility*, 3rd and subsequent eds., (Encino, CA: Dickenson Publishing Co., 1975), pp. 190—208.

This dialogue is an attempt to make the central arguments in Hume's *Enquiry Concerning Human Understanding* — at least sections IV, V, and VII — intelligible to contemporary undergraduates who have no other philosophical background. It translates Hume's concerns into modern language and places them in modern contexts. It relates Hume's problems to matters that the student typically encounters in other college courses, as well as to the experiences of everyday life. Scientific approaches are contrasted with nonscientific methods. The main modern attempts to resolve Hume's problem of induction are considered.

In preparation for writing this piece, I tried to clear my mind of *everything* I had ever read (including Hume's own work) or written about Hume, and to read the *Enquiry* from scratch. I tried to imagine the experience of a very bright student taking — among other courses such as introductory physics and social science — an introductory philosophy course in which Hume's *Enquiry* is being studied. The aim was to make Hume both understandable and interesting to contemporary students.

This writing experience was the most enjoyable I have ever had. The dialogue was written in a single weekend, with time out only to eat and sleep. It seemed to flow almost automatically. For example, as I began it I had no particular intention of writing a dialogue; it just happened that way. As a result of this project, I think I know what novelists and playwrights mean when they refer to the inner necessity of a work.

"Confirmation and Relevance," in Grover Maxwell and Robert M. Anderson, Jr., eds., *Minnesota Studies in the Philosophy of Science*, vol. VI (Minneapolis: University of Minnesota Press, 1975), pp. 3—36.

This essay draws out the consequences of the distinction between two senses of "to confirm": the absolute sense which means "to render highly probable," and the relevance (incremental) sense which means "to render more probable." The force of this distinction, which has been in the literature for many years — the *locus classicus* is chapter 6 of Carnap's *Logical Foundations of Probability* — has been insufficiently understood and appreciated. While the relevance or incremental sense is the one most frequently used in ordinary scientific discourse, logical analyses usually treat the absolute sense.

There are extremely profound disanalogies between these two senses, and this point is reinforced by reference to a variety of paradoxical-sounding examples. (Some, but not all, of these examples were discussed in my *Scientific American* article, "Confirmation" (1973).) Analysis of the paradoxes shows that they arise from an important, but usually unnoticed, aspect of the probability concept on which both senses of confirmation are based.

Reprinted:

"Confirmation and Relevance," in Peter Achinstein, ed., *The Concept of Evidence* (Oxford: Oxford University Press, 1983), pp. 93—123.

"Theoretical Explanation," in Stephan Körner, ed., *Explanation* (Oxford: Basil Blackwell, 1975), pp. 118—145.

This paper was written for presentation at a conference on explanation (not just scientific explanation) held at Bristol in 1973. The title was chosen before the paper was written; had I realized in advance how the paper would come out, I would have entitled it "Causal Explanation," My aim certainly was to deal with theoretical explanation, but every effort to do so became completely entangled with causal considerations.

As it turned out, this essay embodies an extension of the statistical-relevance model of explanation that introduces relations of causal relevance into the explanatory picture. A basic thesis is that relations of statistical relevance between noncontiguous events are to be explained by means of spatio-temporally continuous causal processes. The mark criterion is invoked to distinguish causal processes from pseudo-processes. If neither correlated event is a cause of the other, then a common cause with continuous causal connections to both effects is postulated. Such causal explanations often require postulation of unobserved or unobservable entities. The common cause principle provides the basis for the temporal asymmetry of scientific explanation.

This paper embodies my first serious effort to deal with causality in scientific explanation. I had hoped that it would yield a causal argument in support of scientific realism, but that had to await "Why Ask 'Why?'?" (1978).

Reply to critics:

"Reply to comments," *ibid.*, pp. 160—184.

"Clocks and Simultaneity in Special Relativity *or* Which Twin Has the Timex?" in Peter K. Machamer and Robert G. Turnbull, eds., *Motion and Time, Space and Matter* (Columbus: Ohio State University Press, 1976), pp. 508—545.

See comments on *Space, Time, and Motion: A Philosophical Introduction.*

Reprinted:

Wesley C. Salmon, *Space, Time, and Motion: A Philosophical Introduction* (Encino, CA: Dickenson Publishing Co., 1975), ch. 4.

Foreword to *Laws, Modalities, and Counterfactuals* by Hans Reichenbach (Berkeley & Los Angeles: University of California Press, 1976), pp. vii—xlii.

This book by Reichenbach is a reprint of his 1954 posthumous monograph, *Nomological Statements and Admissible Operations*, which had achieved little notice and had been out of print for many years. Since the original title gave most philosophers no idea of what the book was about, I persuaded the California Press to bring it out under the revised title, *Laws, Modalities, and Counterfactuals*, which gives a clear indication of the subject matter.

My Foreword places the problems to which Reichenbach devoted his attention in this book in historical and contemporary context. It explains the basic concepts and strategy of his extraordinarily complex theory in terms that are, I hope, intelligible to a broader philosophical audience. I argue that, given the absence of any satisfactory theory covering this general area, Reichenbach's theory deserves careful attention. My present view is that Reichenbach's theory is not fully adequate, but that study of his work provides valuable philosophical insight.

Reprinted (with minor revisions):

"Law, Modalities, and Counterfactuals,' *Synthese* **35**, 2 (June, 1977), pp. 191—229.

"Laws, Modalities, and Counterfactuals," in Wesley C. Salmon, ed., *Hans Reichenbach: Logical Empiricist* (Dordrecht: D. Reidel Publishing Co., 1979), pp. 655—696.

Preface to *Hans Reichenbach: Logical Empiricist, Synthese* **34**, 1 (Jan., 1977), pp. 1—2.

"The Philosophy of Hans Reichenbach," *Synthese* **34**, 1 (Jan., 1977), pp. 5—88.

This is essentially a small monograph that presents a comprehensive account of Reichenbach's philosophical achievements. It is designed to show clearly the enormous scope and essential unity of his work. The chief areas discussed are probability and induction; space, time, and relativity; the direction of time, including causality and causal explanation; general epistemology; laws, modalities, and counterfactuals; quantum mechanics; and logic, philosophy of language and philosophy of mathematics. Underlying themes are traced through these various topics in an attempt to exhibit the main purposes and unifying principles of his work. Considerable attention is devoted to demonstrating its relevance to a number of major areas of current philosophical activity. Inasmuch as Reichenbach's work seems generally underappreciated at present, I wish that my essay were more widely read.

It is worth noting that The Archive for Scientific Philosophy in the Twentieth Century at the University of Pittsburgh now contains Reichenbach's papers, as well as those of Carnap and Ramsey. Scholars are invited to make use of this resource.

See also the comments under *Hans Reichenbach: Logical Empiricist.*

Reprinted:

Wesley C. Salmon, ed., *Hans Reichenbach: Logical Empiricist* (Dordrecht: D. Reidel Publishing Co., 1979), pp. 1—84.

German translation with revisions:

"Hans Reichenbachs Leben und die Tragweite seiner Philosophie" (Einleitung zur Gesamtausgabe), in Andreas Kamlah and Maria Reichenbach, eds., *Hans Reichenbach: Gesammelte Werke*, vol. 1 (Wiesbaden: Vieweg, 1977), pp. 5—81. [German translation by Maria Reichenbach.]

"Hempel's Conception of Inductive Inference in Inductive-Statistical Explanation," *Philosophy of Science* **44**, 2 (June, 1977), pp. 180—185.

Carl G. Hempel has often stated that inductive-statistical explanations, as he conceives them, are inductive arguments. This discussion note raises the question of whether such arguments are to be understood as (1) arguments of the traditional sort, containing premises and conclusions, governed by some sort of *inductive acceptance rules*, or (2) something more closely akin to Carnap's degree of confirmation statements, which occur in an inductive logic that entirely eschews inductive acceptance rules. Hempel's writings do not seem unequivocal on this issue. It is suggested that the adoption of construal (2) would remove the need for Hempel's high probability requirement on inductive-statistical explanations.

"Indeterminism and Epistemic Relativization," *ibid.*, pp. 199—202.

Carl G. Hempel's doctrine of *essential epistemic relativization of inductive-statistical explanation* seems to entail the unintelligibility of the notion of *objective homogeneity* of reference classes. This discussion note explores the question of whether, as a consequence, it also entails the unintelligibility of the doctrine of indeterminism.

"An 'At—At' Theory of Causal Influence," *ibid.*, 215—224.

The propagation of causal influences through spacetime seems to play a fundamental role in scientific explanation. Taking as a point of departure a basic distinction between causal interactions (which are localized in spacetime) and causal processes (which may extend throughout vast regions of spacetime), this paper attempts an analysis of the concept of causal propagation on the basis of the ability of causal processes to transmit marks. The analysis rests upon the at—at theory of motion which provided a resolution of Zeno's paradox of the arrow. It is argued that this explication does justice to the concept of the ability of causal processes to transmit causal influence without invoking anti-Humean 'powers' or 'necessary connections.'

Although the formulation in this paper had to be modified in the light of an example given by Nancy Cartwright, the basic idea still seems sound to me. See *Scientific Explanation and the Causal Structure of the World*, chapter 5.

"The Curvature of Physical Space," in John S. Earman, Clark N. Glymour, and John Stachel, eds., *Minnesota Studies in the Philosophy of Science*, vol. VIII (Minneapolis: University of Minnesota Press, 1977).

In a 1972 paper, Clark Glymour takes Adolf Grünbaum to task for claiming that the curvature of space cannot be intrinsic if the metric of space is not intrinsic, for the curvature presupposes (is derivable from) the metric. Glymour makes much of the fact that there is a curvature tensor that can be derived from the affine connection alone, without requiring a metric. There are, as a matter of fact, two rank 4 Riemannian curvature tensors — one covariant, the other mixed. The covariant tensor does presuppose a metric, and that was the one to which Grünbaum obviously had referred. The mixed tensor is derivable from the affine connection alone. On this point of technical detail Glymour was right. However, the point at issue was whether physical space can possess an intrinsic curvature. Glymour offers no hint of an argument to support the notion that the mixed tensor represents an intrinsic type of curvature. In my paper I argue that, if there is no intrinsic metric, still less is there any ground for supposing that the affine connection is intrinsic. Bringing in the curvature represented by the mixed Riemannian tensor provides no aid or comfort to the advocates of geometrodynamics whose theory requires intrinsic curvature if it is to get off the ground. Glymour's observation regarding the affine curvature does nothing to undermine Grünbaum's critique of John A. Wheeler's geometrodynamics.

"The Philosophical Significance of the One-way Speed of Light," *Nous* **XI**, 3 (Sept., 1977), pp. 353—392.

In his initial 1905 paper on special relativity, Einstein offered the following stipulation regarding the synchronization of nonadjacent clocks: given a light signal that travels from A to B, where it is reflected back to A, its speed from A to B is set equal *by definition* (Einstein's italics!) to its speed from B to A. If the equality of the one-way speeds is conventional (as Einstein claims), then it is impossible empirically to determine the one-way speed. The one-way speed is not a matter of fact. This paper reviews several proposed experimental methods for ascertaining the one-way speed of light, and concludes that each fails, for each presupposes some convention that is tantamount to Einstein's convention regarding the one-way speed of light. Since the synchronization of spatially separated clocks involves this convention, the very concept of *any one-way velocity* also contains a nontrivial element of convention.

Ironically, this paper was published in the same issue of *Nous* as a paper by David Malament showing that, given certain minimal assumptions, the relation of simultaneity in any given inertial frame of reference can be defined causally.

I do not claim to have shown that all possible experimental methods for establishing the one-way speed of light must fail; I simply surveyed a rather wide variety of methods. Moreover, the paper by Malament does not yield an experiemental method for ascertaining the one-way velocity. Malament's definition depends on the global features of spacetime. No experiment involves more than local regions of the universe.

"A Third Dogma of Empiricism," in Robert E. Butts and Jaakko Hintikka, eds., *Basic Problems in Methodology and Linguistics*, Part III of the *Proceedings of the Fifth International Congress of Logic, Methodology, and Philosophy of Science* (Dordrecht: D. Reidel Publishing Co., 1977), pp. 149—166.

The dogma to which this paper refers is the widely held thesis that scientific explanations are arguments. This thesis is challenged by posing three questions that seem to raise difficulties for it: (1) Why are irrelevancies harmless to arguments but fatal to explanations? (2a) Can events whose probabilities are low be explained? — or to reformulate essentially the same question — (2b) Is genuine scientific explanation possible if indeterminism is true? (3) Why should requirements of temporal asymmetry be imposed upon explanations but not upon arguments. It is suggested that the statistical-relevance model of scientific explanation can cope with these questions straightforwardly. This paper closes with a strong plea that the time has come to put the "cause" back into "because."

"Objectively Homogeneous Reference Classes," *Synthese* **36**, 4 (Dec., 1977), pp. 399—414.

This paper contains my first attempt to explicate the concept of an *objectively homogeneous reference class*. It is flawed in many ways, and is completely superseded by Chapter 3 of *Scientific Explanation and the Causal Structure of the World.*

"Religion and Science: A New Look at Hume's *Dialogues*," *Philosophical Studies* **33**, (1978), pp. 143—176.

This article was presented orally at a memorial conference at the University of Arizona paying tribute to Hume on the bicentennial of his death. It deals with the design argument for the existence of God as it is discussed in his *Dialogues Concerning Natural Religion*. Using Bayes's theorem — with which, according to Richard Price, Hume had some acquaintance — it shows that the various arguments advanced by Philo and Cleanthes fit neatly into a comprehensive logical structure. The conclusion drawn is that, not only does the empirical evidence fail to support the theistic hypothesis, but also renders the athiestic hypothesis quite highly probable. A postscript speculates on the historical question of Hume's own attitude toward the design argument.

Reply to critic:

"Experimental Atheism," *Philosophical Studies* **35** (1979), pp. 101—104.

This is a brief rejoinder to Nancy Cartwright who served as the commentator on my paper at the Hume conference. She complained that I did not provide enough convincing evidence of order arising out of disorder in the absence of intelligent intervention or planning. She urged appeal to controlled experiments. In this reply I tried to show that *experimentally grounded* scientific results overwhelmingly support experimental atheism. Among other things, I remind her of our trip to the Harvard-Smithsonian Observatory on Mt. Hopkins, just south of Tucson, where we actually observed a site of star formation. The reply might well have been entitled, "A Star is Born."

"Unfinished Business: The Problem of Induction," *Philosophical Studies* **33** (1978), pp. 1—19.

This paper is based on an informal after-dinner speech I had presented a number of years earlier at an Oberlin Philosophy Conference under the title, "What Next, Dammit, David Hume?" Although it was not delivered orally at the above-mentioned Hume Conference, it was written up for inclusion in the published proceedings of that conference. A major part of my motivation was the fact that this conference on the intellectual legacy left to us by Hume included no paper even touching on the problem of justification of induction. I considered that situation intolerable, at least as far as the published record was concerned. The paper retains many of the marks of an after-dinner speech, but it is intended quite seriously.

The article surveys a wide variety of historic and contemporary attempts to deal with the problem of justification of induction. It argues that various attempts to dissolve the problem through linguistic analysis or other means, as well as the attempt to evade it via a deductivist approach, are unsuccessful. Various attempts to provide justifications of induction are also examined and found inadequate. The conclusion is that this deeply significant and highly recalcitrant problem remains with us — *unsolved* — as a valuable part of Hume's philosophical legacy.

"Why Ask, 'Why?'? — An Inquiry Concerning Scientific Explanation," *Proceedings and Addresses of the American Philosophical Association* **51**, 6 (Aug., 1978), pp. 683—705.

This essay, which was my Presidential Address to the Pacific Division of the American Philosophical Association, constitutes an attempt to provide the general outlines of a causal theory of scientific explanation that incorporates the virtues of both the standard inferential view and previous causal accounts. The new theory relies heavily upon the distinction between causal processes and causal interactions, and upon the principle of the common cause. Two types of causal forks — conjunctive and interactive — are distinguished; each is seen to have a distinct function in the account of scientific explanation. The common cause principle is invoked to show how unobservable entities play a fundamental explanatory role, thereby clarifying the explanatory power of theories.

This essay embodies what is, for me, an important extension of the enterprise begun, but not even close to completed, in "Theoretical Explanation" (1975). It was major step toward the theory offered in *Scientific Explanation and the Causal Structure of the World.*

Reprinted:

"Why Ask, 'Why?'? — An Inquiry Concerning Scientific Explanation," in Wesley C. Salmon, ed., *Hans Reichenbach: Logical Empiricist* (Dordrecht: D. Reidel Publishing Co., 1979), pp. 403—425.

"Why Ask, 'Why?'? — An Inquiry Concerning Scientific Explanation," in Janet A. Kourany, ed., *Scientific Knowledge* (Belmont, CA: Wadsworth Publishing Co., 1987), pp. 51—64.

"Alternative Models of Scientific Explanation," *American Anthropologist* **81**, 1 (Mar., 1979), pp. 61–74. [With Merrilee H. Salmon.]

Shortly after my arrival at the University of Arizona in 1973, I was approached by two graduate students in archaeology with an invitation to speak to one of their seminars on scientific explanation. Before that incident I had had no inkling that an important and influential group of archaeologists — practicing an approach known as *the new archaeology* — were endeavoring to put archaeology on a scientific basis, and that, in their view, the hallmark of science is the construction of explanations that conform to Hempel's deductive-nomological model. The students provided me with a few items written by archaeologists and a couple of acerbic responses by philosophers. With my wife Merrilee Salmon — who was also a member of the philosophy faculty and was invited to join us — I met with the seminar for pleasant and useful discussion. Thus began a most fruitful and rewarding interdisciplinary relationship. We were there at their request, not as philosophers attempting to bring unsolicited philosophical wisdom to practitioners in a different discipline. In the ensuing years we met with their classes, visited their field schools, attended their meetings, presented papers at their conferences, and formed firm friendships.

The present paper is a melding of two separate papers, given by M. Salmon and me, at an annual meeting of the Society for American Archaeology. In response to archaeologists' interests in models of scientific explanation, it surveys several 'covering law' models. Primary emphasis is upon a critical comparison of Hempel's deductive-nomological and inductive-statistical models, and Meehan's systems model, with the statistical-relevance model. The crucial difference hinges upon certain relevance conditions. Two advantages of the latter model — of possible interest to anthropologists, and especially to archaeologists — are its ability to incorporate explanations of low-probability events and its potential for furnishing an account of functional explanation. Toward the end, suggestions for supplementation of the statistical-relevance model with factors of causal relevance are advanced.

Although I found archaeology fascinating, and greatly enjoyed our relationships with archaeologists, I never found the time or motivation to take up the study of archaeology in a serious way myself. M. Salmon, in contrast, has studied the subject extensively, and has published many articles on philosophy of archaeology in archaeological journals. Her major work on the subject is *Philosophy and Archaeology* (Academic Press, 1982).

It is, perhaps, regrettable that I have not pursued the subject more avidly, inasmuch as I seem to have a native talent for it. One night, between 9:00 p.m. and 3:00 a.m. (after a full day at another site about 100 miles away with the same crew) we participated in an archaeological survey, by lights from the miners' caps we wore, of an area in the Mammoth Cave in Kentucky. On this survey, I discovered not just one — which might have been sheer luck — but *two* human paleofecal objects, while no one else on the survey found any. Such objects are of considerable archaeological value, for they furnish useful information about prehistoric diets. One might be tempted to say I have a nose for this sort of work, but it should be recalled that, after a few hundred years, they no longer have any odor at all.

"Philosopher in a Physics Course," *Teaching Philosophy* **II**, 2 (1979), pp. 139—146.

During my eight years at the University of Arizona, I participated in "Concepts, Philosophy, and History of Physics," a team-taught two-semester introductory physics course designed to make physical science meaningful to students not planning to major in that field. This course was, I believe, quite unique. During the first year the team consisted of a physicist with a fairly extensive knowledge of the history of science and myself. In subsequent years it was taught by a professional historian, a professional philosopher, and a professional physicist, all of whom were present during virtually every class. This article describes the content and orientation of the course: it discusses pedagogical problems and techniques. It stresses the importance of the involvement of professionals in all three areas. In assessing the role of philosophy in this sort of teaching venture, it concludes that philosophy of science has an important role to play, but that the role history of science is even more significant.

Two different historians and several different physicists participated in the teaching of this course, and I think almost all of us considered it an excellent course in terms of its approach and its content. The regrettable fact was that we were never able to recruit the right student clientele. Since our course did not presuppose calculus, the advisors invariably directed the better students into the standard introductory physics courses with calculus as a prerequisite even if they had no intention of pursuing physical science as a major.

"Propensities: A Discussion Review," *Erkenntnis* **14** (1979), pp. 183–216.

This discussion-review takes, as its point of departure, D. H. Mellor's *The Matter of Chance*, but it contains an extended critique of the so-called propensity interpretation of probability. Popper's initial claim that the propensity interpretation "takes the mystery out of quantum mechanics" is seen to be groundless. Moreover, Popper's treatment of the probabilities of single cases is shown to constitute no advance whatever beyond Reichenbach's explicit treatment of the problem of the single case within the frequency interpretation. It is also shown that the propensity interpretation does not avoid the notoriously difficult problem of selecting an appropriate reference class; exactly the same problem arises in specifying the chance setup that gives rise to the propensity. Finally, as Paul Humphreys pointed out, propensities cannot qualify as an interpretation of the probability calculus, for the probability calculus contains inverse probabilities — e.g., in Bayes's theorem — but it does not make sense to speak of inverse propensities.

"Informal Analytic Approaches to the Philosophy of Science," in Peter Asquith and Henry E. Kyburg, Jr., eds., *Current Research in Philosophy of Science* (East Lansing, Mich: Philosophy of Science Assn., (1979), pp. 3—15.

There is something quite ironic about this paper. At virtually the last minute (because another prospective participant withdrew) I was recruited to represent the informal analytic approach at a conference, sponsored by the National Science Foundation, on current research problems and approaches in philosophy of science. I did what I could, of course, but because I have fairly strong sympathy with formalization and considerable antipathy toward ordinary language analysis, it was rather like leaving the fox to guard the chicken house.

"Probabilistic Causality," *Pacific Philosophical Quarterly* **61**, 1—2 (1980), pp. 50—74.

This article provides a critical analysis and comparison of the theories of probabilistic causality offered by Hans Reichenbach, I. J. Good, and Patrick Suppes. Each of these theories faces certain fundamental difficulties. The most serious is the problem of negative relevance, which is common to all of them, for all require of a probabilistic cause that it raise the probability of its effect. Nevertheless, I maintain, there are bona fide causal chains in which an event renders its successor less probable. In the end I argue that probabilistic causality cannot be explicated in terms of statistical relations among discrete events alone; instead, we must take into account the physical processes that provide causal connections among the events involved. Because of this latter conclusion, Philip Kitcher has recently suggested that my theory of causality could more appropriately be called *indeterministic causality* than *probabilistic causality*. His proposal makes good sense.

"John Venn's *Logic of Chance*", in J. Hintikka, D. Gruender, and E. Agazzi, eds., *Probabilistic Thinking, Thermodynamics, and the Interaction of the History and Philosophy of Science* (Dordrecht: D. Reidel Publishing Co., 1980), pp. 125—138.

This article, which harks back to my doctoral dissertation on Venn, praises his work on probability and explains why I consider him the first thoroughgoing frequentist. It also mentions the fact that I consider the first edition of *The Logic of Chance* considerably superior to the later and much longer editions.

"Robert Leslie Ellis and the Frequency Theory," *ibid.*, pp. 139—143.

In this brief discussion I explain why, in spite of his brilliant insights, Ellis — whose work precedes Venn's — does not qualify as the first bona fide frequentist.

"Rational Prediction," *British Journal for the Philosophy of Science* **32**, 2 (June, 1981), pp. 115—125.

This paper was an invited presentation at a conference on Sir Karl Popper's philosophy, held at the London School of Economics and Political Science in 1980, and attended by Popper. I argued that, whatever its merits in assessing the explanatory value of theories, Popper's deductivism cannot provide any ground for making the kinds of predictions that are required as a basis for rational action. To the dismay of some of his followers who were present on that occasion, Popper apparently accepted the correctness of my analysis.

This paper represents the culmination of a longstanding controversy. My claim about the insufficiency of Popper's deductivism was raised in the paper I had presented at the Bedford College Conference in 1965, the published version of which was "The Justification of Inductive Rules of Inference" (1968). J. W. N. Watkins offered a Popperian response in his published comment on that paper, and I answered him in my "Reply." The same issue was discussed in *The Foundations of Scientific Inference.*

I am not aware of any attempt at rebuttal on behalf of Popper since "Rational Prediction" was presented in 1980. Indeed, at the Seventh International Congress of Logic, Methodology, and Philosophy of Science in 1983 Watkins summed up the outcome in the following terms, "Game, set, match: Salmon." If the point of this paper is correct, it constitutes a refutation of Popper's repeated conjecture that he had solved Hume's problem of induction.

Reprinted:

"Rational Prediction," in Adolf Grünbaum and Wesley C. Salmon, eds., *The Limitations of Deductivism* (Berkeley & Los Angeles: University of California Press, forthcoming).

French translation:

"La Prédiction Rationnelle" in *Karl Popper Cahiers — Science, Technologie, Société*, vol. 8 (Paris: Editions du Centre National de la Recherche Scientifique, 1985), pp. 108—120.

"In Praise of Relevance," *Teaching Philosophy* **4**, 3—4 (July/Oct., 1981), pp. 261—275.

In this article on the teaching of introductory logic I argue that considerable emphasis should be placed upon inductive reasoning, since that is far more *relevant* to the experience and needs of most students than is standard deductive logic. I claim, moreover, that in dealing with inductive logic substantial attention should be paid to causal reasoning, again, because of its high degree of *relevance* to many of our main concerns in life. In attempting to establish causal relations, statistical *relevance* plays a crucial role.

"Comets, Pollen, and Dreams: Some Reflections on Scientific Explanation," in Robert McLaughlin, ed., *What? Where? When? Why? Essays on Induction, Space and Time, Explanation* (Dordrecht: D. Reidel Publishing Co., 1982), pp. 155—178.

This introductory essay on scientific explanation is written at a rather popular level. Three general conceptions of scientific explanation — epistemic, modal, and ontic — are discussed. Three examples of explanation (to which the title of the essay refers) play a central role: (1) explanation in the context of Laplacian determinism, in particular, the Newtonian explanation of Halley's *comet*; (2) irreducibly statistical explanation, in particular, Brownian movement (*pollen*); and (3) functional explanation, in particular, Freudian explanation of *dreams*. This essay foreshadows the more extended and technical treatment of the three basic conceptions in *Scientific Explanation and the Causal Structure of the World*.

"Further Reflections," *ibid.*, pp. 231—280.

Robert McLaughlin's *What? Where? When? Why?* — the first volume of the *Australasian Studies in History and Philosophy of Science* — is a collection of essays "inspired" by my work and "celebrating" my 1978 visit to Australia. In addition to my "Comets, Pollen, and Dreams," it contains nine other philosophical essays. This set of "Further Reflections" contains rather extended discussions of these contributions, including a number of general observations on induction, space and time, and scientific explanation.

"Causality: Production and Propagation," in Peter D. Asquith et al., eds., *PSA 1980* (East Lansing, Mich.: Philosophy of Science Assn., 1982), pp. 49—69.

In this paper a theory of causality based on physical processes is developed. Causal processes are distinguished from pseudo-processes by means of the criterion of mark transmission. Causal interactions are characterized as those intersections of processes in which the intersecting processes are mutually modified in ways that persist beyond the point of intersection. Causal forks of three kinds — conjunctive, interactive, and perfect — are introduced to explicate the principle of the common cause. Causal forks account for the production of order and modifications of order; causal processes account for the propagation of causal influence. The theory adumbrated in this paper is spelled out more fully in chapters 5—6 of *Scientific Explanation and the Causal Structure of the World.*

"Causality in Archaeological Explanation," in Colin Renfrew et al., eds., *Theory and Explanation in Archaeology* (New York: Academic Press, 1982), pp. 45—55.

The theme of this article is that causal considerations always play a major role in explanations in archaeology; consequently, archaeologists need a model in which causality figures prominently. Since the standard Hempelian models pay no attention to causality, they are not especially well-suited to the needs of archaeologists.

"Carl G. Hempel on the Rationality of Science," *Journal of Philosophy* **LXXX**, 10 (Oct., 1983), pp. 555—562.

This is my contribution to an American Philosophical Association Symposium on the Philosophy of Carl G. Hempel. Thomas Kuhn was the other principal symposiast, and Hempel responded to both of us. Inasmuch as scientific rationality was the topic on which Hempel had done his most recent work, and was the subject Kuhn chose to address, I decided to devote my paper to it. In this paper I suggest that it is possible to build a bridge between the views of such logical empiricists as Hempel and such historically oriented philosophers as Kuhn if one adopts a Bayesian conception of scientific confirmation. Although Hempel remained skeptical, Kuhn responded quite positively to the idea.

This paper extends somewhat the ideas broached in "Bayes's Theorem and the History of Science" (1970). It is nevertheless clear that many serious problems are involved in attempting to implement this suggestion. Indeed, during the past year I have been working on a paper, "Rationality and Objectivity in Science," that is growing to monographic dimensions, in which I am trying to work out some of the remaining difficulties. A brief version of this paper will appear in a volume of *Minnesota Studies in the Philosophy of Science* edited by C. Wade Savage. The fate of the larger version has not yet been decided.

I should like to mention another aspect of this APA symposium. It was held in quite a large room, and the room was packed — there was literally standing room only. When Hempel finished delivering his extremely incisive replies to Kuhn and me, the entire audience arose and accorded him the longest and most enthusiastic standing ovation I had ever witnessed for any philosopher. Moreover, at the end of the discussion, after Hempel had made a few concluding remarks, the same kind of standing ovation was repeated — if anything, more enthusiastic than the first. It was an utterly unprecedented spontaneous expression of respect and affection — a truly heartwarming experience.

"Empiricism: The Key Question," in Nicholas Rescher, ed., *The Heritage of Logical Positivism* (Lanham, MD: University Press of America, 1985), pp. 1—21.

From the outset of his provocative book, *The Scientific Image*, Bas van Fraassen presupposes that empiricism and scientific realism are incompatible. This article challenges van Fraassen's assumption. I argue that an empiricist can be a scientific realist. Empiricism demands only that our theoretical claims be supported by sensory experience. The *key question* is whether inductive logic contains the resources to enable us to make inferences from statements about observables to statements about unobservables. I claim that it is possible to make such inductive inferences and I try to show how. If I am correct, empiricism does not preclude scientific realism. In any case, I suggest that philosophers have seldom addressed the key question and that it should be explicitly faced.

"Scientific Explanation: Three Basic Conceptions," in Peter Asquith and Philip Kitcher, eds., *PSA 1984*, (East Lansing, Mich: Philosophy of Science Assn., 1986), pp. 293—305.

This paper was presented at the 1984 meeting of the Philosophy of Science Association just before *Scientific Explanation and the Causal Structure of the World* was published. Comparing the epistemic, modal, and ontic conceptions, I called attention to some of the most outrageous consequences of the theory of explanation offered in that book — for example, acceptance of the fact that in some cases circumstances of type C explain the occurrence of some event of type E, while in other cases circumstances of the same type C explain the non-occurrence of an event of type E. I recommend a change from the epistemic to the ontic conceptions, but point out that this kind of reorientation involves a major gestalt switch — one radical enough to make the outrageous claims seem reasonable.

"Van Fraassen on Scientific Explanation," *Journal of Philosophy*, **LXXXIV** (1987), pp. 315—330. [with Philip Kitcher]

In Chapter 5 of *The Scientific Image* Bas van Fraassen offers an enticing theory of explanation in which pragmatics plays a crucial role. Explanations are answers to why-questions, and why-questions presuppose, at least implicitly, relevance relations. Kitcher and I observe that van Fraassen imposes no limitations whatever on what relations can serve as relevance relations, and without some constraints, we argue — roughly — any fact can explain any other fact. To ascertain what relations are legitimate relevance relations van Fraassen would have to deal with the traditional problems of explanation that he had hoped to avoid through an appeal to pragmatics. We conclude that van Fraassen has offered a valuable theory of the pragmatics of explanation, but we claim that he has not succeeded in constructing a pragmatic theory of explanation.

"Intuitions: Good and Not-So-Good," in Brian Skyrms and William Harper, eds., *Causation, Change, and Credence, Vol. 1*, (Dordrecht: D. Reidel Publishing Co., 1988).

This paper continues a discussion with I. J. Good, initiated in "Probabilistic Causality" (1980), regarding his causal calculus. It takes account of Good's reply to my 1980 paper, and of another paper he published subsequently. I conclude that Good's theory still faces fundamental problems, and that it is very difficult indeed to sort out our intuitions on this subject. Nevertheless, the significance of the project of understanding probabilistic (or indeterministic) causality is sufficient to justify fully further expenditures of intellectual effort.

Italian translation: in Guido Gambetta and Maria Carla Galavotti, eds., *Epistemologia ed Economia* (Bologna: forthcoming). Because the English title is a pun, and thus cannot be translated, a different Italian title will be supplied by the editors.

"Introduction," in Adolf Grünbaum and Wesley C. Salmon, eds., *The Limitations of Deductivism* (Berkeley & Los Angeles: University of California Press, forthcoming).

This essay contains a survey of the main issues surrounding deductivism, including an explanation of Alberto Coffa's intriguing and perceptive phrase, *deductive chauvinism*. It also provides a framework for understanding the essays that make up the body of the book.

"Deductism Visited and Revisited," *ibid.*

This paper, written after the publication of *Scientific Explanation and the Causal Structure of the World*, considers the thesis that all legitimate scientific explanations are deductive, first in deterministic contexts, and second in indeterministic ones. I try to give deductivism the strongest possible defense, but conclude that it is untenable. One argument involves causal considerations not previously noticed; another refers to explanations in applied science.

"Four Decades of Scientific Explanation" in Philip Kitcher and Wesley C. Salmon, eds., *Scientific Explanation* (University of Minnesota Press, forthcoming).

This essay traces the history of discussions of scientific explanation from the Hempel-Oppenheim paper (1948) down to the present (1987). The first decade (1948—57) began with that landmark paper, but contained little discussion of it. Most of the literature deals with explanation in history. The second decade (1958—67) saw much criticism of the Hempel-Oppenheim point of view, and the emergence of theories of statistical explanation. The focus during this decade was on the *received view*, as represented by Hempel's models. The third decade (1968—77) was a period in which new models were developing. It saw the emergence of the statistical-relevance model, and much greater attention was paid to the role of causality. In the fourth decade (1978—87) the previously received view has been superseded, but no single approach has emerged as a clear successor.

"Dynamic Rationality: Propensity, Probability, and Credence," in this volume.

"Rationality and Objectivity in Science," forthcoming in a future volume of *Minnesota Studies in the Philosophy of Science* edited by C. Wade Savage.

BRIEF COMMENTS ON WORKS OF OTHER AUTHORS AND OTHER SMALL ITEMS

"Comments on Barker's 'The Role of Simplicity in Explanation,'" in Herbert Feigl and Grover Maxwell, eds., *Current Issues in the Philosophy of Science* (New York: Holt, Rinehart, and Winston, 1961), pp. 274—276.

Letter to the Editor, *New York Review of Books*, May 14, 1964. A critical response to Stephen Toulmin's review of Adolf Grünbaum, PHILOSOPHICAL PROBLEMS OF SPACE AND TIME, 1st ed.

"Who Needs Inductive Acceptance Rules?" in Imre Lakatos, ed., *The Problem of Inductive Logic* (Amsterdam: North-Holland Publishing Co., 1968), pp. 139—144,

Y. Bar-Hillel, in defending a conception of inductive logic similar to Carnap's, asked rhetorically, "Who needs inductive acceptance rules?" I answer: (1) anyone who adopts the (then standard) covering-law conception of scientific explanations as deductive or inductive arguments; (2) anyone who maintains that analytic degree of confirmation statements cannot constitute an adequate "guide of life." Many philosophers would presumably have one or both of these reasons for attempting to incorporate rules of provisional acceptance of hypotheses into their inductive logics.

"Comment" in Joseph Margolis, ed., *Fact and Existence* (Oxford: Basil Blackwell, 1969), pp. 95—97.

Comment on a paper by R. M. Martin.

"Induction and Intuition: Comments on Ackermann's 'Problems'," in J. W. Davis et al., eds., *Philosophical Logic* (Dordrecht: D. Reidel Publishing Co., 1969) pp. 158—163.

Foreword to the English Edition of Hans Reichenbach, *Axiomatization of the Theory of Relativity* (Berkeley & Los Angeles: University of California Press, 1969), pp. v—x.

Discussion Remarks in Michael Radner and Stephen Winokur, eds., *Minnesota Studies in the Philosophy of Science* vol. IV (Minneapolis: University of Minnesota Press, 1970), pp. 225f, 227, 243, 248f, 251.

Edited *without my approval.* The editors managed to excise the one really interesting exchange involuing N. R. Hanson, Paul Feyerabend, and me.

"A Zenoesque Problem," published without title in Martin Gardner,

"Mathematical Games," *Scientific American* **225**, 6 (Dec., 1971), pp. 97—99.

Shortly after *Zeno's Paradoxes* was published, Martin Gardner sent me a cute puzzle about a boy, a girl, and a dog, asking whether I thought it constituted a Zeno-type paradox. Indeed, it did — and neither it nor anything quite like it had been previously recognized as such. It is analogous to the regressive form of Zeno's dichotomy paradox. The difference is that, whereas the runner in the dichotomy always keeps going in the same direction, the dog in this problem has to change directions infinitely many times. The dog is a kind of infinity machine, but unlike the standard infinity machines, its infinite series of tasks is open toward the past, not toward the future. This puzzle is discussed in *Space, Time, and Motion: A Philosophical Introduction*, chap. 2.

"Explanation and Relevance: Comments on James G. Greeno's 'Theoretical Entities in Statistical Explanation'," in Roger C. Buck and Robert S. Cohen, eds., *PSA 1970* (Dordrecht; D. Reidel Publishing Co., 1971), pp. 27—39.

This paper contains the earliest published account of the statistical-relevance (S-R) model of scientific explanation under that name. It is contrasted with Hempel's inductive-statistical (I-S) model in several basic respects. The important contributions of Richard C. Jeffrey and James G. Greeno to the S-R model are discussed. Particular attention is focused on Greeno's application of information-theoretic concepts to the theory of scientific explanation. This paper is part of a symposium in which Greeno and Jeffrey were the other participants.

"Law, in Science," *Encyclopedia Americana*, 1972, vol. **17**, p. 81.

A brief entry of no particular interest.

Postscript to "Probabilities and the Problem of Individuation" by Bas van Fraassen in Sidney Luckenbach, ed., *Probabilities, Problems, and Paradoxes* (Encino, CA: Dickenson Publishing Co., 1972), pp. 135—138.

These comments on van Fraassen's paper were paraphrased by van Fraassen *with my heartiest approval.*

"Reply to Lehman," *Philosophy of Science* **40**, 3 (Sept., 1973), pp. 397—402.

"Numerals vs. Numbers," *De Pauw Mathnews,* **II**, 1 (Mar., 1973), pp. 8—11.

De Pauw Mathnews was an in-house paper published by a member of the Mathematics Department mainly for the benefit of students and a few other interested parties. (I was living in Greencastle, Indiana, where De Pauw University is located, at the time.) Each issue contained, among other things, problems for students. One such problem had to do with representations of numbers in different notational systems. Because the statement of the problem thoroughly confused numbers with their names, I tried to clarify the situation with a small discussion of the use/mention distinction. The mathematician who published the paper printed my note, including my answer to the problem as suitably restated, but was not particularly pleased to be bothered with such philosophical quibbles.

"Comments on 'Hempel's Ambiguity' by J. Alberto Coffa," *Synthese* **28**, 4 (Oct., 1974), pp. 165—170.

Coffa's paper provided a profound analysis of Hempel's doctrine of *essential epistemic relativization of inductive-statistical explanation.* It made me see an important connection between the doctrine of essential epistemic relativization and determinism in the theory of scientific explanation, and it made me realize the importance of the concept of an objectively homogeneous reference class.

"Dedication to Leonard J. Savage," in Kenneth F. Schaffner and Robert S. Cohen, eds., *PSA 1972* (Dordrecht: D. Reidel Publishing Co., 1974), p. 101.

This brief note is a tribute to the extremely distinguished and influential statistician Jimmy Savage, whose book, *The Foundations of Statistics*, provided the chief initial impetus for contemporary Bayesianism. A thoroughly admirable and likeable person, he was a philosophical thinker of astonishing profundity.

"A Note on Russell's Anticipations," *Russell*, no. 17 (Spring, 1975), p. 29.

In this brief note I suggest that Bertrand Russell had discovered, before Nelson Goodman, a problem that is, in its essentials, just like Goodman's celebrated grue-bleen puzzle. I suggest, moreover, that Russell had provided what seems to me the correct solution.

"Hans Reichenbach," in John A. Garraty, ed., *Dictionary of American Biography, Supplement Five*, 1951—55 (New York: Charles Scribner's Sons, 1977), pp. 562—63.

"An Ontological Argument for the Existence of the Null Set," in Martin Gardner, *Mathematical Magic Show* (New York: Alfred A. Knopf, 1977), pp. 32—33.

This is a joke. Bas van Fraassen once hinted at such an ontological argument: Consider the set than which none emptier can be conceived . . . I completed the argument: Suppose no such set exists. Then the set of such sets has no member; hence, it is the null set. QED

"Hans Reichenbach: A Memoir," in Maria Reichenbach and Robert S. Cohen, eds., *Hans Reichenbach: Selected Writings, 1909—1953*, vol. I (Dordrecht: D. Reidel Publishing Co., 1978), pp. 69—77.

This memoir is a highly personal statement about my relationship to my teacher. It also contains a refutation of a slanderous fabrication about Reichenbach by the novelist Arthur Koestler. In addition, it contains an important clarification of a famous incident involving comments by the illustrious statistician R. A. Fisher on the research results of the celebrated parapsychologist J. B. Rhine.

"Wissenschaft, Grundlagen der," in J. Speck, ed., *Handbuch Wissenschaftstheoretischer Begriff* (Göttingen: Vandenhoeck & Ruprecht, 1980), pp. 252—257. [German translation by the editor.]

A routine exposition of a foundationalist approach to the philosophy of science.

"Autobiographical Note," in *What? Where? When? Why?* (Dordrecht: D. Reidel Publishing Co., 1982), pp. 281—283.

"Probabilistic Explanation: Introduction," in Peter Asquith and Thomas Nickles, eds., *PSA 1982* (East Lansing, Mich.: Philosophy of Science Assn., 1983), pp. 179—180.

This brief note introduces a Philosophy of Science Association Symposium on Probabilistic Explanation. Since it occurred on the 20th anniversary of the publication of Hempel's "Deductive-Nomological vs. Statistical Explanation," I took the opportunity to pay special tribute to that article, which is the first attempt ever made (to the best of my knowledge) to provide a detailed explication of any form of statistical explanation.

"Conflicting Conceptions of Scientific Explanation," *Journal of Philosophy* **LXXXII**, 11 (Nov., 1985), pp. 651—654.

Abstract of contribution to the APA Symposium on Wesley Salmon's *Scientific Explanation and the Causal Structure of the World.*

"Scientific Explanation: Consensus or Progress?" in Barbara C. Klein-Helmuth and David Savold, compilers, *1986 AAAS Annual Meeting: Abstracts of Papers* (Washington, DC: American Association for the Advancement of Science, 1986), p. 90.

Abstract of my contribution to a symposium on "A New Consensus in Philosophy of Science?"

"Eulogy for J. Alberto Coffa," in Adolf Grünbaum and Wesley C. Salmon, eds., *The Limitations of Deductivism* (Berkeley & Los Angeles: University of California Press, forthcoming).

Spanish translation:
Revista Latinoamericana de Filosofia, **XII**, 2 (July, 1986), pp. 245–247.

REVIEWS

Sheldon J. Lachman, THE FOUNDATIONS OF SCIENCE, *Philosophy of Science* **XXIV**, 4 (Oct., 1957), pp. 358–359.

Sir Harold Jeffreys, SCIENTIFIC INFERENCE, *Philosophy of Science* **XXIV**, 4 (Oct., 1957), pp. 364–366.

James T. Culbertson, MATHEMATICS AND LOGIC FOR DIGITAL DEVICES, *Mathematical Reviews* **XIX**, 11 (Dec., 1958), p 1200.

Håkan Törnebohm, "On Two Logical Systems Proposed in the Philosophy of Quantum Mechanics," *Mathematical Reviews* **XX**, 4 (Apr., 1959), # 2278.

Stephen F. Barker, INDUCTION AND HYPOTHESIS, *Philosophical Review* **LXVIII**, 2 (Apr., 1959), pp. 247–253.

G. H. von Wright, THE LOGICAL PROBLEM OF INDUCTION, 2nd ed., *Philosophy of Science* **XXVI**, 2 (Apr., 1959), p. 166.

Richard von Mises, PROBABILITY, STATISTICS, AND TRUTH, 2nd English ed., *Philosophy of Science* **XXVI**, 4 (Oct., 1959), pp. 387–388.

Hans Reichenbach, THE PHILOSOPHY OF SPACE AND TIME, *Mathematical Reviews* **XXI**, 5 (May, 1960), # 3247.

Hans Reichenbach, MODERN PHILOSOPHY OF SCIENCE, *Philosophical Review* **LXIX**, 3 (July, 1960), pp. 409—411.

J. L. Destouches, "Physico-logical Problems," *Mathematical Reviews* **XXI**, 10 (Nov., 1960), # 6319.

Rom Harré, AN INTRODUCTION TO THE LOGIC OF SCIENCE; Greenwood, THE NATURE OF SCIENCE; Smith, THE GENERAL SCIENCE OF NATURE, *Isis* (1962), pp. 234—235.

John Patrick Day, INDUCTIVE PROBABILITY, *Philosophical Review* **LXXII** (1963), pp. 392—396.

In the 1950s and 1960s I was highly critical of ordinary language approaches to the problem of the justification of induction, and I was enormously skeptical about the ability of ordinary language analysis to handle adequately even moderately technical concepts in philosophy of science. This book by Day had been touted to me as a careful and systematic treatment of probability from that standpoint. When I received the review copy and began to read it, I found it full of egregious errors. One of the most memorable was the claim that the notion of "probabilification" represents a transitive relation. I concluded that philosophers who know nothing of the mathematical calculus of probability should not undertake analyses of that concept.

I. J. Good, "A Causal Calculus, I, II," *Mathematical Reviews* **XXVI**, 1 (July, 1963), # 789 a—b.

Robert G. Colodny, ed., FRONTIERS OF SCIENCE AND PHILOSOPHY, *Isis* **LV**, 2 (June, 1964), pp. 215—216.

Adolf Grünbaum, PHILOSOPHICAL PROBLEMS OF SPACE AND TIME, *Science* **147** (12 Feb. 1965), pp. 724—725.

Rudolf Carnap, PHILOSOPHICAL FOUNDATIONS OF PHYSICS, *Science* **155** (10 Mar. 1967), p. 1235.

Henry E. Kyburg, Jr., PROBABILITY AND THE LOGIC OF RA-

TIONAL BELIEF, *Philosophy of Science* **XXXIV**, 3 (Sept., 1965), pp. 283—285.

Adolf Grünbaum, MODERN SCIENCE AND ZENO'S PARADOXES, *Ratio* **XII**, 2 (Dec., 1970), pp. 178—182.

Also published in German edition of *Ratio* in German translation.

Wesley C. Salmon, ZENO'S PARADOXES, *The Monist* **56**, 4 (Oct., 1972), p. 632. Author's abstract-review.

INDEX OF NAMES

INDEX OF SUBJECTS

SYNTHESE LIBRARY

Studies in Epistemology, Logic, Methodology,
and Philosophy of Science

1. J. M. Bochénski, *A Precis of Mathematical Logic.* 1959.
2. P. L. Guiraud, *Problèmes et méthodes de la statistique linguistique.* 1960.
3. Hans Freudenthal (ed.), *The Concept and the Role of the Model in Mathematics and Natural and Social Sciences.* 1961.
4. Evert W. Beth, *Formal Methods. An Introduction to Symbolic Logic and the Study of Effective Operations in Arithmetic and Logic.* 1962.
5. B. H. Kazemier and D. Vuysje (eds.), *Logic and Language. Studies Dedicated to Professor Rudolf Carnap on the Occasion of His Seventieth Birthday.* 1962.
6. Marx W. Wartofsky (ed.), *Proceedings of the Boston Colloquium for the Philosophy of Science 1961–1962.* Boston Studies in the Philosophy of Science, Volume I. 1963.
7. A. A. Zinov'ev, *Philosophical Problems of Many-Valued Logic.* 1963.
8. Georges Gurvitch, *The Spectrum of Social Time.* 1964.
9. Paul Lorenzen, *Formal Logic.* 1965.
10. Robert S. Cohen and Marx W. Wartofsky (eds.), *In Honor of Philipp Frank.* Boston Studies in the Philosophy of Science, Volume II. 1965.
11. Evert W. Beth, *Mathematical Thought. An Introduction to the Philosophy of Mathematics.* 1965.
12. Evert W. Beth and Jean Piaget, *Mathematical Epistemology and Psychology.* 1966.
13. Guido Küng, *Ontology and the Logistic Analysis of Language. An Enquiry into the Contemporary Views on Universals.* 1967.
14. Robert S. Cohen and Marx W. Wartofsky (eds.), *Proceedings of the Boston Colloquium for the Philosophy of Sciences 1964–1966. In Memory of Norwood Russell Hanson.* Boston Studies in the Philosophy of Science, Volume III. 1967.
5. C. D. Broad, *Induction, Probability, and Causation. Selected Papers.* 1968.
6. Günther Patzig, *Aristotle's Theory of the Syllogism. A Logical-Philosophical Study of Book A of the Prior Analytics.* 1968.
7. Nicholas Rescher, *Topics in Philosophical Logic.* 1968.
8. Robert S. Cohen and Marx W. Wartofsky (eds.), *Proceedings of the Boston Colloquium for the Philosophy of Science 1966–1968.* Boston Studies in the Philosophy of Science, Volume IV. 1969

19. Robert S. Cohen and Marx W. Wartofsky (eds.), *Proceedings of the Boston Colloquium for the Philosophy of Science 1966–1968*. Boston Studies in the Philosophy of Science, Volume V. 1969
20. J. W. Davis, D. J. Hockney, and W. K. Wilson (eds.), *Philosophical Logic*. 1969
21. D. Davidson and J. Hintikka (eds.), *Words and Objections. Essays on the Work of W. V. Quine*. 1969.
22. Patrick Suppes. *Studies in the Methodology and Foundations of Science. Selected Papers from 1911 to 1969*. 1969
23. Jaakko Hintikka, *Models for Modalities. Selected Essays*. 1969
24. Nicholas Rescher *et al.* (eds.), *Essays in Honor of Carl G. Hempel. A Tribute on the Occasion of His Sixty-Fifth Birthday*. 1969
25. P. V. Tavanec (ed.), *Problems of the Logic of Scientific Knowledge*. 1969
26. Marshall Swain (ed.), *Induction, Acceptance, and Rational Belief*. 1970.
27. Robert S. Cohen and Raymond J. Seeger (eds.), *Ernst Mach: Physicist and Philosopher*. Boston Studies in the Philosophy of Science, Volume VI. 1970.
28. Jaakko Hintikka and Patrick Suppes, *Information and Inference*. 1970.
29. Karel Lambert, *Philosophical Problems in Logic. Some Recent Developments*. 1970.
30. Rolf A. Eberle, *Nominalistic Systems*. 1970.
31. Paul Weingartner and Gerhard Zecha (eds.), *Induction, Physics, and Ethics*. 1970.
32. Evert W. Beth, *Aspects of Modern Logic*. 1970.
33. Risto Hilpinen (ed.), *Deontic Logic: Introductory and Systematic Readings*. 1971.
34. Jean-Louis Krivine, *Introduction to Axiomatic Set Theory*. 1971.
35. Joseph D. Sneed, *The Logical Structure of Mathematical Physics*. 1971.
36. Carl R. Kordig, *The Justification of Scientific Change*. 1971.
37. Milic Capek, *Bergson and Modern Physics*. Boston Studies in the Philosophy of Science, Volume VII. 1971.
38. Norwood Russell Hanson, *What I Do Not Believe, and Other Essays* (ed. by Stephen Toulmin and Harry Woolf). 1971.
39. Roger C. Buck and Robert S. Cohen (eds.), *PSA 1970. In Memory of Rudolf Carnap*. Boston Studies in the Philosophy of Science, Volume VIII. 1971
40. Donald Davidson and Gilbert Harman (eds.), *Semantics of Natural Language*. 1972.
41. Yehoshua Bar-Hillel (ed.), *Pragmatics of Natural Languages*. 1971.
42. Sören Stenlund, *Combinators, λ-Terms and Proof Theory*. 1972.
43. Martin Strauss, *Modern Physics and Its Philosophy. Selected Paper in the Logic, History, and Philosophy of Science*. 1972.
44. Mario Bunge, *Method, Model and Matter*. 1973.
45. Mario Bunge, *Philosophy of Physics*. 1973.
46. A. A. Zinov'ev, *Foundations of the Logical Theory of Scientific Knowledge (Complex Logic)*. (Revised and enlarged English edition with an appendix by G. A. Smirnov, E. A. Sidorenka, A. M. Fedina, and L. A. Bobrova.) Boston Studies in the Philosophy of Science, Volume IX. 1973.
47. Ladislav Tondl, *Scientific Procedures*. Boston Studies in the Philosophy of Science, Volume X. 1973.
48. Norwood Russell Hanson, *Constellations and Conjectures* (ed. by Willard C. Humphreys, Jr.). 1973
49. K. J. J. Hintikka, J. M. E. Moravcsik, and P. Suppes (eds.), *Approaches to Natural Language*. 1973.

50. Mario Bunge (ed.), *Exact Philosophy – Problems, Tools, and Goals*. 1973.
51. Radu J. Bogdan and Ilkka Niiniluoto (eds.), *Logic, Language, and Probability*. 1973.
52. Glenn Pearce and Patrick Maynard (eds.), *Conceptual Change*. 1973.
53. Ilkka Niiniluoto and Raimo Tuomela, *Theoretical Concepts and Hypothetico-Inductive Inference*. 1973.
54. Roland Fraissé, *Course of Mathematical Logic* – Volume 1: *Relation and Logical Formula*. 1973.
55. Adolf Grünbaum, *Philosophical Problems of Space and Time*. (Second, enlarged edition.) Boston Studies in the Philosophy of Science, Volume XII. 1973.
56. Patrick Suppes (ed.), *Space, Time, and Geometry*. 1973.
57. Hans Kelsen, *Essays in Legal and Moral Philosophy* (selected and introduced by Ota Weinberger). 1973.
58. R. J. Seeger and Robert S. Cohen (eds.), *Philosophical Foundations of Science*. Boston Studies in the Philosophy of Science, Volume XI. 1974.
59. Robert S. Cohen and Marx W. Wartofsky (eds.), *Logical and Epistemological Studies in Contemporary Physics*. Boston Studies in the Philosophy of Science, Volume XIII. 1973.
60. Robert S. Cohen and Marx W. Wartofsky (eds.), *Methodological and Historical Essays in the Natural and Social Sciences. Proceedings of the Boston Colloquium for the Philosophy of Science 1969–1972*. Boston Studies in the Philosophy of Science, Volume XIV. 1974.
61. Robert S. Cohen, J. J. Stachel, and Marx W. Wartofsky (eds.), *For Dirk Struik. Scientific, Historical and Political Essays in Honor of Dirk J. Struik*. Boston Studies in the Philosophy of Science, Volume XV. 1974.
62. Kazimierz Ajdukiewicz, *Pragmatic Logic* (transl. from the Polish by Olgierd Wojtasiewicz). 1974.
63. Sören Stenlund (ed.), *Logical Theory and Semantic Analysis. Essays Dedicated to Stig Kanger on His Fiftieth Birthday*. 1974.
64. Kenneth F. Schaffner and Robert S. Cohen (eds.), *Proceedings of the 1972 Biennial Meeting, Philosophy of Science Association*. Boston Studies in the Philosophy of Science, Volume XX. 1974.
65. Henry E. Kyburg, Jr., *The Logical Foundations of Statistical Inference*. 1974.
66. Marjorie Grene, *The Understanding of Nature. Essays in the Philosophy of Biology*. Boston Studies in the Philosophy of Science, Volume XXIII. 1974.
67. Jan M. Broekman, *Structuralism: Moscow, Prague, Paris*. 1974.
68. Norman Geschwind, *Selected Papers on Language and the Brain*, Boston Studies in the Philosophy of Science, Volume XVI. 1974.
69. Roland Fraissé, *Course of Mathematical Logic* – Volume 2: *Model Theory*. 1974.
70. Andrzej Grzegorczyk, *An Outline of Mathematical Logic. Fundamental Results and Notions Explained with All Details*. 1974.
71. Franz von Kutschera, *Philosophy of Language*. 1975.
72. Juha Manninen and Raimo Tuomela (eds.), *Essays on Explanation and Understanding. Studies in the Foundations of Humanities and Social Sciences*. 1976.
73. Jaakko Hintikka (ed.), *Rudolf Carnap, Logical Empiricist. Materials and Perspectives*. 1975.
74. Milic Capek (ed.), *The Concepts of Space and Time. Their Structure and Their Development*. Boston Studies in the Philosophy of Science, Volume XXII. 1976.

75. Jaakko Hintikka and Unto Remes, *The Method of Analysis. Its Geometrical Origin and Its General Significance.* Boston Studies in the Philosophy of Science, Volume XXV. 1974.
76. John Emery Murdoch and Edith Dudley Sylla, *The Cultural Context of Medieval Learning.* Boston Studies in the Philosophy of Science, Volume XXVI. 1975.
77. Stefan Amsterdamski, *Between Experience and Metaphysics. Philosophical Problems of the Evolution of Science.* Boston Studies in the Philosophy of Science, Volume XXXV. 1975.
78. Patrick Suppes (ed.), *Logic and Probability in Quantum Mechanics.* 1976.
79. Hermann von Helmholtz: *Epistemological Writings. The Paul Hertz/Moritz Schlick Centenary Edition of 1921 with Notes and Commentary by the Editors.* (Newly translated by Malcolm F. Lowe. Edited, with an Introduction and Bibliography, by Robert S. Cohen and Yehuda Elkana.) Boston Studies in the Philosophy of Science, Volume XXXVII. 1975.
80. Joseph Agassi, *Science in Flux.* Boston Studies in the Philosophy of Science, Volume XXVIII. 1975.
81. Sandra G. Harding (ed.), *Can Theories Be Refuted? Essays on the Duhem-Quine Thesis.* 1976.
82. Stefan Nowak, *Methodology of Sociological Research. General Problems.* 1977.
83. Jean Piaget, Jean-Blaise Grize, Alina Szeminska, and Vinh Bang, *Epistemology and Psychology of Functions.* 1977.
84. Marjorie Grene and Everett Mendelsohn (eds.), *Topics in the Philosophy of Biology.* Boston Studies in the Philosophy of Science, Volume XXVII. 1976.
85. E. Fischbein, *The Intuitive Sources of Probabilistic Thinking in Children.* 1975.
86. Ernest W. Adams, *The Logic of Conditionals. An Application of Probability to Deductive Logic.* 1975.
87. Marian Przelecki and Ryszard Wójcicki (eds.), *Twenty-Five Years of Logical Methodology in Poland.* 1976.
88. J. Topolski, *The Methodology of History.* 1976.
89. A. Kasher (ed.), *Language in Focus: Foundations, Methods and Systems. Essays Dedicated to Yehoshua Bar-Hillel.* Boston Studies in the Philosophy of Science, Volume XLIII. 1976.
90. Jaakko Hintikka, *The Intentions of Intentionality and Other New Models for Modalities.* 1975.
91. Wolfgang Stegmüller, *Collected Papers on Epistemology, Philosophy of Science and History of Philosophy.* 2 Volumes. 1977.
92. Dov M. Gabbay, *Investigations in Modal and Tense Logics with Applications to Problems in Philosophy and Linguistics.* 1976.
93. Radu J. Bodgan, *Local Induction.* 1976.
94. Stefan Nowak, *Understanding and Prediction. Essays in the Methodology of Social and Behavioral Theories.* 1976.
95. Peter Mittelstaedt, *Philosophical Problems of Modern Physics.* Boston Studies in the Philosophy of Science, Volume XVIII. 1976.
96. Gerald Holton and William Blanpied (eds.), *Science and Its Public: The Changing Relationship.* Boston Studies in the Philosophy of Science, Volume XXXIII. 1976.
97. Myles Brand and Douglas Walton (eds.), *Action Theory.* 1976.
98. Paul Gochet, *Outline of a Nominalist Theory of Proposition. An Essay in the Theory of Meaning.* 1980.

99. R. S. Cohen, P. K. Feyerabend, and M. W. Wartofsky (eds.), *Essays in Memory of Imre Lakatos.* Boston Studies in the Philosophy of Science, Volume XXXIX. 1976.
100. R. S. Cohen and J. J. Stachel (eds.), *Selected Papers of Léon Rosenfield.* Boston Studies in the Philosophy of Science, Volume XXI. 1978.
101. R. S. Cohen, C. A. Hooker, A. C. Michalos, and J. W. van Evra (eds.), *PSA 1974: Proceedings of the 1974 Biennial Meeting of the Philosophy of Science Association.* Boston Studies in the Philosophy of Science, Volume XXXII. 1976.
102. Yehuda Fried and Joseph Agassi, *Paranoia: A Study in Diagnosis.* Boston Studies in the Philosophy of Science, Volume L. 1976.
103. Marian Przelecki, Klemens Szaniawski, and Ryszard Wójcicki (eds.), *Formal Methods in the Methodology of Empirical Sciences.* 1976.
104. John M. Vickers, *Belief and Probability.* 1976.
105. Kurt H. Wolff, *Surrender and Catch: Experience and Inquiry Today.* Boston Studies in the Philosophy of Science, Volume LI. 1976.
106. Karel Kosík, *Dialectics of the Concrete.* Boston Studies in the Philosophy of Science, Volume LII. 1976.
107. Nelson Goodman, *The Structure of Appearance* (Third edition.) Boston Studies in the Philosophy of Science, Volume LIII. 1977.
108. Jerzy Giedymin (ed.), *Kazimierz Ajdukiewicz: The Scientific World-Perspective and Other Essays, 1931-1963.* 1978.
109. Robert L. Causey, *Unity of Science.* 1977.
110. Richard E. Grandy, *Advanced Logic for Applications.* 1977.
111. Robert P. McArthur, *Tense Logic.* 1976.
112. Lars Lindahl, *Position and Change. A Study in Law and Logic.* 1977.
113. Raimo Tuomela, *Dispositions.* 1978.
114. Herbert A. Simon, *Models of Discovery and Other Topics in the Methods of Science.* Boston Studies in the Philosophy of Science, Volume LIV. 1977.
115. Roger D. Rosenkrantz, *Inference, Method and Decision.* 1977.
116. Raimo Tuomela, *Human Action and Its Explanation. A Study on the Philosophical Foundations of Psychology.* 1977.
117. Morris Lazerowitz, *The Language of Philosophy. Freud and Wittgenstein.* Boston Studies in the Philosophy of Science, Volume LV. 1977.
119. Jerzy Pelc, *Semiotics in Poland, 1894-1969.* 1978.
120. Ingmar Pörn, *Action Theory and Social Science. Some Formal Models.* 1977.
121. Joseph Margolis, *Persons and Mind. The Prospects of Nonreductive Materialism.* Boston Studies in the Philosophy of Science, Volume LVII. 1977.
122. Jaakko Hintikka, Ilkka Niiniluoto, and Esa Saarinen (eds.), *Essays on Mathematical and Philosophical Logic.* 1978.
123. Theo A. F. Kuipers, *Studies in Inductive Probability and Rational Expectation.* 1978.
124. Esa Saarinen, Risto Hilpinen, Ilkka Niiniluoto, and Merrill Provence Hintikka (eds.), *Essays in Honour of Jaakko Hintikka on the Occasion of His Fiftieth Birthday.* 1978.
125. Gerard Radnitzky and Gunnar Andersson (eds.), *Progress and Rationality in Science.* Boston Studies in the Philosophy of Science, Volume LVIII. 1978.
126. Peter Mittelstaedt, *Quantum Logic.* 1978.
127. Kenneth A. Bowen, *Model Theory for Modal Logic. Kripke Models for Modal Predicate Calculi.* 1978.

128. Howard Alexander Bursen, *Dismantling the Memory Machine. A Philosophical Investigation of Machine Theories of Memory.* 1978.
129. Marx W. Wartofsky, *Models: Representation and the Scientific Understanding.* Boston Studies in the Philosophy of Science, Volume XLVIII. 1979.
130. Don Ihde, *Technics and Praxis. A Philosophy of Technology.* Boston Studies in the Philosophy of Science, Volume XXIV. 1978.
131. Jerzy J. Wiatr (ed.), *Polish Essays in the Methodology of the Social Sciences.* Boston Studies in the Philosophy of Science, Volume XXIX. 1979.
132. Wesley C. Salmon (ed.), *Hans Reichenbach: Logical Empiricist.* 1979.
133. Peter Bieri, Rolf-P. Horstmann, and Lorenz Krüger (eds.), *Transcendental Arguments in Science Essays in Epistemology.* 1979.
134. Mihailo Marković and Gajo Petrović (eds.), *Praxis, Yugoslav Essays in the Philosophy and Methodology of the Social Sciences.* Boston Studies in the Philosophy of Science, Volume XXXVI. 1979.
135. Ryszard Wójcicki, *Topics in the Formal Methodology of Empirical Sciences.* 1979.
136. Gerard Radnitzky and Gunnar Andersson (eds.), *The Structure and Development of Science.* Boston Studies in the Philosophy of Science, Volume LIX. 1979.
137. Judson Chambers Webb, *Mechanism, Mentalism, and Metamathematics. An Essay on Finitism.* 1980.
138. D. F. Gustafson and B. L. Tapscott (eds.), *Body, Mind, and Method. Essays in Honor of Virgil C. Aldrich.* 1979.
139. Leszek Nowak, *The Structure of Idealization. Towards a Systematic Interpretation of the Marxian Idea of Science.* 1979.
140. Chaim Perelman, *The New Rhetoric and the Humanities. Essays on Rhetoric and Its Applications.* 1979.
141. Wlodzimierz Rabinowicz, *Universalizability. A Study in Morals and Metaphysics.* 1979.
142. Chaim Perelman, *Justice, Law, and Argument. Essays on Moral and Legal Reasoning.* 1980.
143. Stig Kanger and Sven Öhman (eds.), *Philosophy and Grammar. Papers on the Occasion of the Quincentennial of Uppsala University.* 1980.
144. Tadeusz Pawlowski, *Concept Formation in the Humanities and the Social Sciences.* 1980.
145. Jaakko Hintikka, David Gruender, and Evandro Agazzi (eds.), *Theory Change, Ancient Axiomatics, and Galileo's Methodology. Proceedings of the 1978 Pisa Conference on the History and Philosophy of Science,* Volume I. 1981.
146. Jaakko Hintikka, David Gruender, and Evandro Agazzi, *Probabilistic Thinking, Thermodynamics, and the Interaction of the History and Philosophy of Science. Proceedings of the 1978 Pisa Conference on the History and Philosophy of Science,* Volume II. 1981.
147. Uwe Mönnich (ed.), *Aspects of Philosophical Logic. Some Logical Forays into Central Notions of Linguistics and Philosophy.* 1981.
148. Dov M. Gabbay, *Semantical Investigations in Heyting's Intuitionistic Logic.* 1981.
149. Evandro Agazzi (ed.), *Modern Logic – A Survey. Historical, Philosophical, and Mathematical Aspects of Modern Logic and Its Applications.* 1981.
150. A. F. Parker-Rhodes, *The Theory of Indistinguishables. A Search for Explanatory Principles below the Level of Physics.* 1981.
151. J. C. Pitt, *Pictures, Images, and Conceptual Change. An Analysis of Wilfrid Sellars' Philosophy of Science.* 1981.

152. R. Hilpinen (ed.), *New Studies in Deontic Logic. Norms, Actions, and the Foundations of Ethics.* 1981.
153. C. Dilworth, *Scientific Progress. A Study Concerning the Nature of the Relation Between Successive Scientific Theories.* 1981.
154. D. W. Smith and R. McIntyre, *Husserl and Intentionality. A Study of Mind, Meaning, and Language.* 1982.
155. R. J. Nelson, *The Logic of Mind.* 1982.
156. J. F. A. K. van Benthem, *The Logic of Time. A Model-Theoretic Investigation into the Varieties of Temporal Ontology, and Temporal Discourse.* 1982.
157. R. Swinburne (ed.), *Space, Time and Causality.* 1982.
158. R. D. Rozenkrantz, *E. T. Jaynes: Papers on Probability, Statistics and Statistical Physics.* 1983.
159. T. Chapman, *Time: A Philosophical Analysis.* 1982.
160. E. N. Zalta, *Abstract Objects. An Introduction to Axiomatic Metaphysics.* 1983.
161. S. Harding and M. B. Hintikka (eds.), *Discovering Reality. Feminist Perspectives on Epistemology, Metaphysics, Methodology, and Philosophy of Science.* 1983.
162. M. A. Stewart (ed.), *Law, Morality and Rights.* 1983.
163. D. Mayr and G. Süssmann (eds.), *Space, Time, and Mechanics. Basic Structure of a Physical Theory.* 1983.
164. D. Gabbay and F. Guenthner (eds.), *Handbook of Philosophical Logic.* Vol. I. 1983.
165. D. Gabbay and F. Guenthner (eds.), *Handbook of Philosophical Logic.* Vol. II. 1984.
166. D. Gabbay and F. Guenthner (eds.), *Handbook of Philosophical Logic.* Vol. III. 1985.
167. D. Gabbay and F. Guenthner (eds.), *Handbook of Philosophical Logic.* Vol. IV, forthcoming.
168. Andrew, J. I. Jones, *Communication and Meaning.* 1983.
169. Melvin Fitting, *Proof Methods for Modal and Intuitionistic Logics.* 1983.
170. Joseph Margolis, *Culture and Cultural Entities.* 1984.
171. Raimo Tuomela, *A Theory of Social Action.* 1984.
172. Jorge J. E. Gracia, Eduardo Rabossi, Enrique Villanueva, and Marcelo Dascal (eds.), *Philosophical Analysis in Latin America.* 1984.
173. Paul Ziff, *Epistemic Analysis. A Coherence Theory of Knowledge.* 1984.
174. Paul Ziff, *Antiaesthetics. An Appreciation of the Cow with the Subtile Nose.* 1984.
175. Wolfgang Balzer, David A. Pearce, and Heinz-Jürgen Schmidt (eds.), *Reduction in Science. Structure, Examples, Philosophical Problems.* 1984.
176. Aleksander Peczenik, Lars Lindahl, and Bert van Roermund (eds.), *Theory of Legal Science. Proceedings of the Conference on Legal Theory and Philosophy of Science, Lund, Sweden, December 11–14, 1983.* 1984.
177. Ilkka Niiniluoto, *Is Science Progressive?* 1984.
178. Binal Matilal and Jaysankar Lal Shaw (eds.), *Exploratory Essays in Current Theories and Classical Indian Theories of Meaning and Reference.* 1985.
179. Peter Kroes, *Time: Its Structure and Role in Physical Theories.* 1985.
180. James H. Fetzer, *Sociobiology and Epistemology,* 1985.
181. L. Haaparanta and J. Hintikka, *Frege Synthesized. Essays on the Philosophical and Foundational Work of Gottlob Frege.* 1986.
182. Michael Detlefsen, *Hilbert's Program. An Essay on Mathematical Instrumentalism.* 1986.

183. James L. Golden and Joseph J. Pilotta (eds.), *Practical Reasoning in Human Affairs. Studies in Honor of Chaim Perelman*. 1986.

184. Henk Zandvoort, *Models of Scientific Development and the Case of Nuclear Magnetic Resonance*. 1986.

185. Ilkka Niiniluoto, *Truthlikeness*. 1987

186. Wolfgang Balzer, C. Ulises Moulines, and Joseph D. Sneed, *An Architectonic for Science*. 1987